A Handbook for Construction Planning and Scheduling

The book's companion website is at

www.wiley.com/go/baldwin/constructionplansched

You will find here freely downloadable support materials.

A Handbook for Construction Planning and Scheduling

Andrew Baldwin

and

David Bordoli

WILEY Blackwell

Registered Office
John Wiley & Sons, Ltd, The Atrium, Southern Gate, Chichester, West Sussex, PO19 8SQ, UK

Editorial Offices
9600 Garsington Road, Oxford, OX4 2DQ, UK
The Atrium, Southern Gate, Chichester, West Sussex, PO19 8SQ, UK

For details of our global editorial offices, for customer services and for information about how
to apply for permission to reuse the copyright material in this book please see our website at
www.wiley.com/wiley-blackwell.

Library of Congress Cataloging-in-Publication Data

Baldwin, Andrew, 1950–
  A handbook for construction planning and scheduling / Andrew Baldwin, David Bordoli.
    pages   cm
  Includes bibliographical references and index.
  ISBN 978-0-470-67032-3 (paperback)
1. Building–Superintendence. 2. Production scheduling. I. Bordoli, David. II. Title.
  TH438.4.B35 2014
  624.068′4–dc23
                                        2013043938

A catalogue record for this book is available from the British Library.

Contents

Appendices

Notes on Contributors

Simon Austin BSc, PhD, CEng, MICE

Simon Austin is Professor of Structural Engineering in the School of Civil and Building Engineering at Loughborough University. Prior to this, he worked for Scott Wilson Kirkpatrick & Partners and Tarmac Construction. He has undertaken industry-focused research for over 30 years into the design process, integrated working, value management, structural materials and their design. The latter includes the behaviour and design of structural elements, sprayed, cast and, most recently, 3D printed concretes. Most of this research has been funded by the EPSRC with collaboration from industry and the findings disseminated in over 200 publications. A strong believer in extending academic research into practice, Simon has served on various BSi and CEN standardisation committees and is a consultant member of two trade associations. In 2001, he co-founded Adept Management, a management consultancy specialising in design, development and engineering management. The company works with many large construction clients, designers and contractors, particularly helping in planning and process improvement.

Andrew Baldwin PhD, MSc, BSc (Hons), FICE, Eur Ing

Andrew Baldwin is an Emeritus Professor of Loughborough University where he was previously Professor of Construction Management in the School of Civil and Building Engineering. His background is Civil Engineering, and he worked extensively in the UK construction industry on major capital projects before embarking on an academic career. These capital projects included major roadworks, offshore engineering projects and major flood defence systems where he gained extensive planning and scheduling experience. His research interests have focused on construction planning, information modelling, process improvement and the development of new ways of working for both design and construction. He has worked in the United Kingdom, Hong Kong and mainland China. His last research management position at Loughborough University was as Director of the Innovative Manufacturing and Construction Research Centre (IMCRC), a major research centre which comprised some 50 academic staff engaged on a range of innovative research projects. He is currently a Co-Director of the National Centre for International Research of Low-Carbon and Green Buildings at Chongqing University, China.

David Bordoli BSc, MSc, FCIOB, MAPM, ACIArb
David Bordoli is an extremely experienced planning professional who began his career as a planning engineer with construction contractors following graduating in Construction Engineering in 1978. His first appointment as an expert witness was in 1989, where he used innovative network techniques to analyse project delays. In 1994 he returned to academic studies where he first met Professor Andrew Baldwin with whom he subsequently authored a number of articles and papers including 'A methodology for assessing construction project delays' which developed the analysis technique now known as 'Time Impact Analysis'.

In 2001 David left general contracting to work as a consultant, providing contractual advice, preparing time delay claims, reports for adjudications, arbitrations and litigation, and undertaking expert witness appointments in delay and disruption disputes in construction and engineering. In 2012 he was appointed a Director of Driver Consult and has recently spent most of his time working on overseas projects, particularly in South Africa.

Sam Ewuosho BSc (Hons)
Sam Ewuosho inherited an interest in engineering from his father who gained a BEng in Mechanical Engineering. However, a brief period with a local architectural firm at age 16 led him to undertake an undergraduate programme in Construction Management at the School of Civil and Building Engineering at Loughborough University. This programme included construction site experience with a leading UK construction organisation and a period of study in Hong Kong where he studied international real estate and was part of research task force that sought ways to transform a valuable but disused coastal stretch of the Hong Kong Special Authority region. He graduated from Loughborough in 2012 with a First Class Honours degree and is currently undergoing professional development with an international financial services and consulting organisation.

Alistair Gibb PhD, BSc, CEng, MICE, MCIOB
Alistair Gibb is the European Construction Institute (ECI) Royal Academy of Engineering Chair of Complex Project Management. He leads the ECI, a pan-European, evidence-based organisation, providing pragmatic, industry-focused evidence through research collaboration between industry and academia. The ECI, based at Loughborough, provides a knowledge network, with processes and programmes for disseminating, assessing and managing knowledge. Alistair joined Loughborough University in 1993, following a career in engineering and project management with John Laing, Taylor Woodrow and Sir Robert McAlpine. He leads Loughborough's Construction Health and Safety Research Unit, working closely with the Health and Safety Executive (HSE) and industrialists, maintaining a leading role in UK, European and international networks in health and safety.

Sarah-Jane Holmes BSc (Hons)

Since graduating from Loughborough University with a degree in Architectural Engineering and Design Management, Sarah-Jane Holmes has undertaken the role as an Environmental Advisor for a major contractor, Keepmoat Ltd. Her current role within the Environmental Team focuses on the implementation of the environmental management system, policies and procedures throughout a range of new-build and refurbishment projects. In particular, this focuses on waste management, more specifically site waste management plans, and broader environmental compliance issues on-site, through the creation and delivery of best practice guidance, environmental training and on-site auditing. Currently, she is working towards chartered membership of the Chartered Institute of Building (CIOB) and full IEMA membership of the Institute of Environmental Management & Assessment IEMA.

Baiyi Li PhD, BSc (Hons)

Baiyi Li graduated from Chongqing Jianzhu University (Chongqing University), China, in 1999. After a period of working in the local construction industry, he decided to secure a postgraduate degree. He completed his PhD at Loughborough University, UK, in 2008 under the supervision of Professor Simon Austin and Professor Tony Thorpe. In this research, a generic preconstruction planning process model with a method to support the management of preconstruction planning was developed and validated. With extensive construction experience, Baiyi Li is recognised as a leading expert in innovative construction planning techniques and their use on large capital projects including commercial centres, airport and new town development.

Mohamed Osmani BA (Hons), Dip Arch, MSc, RIBA, HEA

Mohamed Osmani is a Senior Lecturer in Architecture and Sustainable Construction at Loughborough University. He teaches on undergraduate programmes and postgraduate courses in the areas of architecture, sustainable building design and construction and CAD modelling and rendering. He has more than 10 years industrial experience as an architect and over 15 years as an academic. Mohamed has developed a significant portfolio of research projects and has been a member of numerous committees and task groups including the CIRIA Sustainability Advisory Panel, House of Lords Waste Enquiry, the UK Green Building Council Vision for Sustainable Built Environment, the Office of Government Commerce Construction and Refurbishment: Building a Future and the British Standards Institution (BSI).

Stacy Sinclair BA (Hons), MSc, RIBA, SCL, AS, DRBF, DBF

Stacy Sinclair, a solicitor at Fenwick Elliott LLP, advises on a broad range of construction and engineering issues. Before qualifying as a solicitor, Stacy practised as an architect, principally designing large-scale projects such as stadiums, hospitals and education buildings both in the United Kingdom and the United States. Stacy has a particular interest in Building Information Management (BIM) and its impact on the construction industry and regularly writes for *Building Magazine* and the *RIBA Journal*. She is the co-editor of the *Dictionary of Construction Terms* and is also a lecturer and oral examiner on the RIBA Part III postgraduate course at a number of universities.

Foreword

This is an excellent publication that will be welcomed by both practitioners and students.

Although the subject of planning and scheduling is a 'mature' academic subject and the basics well established, as with all aspects of construction practice, the requirements of the construction client and demands of the industry continually require a re-assessment of current practice.

This publication is timely. It reviews current practice, returning to the basics of the topics and reiterating the fundamentals. It then examines current planning and scheduling methods including the new methods of working that are emerging to meet the demands of both contractors and design managers. It also considers Building Information Management, (BIM) and its impact on planning and scheduling. Other additional topics relate to the need for sustainable construction and planning to meet the requirements of health and safety.

Regrettably the construction industry still regularly fails to meet the targets for the completion of projects on time and at cost. Section IV by David Bordoli is an excellent summary of how delay and disruption may be assessed both from the perspective of assessing the impact of delays and seeking compensation.

Andrew Baldwin and David Bordoli have a wealth of experience that is founded in management thinking and industry based. This means that the text focuses on the requirements of practitioners. The style of the text ensures that the detail required by the reader is easily accessible. The book may therefore be either a supporting text for an academic course or the reference book for the construction planner in industry. In addition to the knowledge of the main authors it includes contributions from a number of colleagues within the School of Civil and Building Engineering at Loughborough University, one of the leading universities in the United Kingdom.

I strongly recommend it to you.

Professor Li Baizhan
Director, National Centre for International Research
of Low-carbon and Green Buildings
Chongqing University, Chongqing
P.R. China

CHONGQING UNIVERSITY

Preface

Why another book about Construction Planning and Scheduling? Planning and scheduling is a 'mature' subject. The knowledge base is long established; there are many excellent texts specifically on the subject. Planning and scheduling is covered in many excellent project management and business management textbooks. Why another book?

Although the basis for construction planning and scheduling is long established and the subject firmly embedded in university and college teaching programmes, there is substantial evidence that most planners and schedulers are trained by experience 'on the job'. Planners and schedulers are, in the main, self-taught. This learning extends over time and like all industry-based learning needs to be supported by the knowledge and experience of others.

The aim of this handbook is therefore to present the key issues of planning and scheduling in a clear, concise and practical way in a readily acceptable format whereby individual chapters and sections can be accessed and read in isolation to provide a guide to good practice. Our objective was to provide a text to accompany learning, a reference document which, supported by web-based information, would provide information on the background to planning and scheduling together with guidance on best practice and practical methods for the application of construction planning and scheduling on different types of construction work. In addition to revisiting the basic elements of planning and scheduling, we have included chapters on current topics that are demanding consideration by all those within the construction industry. These include planning for sustainability, waste, health and safety and Building Information Modelling (BIM).

The book is divided into four sections.

The first section looks at planning and scheduling within the construction context. It provides both an outline of the evolution of planning and scheduling and a review of the basics: who plans, when and why. We consider the overall project cycle and then explore what the construction planner actually does and how the form of procurement adopted by the client impacts both the type of planning undertaken and when planning takes place. We complete the first section by looking at different construction management schools of thought and how these approaches influence how the managers of construction organisations plan, monitor and control construction projects.

The second section looks at planning and scheduling techniques and practice. There are numerous planning and scheduling techniques available to assist the construction planner. These have been developed over extended periods of time. We provide details of the basis of these techniques and then look at how they are used in practice and how they are adopted, adapted and utilised in practical situations. This section also looks at other aspects of planning such as how the cash flow for the contract may be calculated, the method statements that need to be produced and the uncertainty and the risks that may arise due to insufficient information.

The third section considers planning and scheduling methods and how the techniques described and discussed in Section II are incorporated into current ways of working including Critical Chain Project Management, Earned Value Analysis, Last Planner, ADePT (for planning the design process), BIM, Planning for Sustainability, Planning for Waste Management and Planning for Health Safety and the Environment.

Delays and disruption are an inevitable part of most construction projects. The fourth section, Delay and Forensic Analysis, looks at delay and disruption, their differences and how their impact on the original production schedule may be assessed. We look at the different approaches used and the information required in order that the analysis may be undertaken. Practitioners who specialise in this 'forensic analysis' have established protocols on how to approach their modelling and analysis. We look at the approach adopted by the Society of Construction Law Delay and Disruption Protocol and the Recommended Practice for Forensic Schedule Analysis produced by the Association for the Advancement of Cost Engineering International (AACEI). We outline both these protocols, their background and their guidance on method implementation, analysis evaluation and method analysis selection.

Analysing delays and disruption is seldom straightforward; a number of other issues may need to be taken into consideration. We look at issues including out-of-sequence progress; the effect of different types of calendars; the impact of abnormal weather; concurrent delays; the relatively new concept of pacing, mitigation, acceleration and the impact of different employer, contractor and subcontractor schedules. We define each of these and provide guidance on how to go about assessing the implications of each of these on the planner's analysis on the construction schedule.

Together, each of these sections provides a basis for the understanding of both the basics of planning and scheduling techniques and how they may be used in practice. We define planning and scheduling and differentiate between these two terms. Our research for the book identified that whilst there is no confusion over the meaning of 'planning', there is frequently discussion with respect to the terms 'scheduling' and 'programming'. Throughout the text, we have adopted the term 'schedule' in preference to 'programme'. In the United Kingdom and current and former Commonwealth countries, 'programme' was generally the preferred term. However, increasingly the original American term 'schedule' is being adopted throughout the world. We also note that in the United Kingdom, the term 'schedule' may also refer to a tabular list of information. For example, an 'information required schedule' is a tabular list of information items and dates by which the information is required by the project team. Terminology is always important.

To assist the practitioner, there is an extended glossary of terms in which the terminology used by practitioners is explained.

In writing the book, we have considered not only our own knowledge gleaned from industry experience and academic study but also the experience of many other industry practitioners and leading academics. We have reviewed conference and journal papers and considered recent research findings. It was never our intention to 're-write the subject' but rather to provide a handbook that included links to the important works of others. Here, we have revisited standard texts such as those of Frank Harris and Ron McCaffer and recognised highly rated works such as those by Michael Mawdesley, William Askew and Michael O'Reilly, Thomas Uher and Adam Zantis, and Brian Cooke and Peter Williams. We have also incorporated information from the current guides and best practice produced by professional institutions. These include several publications by the Chartered Institue of Building CIOB: the CIOB Guide to Estimating; the CIOB Guide to Good Practice in the Management of Time in Complex Projects; and the CIOB Code of Practice for Project Management for Construction and Development; all of which we believe the practitioner should always keep readily available for reference and guidance.

Acknowledgements

The handbook could not have been produced without the help and assistance of others. We should like to thank all those who have assisted in the preparation of material and the production of the book. First are the contributors who have provided contributions in the form of individual chapters: Simon Austin, Alistair Gibb, Baiyi Li, Mohamed Osmani, Sam Ewuosho, Sarah-Jane Holmes and Stacy Sinclair. Some are long-standing friends at Loughborough University, others are more recent colleagues. We are delighted that Stacy Sinclair of Fenwick Elliott LLP was willing to assist us with consideration of the legal issues relating to BIM and how this new way of working impacts the industry. For this important perspective we are extremely grateful.

Lean Construction is a way of working that is now firmly established in the construction industry. We should like to thank Glenn Ballard and Ian Mossmann for their assistance in providing background material for us to use within the text and also their time in reviewing drafts of the text. Their contribution has been invaluable to our understanding of not only how Lean Construction thinking has evolved but also the current perspectives.

More difficult to identify by name but no less valuable are the academic colleagues and industry practitioners who over the years have extended our knowledge and improved our thinking around the subject of planning and scheduling for construction. They too have all contributed to this text, even though they may not have been aware of their potential contribution when we discussed issues and problems in the context of the projects on which we were working. We nevertheless thank them for their time and perspectives on the problems under consideration. We should also like to thank Driver Trett, Loughborough University and the 'One Thousand Experts' programme for their support for the production of the book. Finally, we should like to thank Andy Mathers and Christine O'Mahony for their artwork and Madeleine Metcalfe and all the editorial and production team at Wiley-Blackwell for their time, patience and understanding.

Andrew Baldwin and David Bordoli
January 2014

About the Companion Website

This book's companion website is at

www.wiley.com/go/baldwin/constructionplansched

where you will find freely downloadable support materials.

Section I
Planning and Scheduling within the Construction Context

Introduction

This section comprises two chapters. Chapter 1 looks at why and when we plan. It considers different types of planning within the context of a construction project. It looks at the differences between Planning, Programming and Scheduling and the costs and benefits of undertaking these tasks on a project. We look at the planning process with the construction project cycle within the context of several different frameworks. Production in all industries requires planning. What are the distinct characteristics of planning in the construction industry? We look at how the clients of construction work procure the new assets and services that they require, what is unique to construction, what is similar to other industries, what has happened in the past and what may happen in the future. It is widely accepted that there are four functions of management – Planning, Organising, Control and Leadership – and that good planning is imperative for successful management. Chapter 2 looks at how planning is considered by the leading construction management schools of thought. We examine the position of planning in each of these perspectives and at the 'root theory' of each school, their current thinking, how they compare when considering the demands of complex construction projects and their perspective on the role of planning in the project process.

A Handbook for Construction Planning and Scheduling, First Edition. Andrew Baldwin and David Bordoli.
© 2014 John Wiley & Sons, Ltd. Published 2014 by John Wiley & Sons, Ltd.

Chapter 1
An Introduction to Planning and Scheduling

A brief history of planning and scheduling

Frederick Winslow Taylor was the founder of modern scientific management. His studies in the latter part of the 19th century formed the basis for management thinking in the 20th century and continue through to the present. Currie (1977) states that Taylor's work and philosophy may be seen in three major phases. First, he made improvements in the management of production. These sprang from his application of scientific methods. Second, he introduced systems of pay designed to produce 'a fair day's work for a fair day's pay'. Then, moving from the individual scale to the overall scale, he produced his 'grand design' for an industrial society. He hoped that this 'grand design' would lead to improved standards of living. His detailed, careful analysis of production tasks and functions led to new machines and tools, new methods of production control and stock control and new office procedures. Taylor's contribution to manufacturing production scheduling was establishing the planning office in a separate location away from the production area and the recognition that planning was a decision making process that required sharing of information. This and his other works attracted the attention of many other industrialists and professionals.

Henry L. Gantt (1861–1919) was a teacher, draughtsman, engineer and management consultant. He was contemporary and protégé of Taylor, and between 1887 and 1893 he worked with him in his experiments at the Midvale Steel Works (Currie, 1977). His contribution to manufacturing production management includes the application of scientific analysis to all aspects of production, the introduction of tasks and bonus systems where the bonus was linked to how well managers taught employees to improve performance and the social responsibility of business. Gantt focused on the motivation of workers and the application of knowledge to the advantage of all concerned with a

A Handbook for Construction Planning and Scheduling, First Edition. Andrew Baldwin and David Bordoli.
© 2014 John Wiley & Sons, Ltd. Published 2014 by John Wiley & Sons, Ltd.

business. He believed that business organisations had an obligation to the welfare of the society in which they operate and this directed much of his thinking. He developed the Gantt chart, a chart that allowed supervisors to identify and schedule the work of each worker and then review and assess the actual production. Gantt did *not* invent the bar chart, the concept of bar charts pre-date Gantt's work by at least a century. Gantt took existing methods of visually displaying work tasks and developed them to produce a new chart to form a visual statement of productivity. He also recognised the advantages of reducing inventory and clean, well laid out workspace and developed other management techniques (Weaver, 2012).

Critical path methods

For the first half of the twentieth century the bar chart was the dominant technique for planning and scheduling on projects of all sizes. This changed in the 1950s. A bar chart is excellent at showing *when* activities are scheduled to take place. However, it fails to show the *inter-relationships* between activities and the effect of delay in individual activities on the overall project. The decade of the 1950s included many major military, industrial and infrastructure projects both in the United Kingdom and the United States, and new systems were sought to manage these complex projects. Within the operational research community there was widespread interest in solving the problem of modelling the inter-relationship between the activities within a project.

By 1957 the Central Electricity Generating Board (CEGB) in the United Kingdom had developed a technique for 'identifying the longest irreducible sequence of events'. At the same time work in the United States, the U.S. Navy Special Projects Office was devising a means of planning and controlling complex projects. July 1958 saw the publication of a report entitled PERT, Summary Report, Phase 1 in which the technique entitled the 'Programme Evaluation Review Technique' (PERT) was proposed. In October 1958 it was decided to apply PERT to the Fleet Ballistic Missiles Programme. Meanwhile at the U.S. company E.I. du Pont de Nemours a technique called the Critical Path Method was under development. (For more details see Lockyer, 1974.)

Early successes of these techniques led to their widespread adoption by project managers, and the next decade saw the development of the techniques by researchers in academia and industry and their use across a range of projects. Originally the calculations were undertaken manually, then using computer programs operating on large mainframe computers. The generic term 'Critical Path Analysis' (CPA) arose emphasising the ability of the technique to identify the key activities that form the shortest duration for the project.

Two forms of the technique emerged: activity-on-arrow and the precedence method. By the early 1970s CPA was the *de facto* standard for planning and scheduling major projects and was adopted by both clients and contractors for project planning, monitoring and control. However, project managers soon discovered that adopting the technique did not guarantee the success of the project. For some projects the technique simply highlighted the problems. On others the technique (or rather its use) became the problem.

A report by NEDO compared construction performance on major engineering projects in the United Kingdom, Europe and the United States. It found 'That there was no correlation across the case studies between the sophistication with which programming was done and the end result in terms of successful completion on time'.

It became clear that CPA-based planning and scheduling systems were only an aid, albeit an important one, to the project manager and not a panacea for poor management. 'There was a general feeling that project planning was generally unsuccessful, that project planning using network planning was even more unsuccessful and that network planning using computers was the least successful of [all] techniques' (NEDO Report, 1983). Enthusiasm for the technique waned. Many project managers who were required contractually or by their organisations to use the technique paid only lip service to it.

Its use was resurrected by the introduction of the micro-computer, now generally known as the PC.

The impact of the PC

The introduction of the personal computer/micro computer provided cheap 'local' computing power for every office and every construction site. This meant that the time required for the preparation of plans and the production of bar chart schedules could be significantly reduced and they could become readily available to the construction team. 'The bar chart was no longer out of date before you pinned it to the wall' (Reiss, 1995).

The success of the IBM PC (introduced in 1981), its subsequent models and alternative computer products resulted in the production of many new software applications including new software for CPA. Soon, CPA software was re-packaged and marketed to all industries as a 'Project Management System'. New features were added. Some of these products adopted a new approach to inputting and displaying project data that was based on a bar chart format. This combined with improved facilities for producing and printing the output of schedules led to the 'linked bar chart' software product. This linked bar chart format became the preferred form of planning and scheduling for many planning engineers and led to a resurgence in the use of project management software.

With the opportunities of new computer systems came an awareness that the successful adoption of computer systems requires more than just hardware and software but consideration of data, procedures and people. There came a wider appreciation of the need to plan the implementation of systems around the users, not the computer hardware and related equipment. Moreover, there was a clear need to develop and work with collaborative systems whereby all parties involved in the project may contribute to the project planning scheduling and monitoring process. The boundaries between the technical innovations of information and communication technologies and the human aspects of systems adoption and performance became less distinctive. However, the success of project management systems in the overall management of construction projects remained inconsistent. This led to new systems and new thinking.

New systems and new thinking

The decade commencing 1990 saw the development of two important developments in planning and scheduling: Critical Chain Project Management (CCPM) and Last Planner. Both were the result of the realisation that, even with the cheap computing power and many additional features, the adoption of existing project management systems could not ensure project success.

CCPM focuses on the uncertainty in schedule activities and identifies the key activities that, based on time and resource constraints, form the 'critical chain' for the construction work. Rather than adopt traditional critical path methods that allow individual managers to create and use up buffer time relating to 'their' activities, CCPM creates a 'project buffer' and argues that production should monitor this buffer time on an on-going basis, always allocating resources to critical chain tasks. CCPM stresses the importance of focusing on the critical activities and the resources required to complete these activities. It argues that by monitoring the project buffer time you will ensure successful project completion. Advocates of CCPM claim that the introduction of the CCPM methodology ensures project success, reduces project durations, enables increased project throughput with no resource increases, and reduces manager and worker stress, all with minimal investment.

Last Planner was developed from research that concluded that even with the strict adoption of critical path planning techniques only 50% of the activities on a typical construction project were completed to schedule. One major shortcoming of CPM is that it is ill suited to direct production on site. Ballard and Howell (1992) argued that the CPM approach as a basis for production planning was fundamentally flawed and that production should only commence if all the resources required for the completion of an activity are available, that is, you should consider not only what *should* be done but what *can* be done. Introduced in 1992 the Last Planner System has become the platform for Lean Construction and is now fully recognised as a proven approach to production-based construction management.

New information and communication technologies

The last decade has seen the emergence and acceptance of Building Information Modelling (BIM) and Virtual Prototyping as the basis for the design, production and maintenance of many new buildings. These technologies together with a focus on sustainable building developments and new procurement requirements are influencing the thinking of both public and private clients who are demanding new standards and new ways of working.

The ability to model the building product and link the contents of the building model to other systems was first developed in the 1980s. With respect to construction planning this became known as 4D Planning and typically comprised the ability to link the elements and quantities from the computer model to project management software to introduce the dimension of time and generate simulations showing how the construction would proceed throughout the duration of the project. (Similarly, using product model data to analyse cost has become known as 5D planning.)

The use of digital product models for all aspects of building design and management is now known as building information modelling or BIM. The development of realistic graphical simulations for planning the broader aspects of physical and operational aspects of the building or facility is known as virtual prototyping or virtual construction. The benefits of these tools and techniques are already proven on large commercial buildings and infrastructure works, and they are increasingly being adopted on medium and smaller projects. In some countries they are a recognised part of the government procurement process. In 2011 the British Government announced their intention that, by 2016, all public sector contracts would be procured using BIM.

It is within this context of planning and scheduling in the twenty-first century that we need to consider fundamentals of planning and scheduling.

Planning

A plan is 'a formulated and especially detailed method by which a thing is to be done' (Oxford, 2002). This definition indicates the importance of working towards an objective and identifying how that objective will be achieved. Within a construction context the objective may be simply stated as the successful completion of the design and construction of the building or infrastructure. The 'plan' for the completion of any building, let alone a large commercial building or complex infrastructure works, will comprise a number of linked plans.

It is easy to think of planning as the production of a time schedule but this is only one aspect of successful project planning. There is always a need to consider planning in a wider context. Planning for a project must include not only consideration of time but also consideration of cost, quality, health and safety and other aspects such as design and production. Many construction writers and construction researchers have considered what constitutes planning from this broader perspective context and provided appropriate definitions. Here are some:

> Planning is 'the determination and communication of an intended course of action incorporating detailed methods showing time, place and the resources required' (CIOB, 2011).

> Planning is 'the creative and demanding mental activity of working out what has to be done, how, and when, by whom, and with what, i.e. doing the job in the mind' (Neale and Neale, 1989).

> Planning is 'a decision making process performed in advance of action which endeavours to design a desired future and effective ways of bringing it about' (Ackoff, 1970).

> Planning is 'the production of budgets, schedules, and other detailed specifications of the steps to be followed and the constraints to be obeyed in project execution' (Ballard and Howell, 1998b; Ballard, 2000).

There are many more. Whatever definition is chosen it is clear that a number of factors emerge:

- Planning precedes execution. (You plan before you commence work.)
- Planning is a process and it is important to complete all the stages in the process.

- Planning is more than an aid to the successful completion of the project; it is an essential part of the project.
- Planning is a creative and demanding mental activity.
- To plan you need to make decisions.
- The objective of planning is to ensure that things happen successfully.
- The output of planning comprises schedules and budgets and information for others to use.
- The results of planning have to be communicated to others.
- Having set in place plans it is necessary to monitor progress and, in the event of the unexpected or failing to achieve expected performance, re-plan.

Who plans?

Everyone should plan the work for which they are responsible. All those responsible for the management of construction work, client, designer, contractor and subcontractors, need to plan. Each party within a project will be working to a different schedule that reflects their own requirements. Planning takes place within different parts of the organisation and at all levels within the organisation. Within the client, designer or contractor's organisation there is the likelihood that different programmes for the same project will exist. Within the contractor's organisation the estimator, project manager, site agent, site engineer and subcontractor/gang leader will have different plans. What is important is that all parties and all within the organisation are working to integrated plans that contribute to the overall objectives of the project master plan.

Planning, programming and scheduling

The term programming is based on the traditional production of graphical, paper-based schedules. The production of such schedules requires the planner to identify and state the activities involved, the timing of activities and their durations. From this information a chart showing the activities involved in the project, key dates, material and equipment delivery dates, manpower requirements and when subcontractors would be involved in the site production, would be produced.

The term scheduling has now largely replaced programming. Scheduling has been defined as the process of 'quantifying the programme' and 'the production of computerised calculated dates and logic' (see Uher and Zantis, 2011; CIOB, 2011). Within this text we shall use the term scheduling to cover all aspects of the production of bar charts, networks, method statements and other material relating to the project.

Although the terms planning and scheduling are often used synonymously they are separate activities. Here the Chartered Institute of Building (CIOB), definition is useful: 'Project planning is an experienced-based art, a group process requiring contribution from all affected parties for its success. Scheduling is the science of using mathematical calculations and logic to predict when and where work is to be carried out in an efficient and time-effective sequence' (CIOB, 2011).

Planning requires decisions concerning:

- The overall strategy of how the work process is to be broken down for control
- How control is to be managed
- How design will be undertaken and by whom
- The methods to be used for construction
- The strategy for subcontracting and procurement
- The interfaces between the various participants
- The zones of operation and their interface
- Maximising efficiency of the project strategy with respect to cost and time
- The management of risk and opportunity (CIOB, 2011)

Having reached these decisions, scheduling is the *process* by which plans are prepared and presented to all those involved in the project. Scheduling involves answering new questions and making new decisions such as:

- When will the work be carried out?
- How long will it take?
- What level of resources will be required?

Scheduling is concerned with sequencing and timing. Sequencing and timing leads to considerations of time and cost. As more information becomes available and the project progresses, there will inevitably be a need to revise and amend such forecasts. Planning and scheduling is therefore an iterative process. A good schedule is more than simply a good graphical representation of activities and events. It must provide the basis for analysis and production. It must expose difficulties likely to occur in the future and facilitate re-organisation to overcome them. A good schedule must enable the unproductive time of both labour and machines to be minimised. It must be suitable for use as a control tool against which progress may be measured. The schedule must be sufficiently accurate to enable its use for forecasting material, manpower, machines and money requirements. It must show an efficient work method based on an optimal cost, bearing in mind the availability of the resources (Pilcher, 1992).

Having identified the differences between planning and scheduling, it is important to look at how and when these tasks are undertaken.

Planning may be an iterative process but the tasks of planning and scheduling should not be attempted concurrently. Planning should precede scheduling. Scheduling should never precede planning. It is not good practice to plan whilst scheduling. It is not a good practice to schedule whilst planning. Planning and scheduling therefore requires timing, organisation and discipline. On larger projects, where planning and scheduling will be separate tasks undertaken by different people, it is easier to differentiate between the two tasks, and the tendency to confuse the roles of planning and scheduling tasks is less likely to arise. However, for all of us it is always tempting, particularly when using computer-based tools, to start drawing up the schedule before having fully thought through the key elements, the relationships between them and the information that you wish to communicate. Avoiding this temptation will enable you to plan faster and produce better schedules.

The cost and benefits of planning

Mawdesley et al. (1997) review the costs and benefits of planning. They stress that all parties to the project can benefit from planning. The benefits for the construction client and architect/designer include:

- Established deadline dates for the release of information on the project
- The ability to forecast resource requirements and resource costs
- The ability to forecast the expenditure and payment schedules
- The ability to forecast the staffing levels
- The ability to provide information to the public and other third-parties
- Improved co-ordination of the work of the project team
- Co-ordination of the project with work on other projects within the client's/architect's portfolio

For the construction contractor the benefits include:

- Predicting the timing of activities and their sequence
- Predicting the total construction period
- Full consideration of the safety, quality and environmental impact of the construction work
- Evaluating risks and opportunities
- Providing a basis for the estimate
- Providing a basis for monitoring and control
- Predicting the contract cash flow and return on capital
- Providing a basis for claims
- Identifying when materials are required
- Minimising materials wastage
- Determining average and peak levels of materials demand
- Predicting labour, staff and plant resource levels (Mawdesley et al., 1997)

These two lists of benefits show the range of outcomes from planning that assist the main parties to the typical construction project. When committing resources to planning activities the anticipated benefits need to be carefully considered to ensure expectations meet reality. Both direct and indirect benefits need to be considered.

However, the benefits of planning may be many but they can only be achieved at a cost.

Planning requires time. Planning requires experience. Experienced planners are an expensive overhead to construction costs. To plan and communicate effectively requires a minimum commitment to computer equipment, software and technical support. (On larger projects this may extend to the provision of comprehensive information systems.) Good planning requires communication (via methods of working etc.), schedules, report, estimates, monitoring and control. These then are some of the costs of planning. They are not insignificant. Whilst the benefits of planning are widely recognised and the cost of little or no planning may be disruption, delay and late completion the issue of how much planning (and hence the cost of planning) is always a question for the project manager. Therefore, when committing resources to planning it is essential to have a clear understanding of the benefits, both direct and indirect.

Types of plans

There are three different types of plans: strategic plans, operational plans, and co-ordinating plans.

Any successful business operates under a corporate strategic plan. Many organisations have preferred suppliers and entered into strategic alliances with other organisations to operate in specific market areas or geographical regions. However, this type of planning is not the focus of this book. In this text we are concerned with project-based planning, the operational aspects of design and construction, how the design team plans and, in particular, how the construction organisation plans construction work at the pre-tender stage; the pre-contract stage; and the contract stage, including short-term planning and the monitoring of progress.

An activity of the mind

Planning is an activity of the mind. The implications of this are not always fully recognised.

It is widely accepted that planning is a task best undertaken by experienced practitioners. In the context of construction this means experienced construction professionals who have worked on a number of similar types of construction projects, ideally under different forms of contract in different geographical locations. Experts' work patterns reflect their experience. Studies of expertise in planning and scheduling reveal that experienced planners and experienced estimators work patterns are different from novices. Novices work systematically through the documentation provided. Experts work randomly in a peripatetic manner, they seek out the differences between the project under review and previous projects. Experience leads to faster conclusions but sometimes at a cost. Sometimes the solutions presented are not the optimal solution. Sometimes significant errors result and substantial additional costs are incurred when construction commences.

Ben Goldacre (2009) examined why clever people believe stupid things. He lists several traits that lead us all to wrong conclusions. These traits include randomness, regression to mean, the bias towards positive evidence, being biased by our prior beliefs, availability of information and social influences.

'Randomness' is the name given to the propensity of human beings to spot patterns where none exist. 'Regression to the mean' refers to the tendency when faced with extremes to revert to the middle view. When considering a new situation we all have a tendency to be biased towards the positive. We are also biased by our prior beliefs. (If the proposed method of working went well on the last project then inevitably we think the same method should be used on the next project.) Planners frequently suffer from lack of information and have to make assumptions. This may present changes later when new information emerges. New or different information may also present problems. All of us are often unduly influenced when information is made more 'available'. When we spot something different it tends to be given greater importance than existing information. This may unduly affect our decisions. We are all subject to social

Section I

influence of the company that we keep; ideas for new ways of working are easily influenced by the views of our peers.

Ben Goldacre (2009) sums this up as follows:

"We see patterns when there is only random noise" (p. 247)

"We identify causal relationships when there are none" (p. 247)

"We overvalue confirmatory information for any given hypothesis" (p. 248)

"We seek out confirmatory information for any given hypothesis" (p. 248)

"Our assessment of the quality of new information is biased by our previous beliefs" (p. 250)

We all need to be aware that these traits lead not only to 'Bad Science' but also to bad planning.

Planning for construction

Whatever the type of project under consideration, there is a need to plan. Whatever the type of project under consideration, there is a need to produce a schedule. Different industries have different characteristics. It is generally accepted that construction is 'different' from other industries, but how different? What is construction and what are the characteristics of construction projects?

The Oxford Dictionary (Thompson, 1996) defines construction as 'the act or mode of constructing, a thing constructed'. In the context of the built environment and the construction industry, the term construction is widely used to include all types of work ranging from house building to commercial building, infrastructure work, civil engineering and heavy industrial engineering. The importance of the construction industry to the economy is widely recognised. In a typical developed economy the industry will contribute around 10% to the Gross Domestic Product. Within the context of planning and scheduling what is included within the term construction is not of major importance. What is important is to consider the characteristics of construction projects and how these characteristics impact the planning and scheduling process.

A construction project is widely accepted as complex in nature. This is normally due to one or all of the following factors: the physical constraints, the size of the project, technical complexity, contractual arrangements, the range of client–consultant contractor relationships and the general 'one-off' nature of the project. Each project is different from other projects, and the environment in which a project occurs is always different and constantly changing. Construction projects are characterised by uncertainty. Winch (2002) focuses on this aspect of uncertainty and argues that this is central to the management of construction projects. It is this uncertainty that leads to problems with respect to information and information flow:

"The management of construction projects is a problem in information, or rather, a problem in the lack of information required for decision making. In order to keep the project roiling, decisions have to be made before all the information required for the decision is available" (p. 32).

It is in this context construction that planning and the planning process need to be seen.

The planning process in the project cycle

The main objective of planning is to ensure that things happen successfully. This requires objectives to be established, tasks to be identified and progress to be monitored. The project schedule provides the basis for measuring progress, the basis for regular review and an updating of the plan.

All projects have a life-cycle. This is commonly known as the Project Cycle. This cycle comprises a number of phases (or stages) from inception through completion to operation. The exact nature of these phases, the time span of each phase and hence the total time of the life cycle will vary depending on the type of project and the industry. Compare, for example, the phases of a project to implement a new computer system: feasibility study, system selection, training and implementation, with a project for planning, designing, constructing, commissioning, operating and decommissioning a nuclear power station. The former may be completed in a few months; the latter may take 60 years.

The CIOB Code of Practice for Project Management for Construction and Development (2010a) recognises eight stages in the lifetime of a project:

- Inception
- Feasibility
- Strategy
- Pre-construction
- Construction
- Engineering services commissioning
- Completion, handover and occupation
- Post-completion review/project close-out report

Each stage represents a point where a key decision must be made.

This is not the only framework for the project process. Figure 1.1 shows the CIOB framework together with that of the:

- Office of Government Commerce (OGC)
- British Standards BS6079-1: 2000
- British Property Federation (BPF)
- Royal Institute of British Architects (RIBA)

Reviewing Figure 1.1, it is evident that each project framework contains a number of stages that commence with inception and conclude with completion. The exact number of phases and the terminology used vary with the organisation that has developed the framework.

Within this text we shall consider in more detail two of the frameworks shown in Figure 1.1: the CIOB Code of Practice and The RIBA Plan of Work. To these we have added the Process Protocol, a framework that was produced following extensive research and development with contributors from academia and industry that provides a detailed map of information flow and information requirements. We also consider the PRINCE2 methodology.

Chartered Institute of Building (CIOB)	Office of Government Commerce (OGC)	British Standards BS6079–1:2000	Construction Industry Council (CIC) Work Stages	British Property Federation (BPF)	Royal Institute of British Architects (RIBA)
1. Inception	Gate 0 strategic assessment	1. Conception	1. Brief	1. Concept	1. Preparation
2. Feasability	Gate 1 business justification	2. Feasability	2. Feasability	2. Preparation of the brief	2. Concept design
3. Strategy	Gate 2 Procurement strategy Gate 3 Investment decision		3. Developed design	3. Design development	3. Developed design
4. Pre-construction			4. Production	4. Tender documentation and tendering	4. Production
5. Construction		3. Realisation	5. Installation	5. Construction	5. Specialist design
6. Engineering services commissioning	Gate 4 readiness for service				
7. Completion handover and occupation		4. Operation	6. As constructed		6. Construction (on and off-site)
8. Post-completion review / project close-out report	Gate 5 benefits evaluation	5. Termination	7. In use		7. Use and aftercare

Figure 1.1 The life-cycle of a project – six frameworks.

PRINCE2

PRINCE2 is a project management methodology. The name stands for: Projects in Controlled Environments. The number 2 highlights the second version of the methodology which has become the basis for all subsequent versions.

PRINCE2 was originally developed by a UK government agency for the management of Information Technology (IT) projects. As a major client for the procurement of a wide range of goods and services, the UK government clearly has a need to adopt a consistent approach and framework to project delivery, and PRINCE2 is now mandated for the management of many different types of UK government projects. It provides a structured approach to project management aimed at the efficient control of resources. The methods and their terminology provide a common language for all involved in the project. PRINCE2 adopts the following basic principles: continued business justification, manage by stages, management by exception, focus on products and tailor to suit the context of the project environment. Plans (Planning) is one of the seven key themes, along with Providing and Maintaining the Business Case, Organisation, Quality, Risk, Change, and Monitoring Progress. The methodology is based on the key stages of Starting Up a Project, Initiating a Project, Managing Stage Boundaries, Managing Product Delivery and Closing a Project. These are supported by the processes of Directing a Project and Controlling a Stage. Planning supports all of these processes but is particularly important for Initiating a Project, Managing Product Delivery by Controlling each Stage and for Managing Stage Boundaries. Planning and producing a Project Schedule are also integral parts of Managing Product Delivery.

PRINCE2 adopts a Product-Based Planning approach requiring the user to identify all the products or project deliverables that are required if the project is to meet its objectives. The project team is required to produce: a product-breakdown structure, a product flow diagram and a work-breakdown structure (WBS). The product flow diagram typically includes multiple and complex parallel paths. This is essentially the same concept as a Precedence chart or PERT chart used for critical path scheduling. From this flow diagram the project schedule is produced. The PRINCE2 Methodology emphasises project control: an 'organised and controlled' start, an 'organised and controlled' middle and an 'organised and controlled' end. Each project has a Project Board comprising representatives of the 'customers', the 'suppliers', and 'service suppliers'. The Project Manager for the project reports regularly to this Board and the main focus is the delivery of the product on time and to cost.

Further details of PRINCE2 may be found in the sources of information provided at the website for this book: www.wiley.com/go/baldwin/constructionplansched

CIOB code of practice for project management for construction and development

Within this framework there are a number of key actions. These are summarised in Table 1.1.

With respect to planning and scheduling the CIOB Code of Practice highlights the need for a project execution plan (PEP) which it identifies as 'the core document for the management of the project' (CIOB, 2010b).

The project execution plan, PEP

The Code of Practice emphasises that the PEP should:

'Include plans, procedures and control processes for project implementation and for monitoring and reporting progress'. It should 'define the role and responsibilities of all project participants', and provide 'a means of ensuring that everyone understands, accepts and carries out their responsibilities'. It should 'set out the mechanisms for quality control, audit, review and feedback, by defining the reporting and meeting requirements, and, where appropriate, the criteria for independent external review (CIOB, 2010).

Essential contents of the PEP include:

- The project definition and brief
- The statement of objectives
- The business plan: with costs, revenues and cash flow projections including borrowings, interest and tax calculations
- Details of market predictions and assumptions with respect to the likely revenue and return

Section I

Table 1.1 CIOB code of practice for project management for construction and development: stages and key actions.

Stage	Key actions
1. Inception	The business decision by the client that confirms that a new facility may be required. The commissioning of a project manager to examine the feasibility of the project.
2. Feasibility	A broad ranging assessment of the feasibility of the project. This is undertaken with input from a number of experts who examine all aspects of the proposed facility.
3. Strategy	Establish the project objectives, approach and procedures. Select key team members. Check procedures for ensuring sustainability and environmental issues. Determine the overall procurement approach. Establish all necessary control systems and the means for controlling project value.
4. Pre-construction	Design Development. Principle decisions relating to time, quality and cost management. Secure statutory approvals and consents and the provision of all utilities. Provide all the necessary information for construction to begin.
5. Construction	Construct the building and/or facilities required. Control cost and time within the parameters of the project objectives. Meet environmental performance targets.
6. Engineering services commissioning	Ensure that all operational and statutory inspections and approvals have been satisfactorily completed. Ensure the provision of proper records, test results, certification etc. Arrange for advice on maintenance staff training.
7. Completion handover and occupation	Handover the building and/or facility to the client. Facilitate occupation of the building.
8. Post-completion review/ project close-out report	Evaluate the performance of the project team. Identify lessons learned. A careful, objective review.

Adapted from the Charted Institute of Building 'Code of Practice for Project Management and Construction Development' (2010).

- Functional and aesthetic brief
- Client management and limits of authority including the project manager
- Financial procedures and details of delegated authority to place orders
- A full development strategy and procurement route
- Statutory proposals
- Risk assessment
- Project planning and phasing
- The scope of content of each consultant appointment
- Reconciled concept design and budget
- Method statement for design development, package design and tendering, construction, commissioning and handover and operation
- Safety and environmental issues such as the construction design and management regulations, carbon dioxide emissions and energy targets
- Management of information systems including document management systems

- Quality assurance
- Post-project evaluation

The PEP will change during the design and construction process and therefore should be viewed as an on-going tool for review and communication. The code identifies the key role of the project master schedule (the code prefers the term 'programme') that needs to be developed and agreed by the client and consultants. 'It is against this document that the project manager must monitor the progress of the project, assess risks to progress and initiate necessary action to rectify potential or actual non-compliance' (CIOB, 2010).

Figure 1.2 summarises the code of practice perspective of the project planning required for a project.

The RIBA plan of work

The RIBA Plan of Work, first published in 1963, provides a model for the design team and a basis for managing the design and administration of the building project. It is an established framework within which the client, designer and contractor can plan and schedule their contribution to the project. Since its initial publication it has been revised and amended on several occasions.

The 2007 Plan of Work comprised five phases: preparation, design, pre-construction, construction and use. Within each of these phases there are 11 work stages (A–L). For each of these stages guidance is given as to the key tasks to be undertaken by the design team together with the outputs and the controls required (RIBA, 2007).

In 2013 the RIBA introduced a new Plan of Work. The new plan reflects the increasingly complex construction landscape including:

- UK Government Construction Strategy
- Changing procurement processes
- Need for earlier collaboration within the project team
- Improved Client Briefing
- The importance of project handover and post occupancy
- Increasing use of information management including BIM

The new plan of work comprises eight stages: Strategic Definition; Preparation and Brief; Concept Design; Developed Design; Technical Design; Construction; Handover and Closeout; and In Use. These stages are numbered 0–7.

The tasks within each stage are identified within eight task bars: Core Objectives; Procurement; Programme; (Town) Planning; Suggested Key Support Tasks; Sustainability Checkpoints; Information Exchange; and UK Government Information Exchanges.

Table 1.2 and 1.3 identify the core objectives and key support tasks at each stage. For the full details of the tasks within the eight task bars see RIBA Plan of Work (2013).

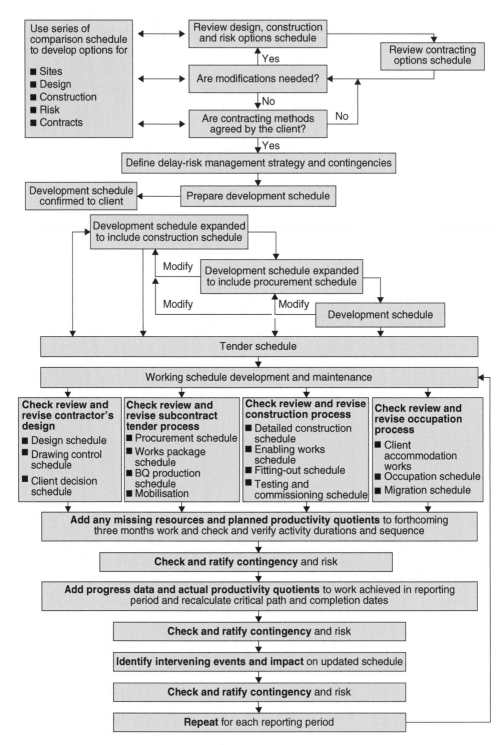

Figure 1.2 The project planning required for a project (CIOB, 2010).

Table 1.2   RIBA Plan of Work 2013, Stages 0–3.

Stage 0 – Strategic definition	Stage 1 – Preparation and Brief	Stage 2 – Concept Design	Stage 3 – Developed Design
Identify client's Business Case and Strategic Brief and other core project requirements.	Develop Project Objectives, including Quality Objectives and Project Outcomes, Sustainability Aspirations, Project Budget, and Initial Project Brief. Undertake Feasibility Studies and a review of Site Information.	Prepare Concept Design, and preliminary Cost Information along with relevant Project Strategies in accordance with the Design Programme. Agree alterations to the brief and issue a Final Project Brief.	Prepare Developed Design, Cost Information and Project Strategies in accordance with the Design Programme.
Review Feedback from previous projects.	Prepare Handover Strategy and Risk Assessments. Agree Schedule of Services, Design Responsibility Matrix and Information Exchanges and prepare a Project Execution Plan including Technology and Communication Strategies and Standards.	Prepare Sustainability Strategy, Maintenance and Operational Strategy and review Handover Strategy and Risk Assessments. Identify any Research and Development aspects. Review and update the Project Execution Plan. Consider Construction Strategy, including offsite fabrication, and develop the Health and Safety Strategy.	Review and update Sustainability, Maintenance and Operational and Handover Strategies and Risk Assessments. Conclude Research and Development aspects. Review and update the Project Execution Plan, including Change Control Procedures. Review and update Construction and Health and Safety Strategies.

(Data Summary from RIBA, 2013).

The exact nature of these tasks and their sequencing will vary depending upon the procurement method selected. It is the aim of the new plan of work to provide a straight forward mapping and flexibility for all forms of procurement. Sinclair has produced a BIM Overlay to the RIBA Plan of Work (Sinclair, 2012).

The process protocol map

The Process Protocol Map illustrates the design and construction process in terms of the various phases of development, the main participants, the deliverables to be produced and how the process may be/is managed through a series of phase reviews. The map is applicable for a wide range of construction projects. The process is divided into ten phases covering aspects of the project lifecycle from the demonstration and conception of need to the operation and maintenance of the

Section I

Table 1.3   RIBA Plan of Work 2013, Stages 4–7.

Stage 4 – Technical Design	Stage 5 – Construction	Stage 6 – Handover and Close Out	Stage 7 – In Use
Prepare Technical Design in accordance with Design Responsibility Matrix and Project Strategies to include all architectural, structural and building services information, specialist subcontractor design and specifications, in accordance with the Design Programme.	Offsite manufacturing and onsite Construction in accordance with the Construction Programme and resolution of Design Queries from site.	Handover of building and conclusion of Building Contract.	Undertake In Use services in accordance with the Schedule of Services
Review and update Sustainability, Maintenance and Operational and Handover Strategies and Risk Assessments.	Review and update Sustainability Strategy and implement Handover Strategy, including agreement of information required for commissioning, training, handover, asset management, future monitoring and maintenance and the ongoing compilation of 'As-constructed' Information.	Carry out activities listed in Handover Strategy including Feedback for use during the future life of the building or on future projects.	Conclude activities listed in Handover Strategy including Post-occupancy Evaluation, review of Project Performance, Project Outcomes and Research and Development aspects.
Prepare and submit Building Regulations and other third party consent submissions.		Update the Project Information as required.	
Review and update the Project Execution Plan.	Update Construction and Health and Safety Strategies.		Update the Project Information, as required.
Review the Construction Strategy, including sequencing, and update Health and Safety Strategy.			

(Data summary from RIBA, 2013).

constructed facility. The role of the project participants is integrated into a number of teams covering the whole supply chain. These teams are responsible for producing the deliverables of the process in terms of documents, designs or simply pieces of information. The 'gates' between the phases are categorised as 'soft' or 'hard', to distinguish when phases may commence with incomplete information and when information must be complete before the next phase may commence.

Table 1.4 shows the structure of the Process Protocol Map.

Figure 1.3 shows an extract from the Process Protocol Map.

The processes within the map are sub-divided into two additional levels of detail. These provide a basis for planning and management. A key principle of the process protocol is that it may be customised to meet the requirements of specific projects and the teams involved – it is a flexible system that enables the alignment of project process with existing and new business and operational procedures (Cooper et al., 1998).

Table 1.4 The process protocol map – Overall framework.

	Phase zero	Phase one	Phase two	Phase three	Phase four	Phase five	Phase six	Phase seven	Phase eight	Phase nine
	Demonstrating the need	Conception of need	Outline feasibility	Substantial feasibility study and outline financial authority	Outline conceptual design	Full conceptual design	Detailed design, procurement and full financial authority	Production information	Construction	Operation and maintenance
Development management										
Resource management										
Design management										
Production management										
Facilities management										
Health, safety, statutory Legal management										
Process management										

The columns represent the phases of the project. The rows represent the management themes.

Section I

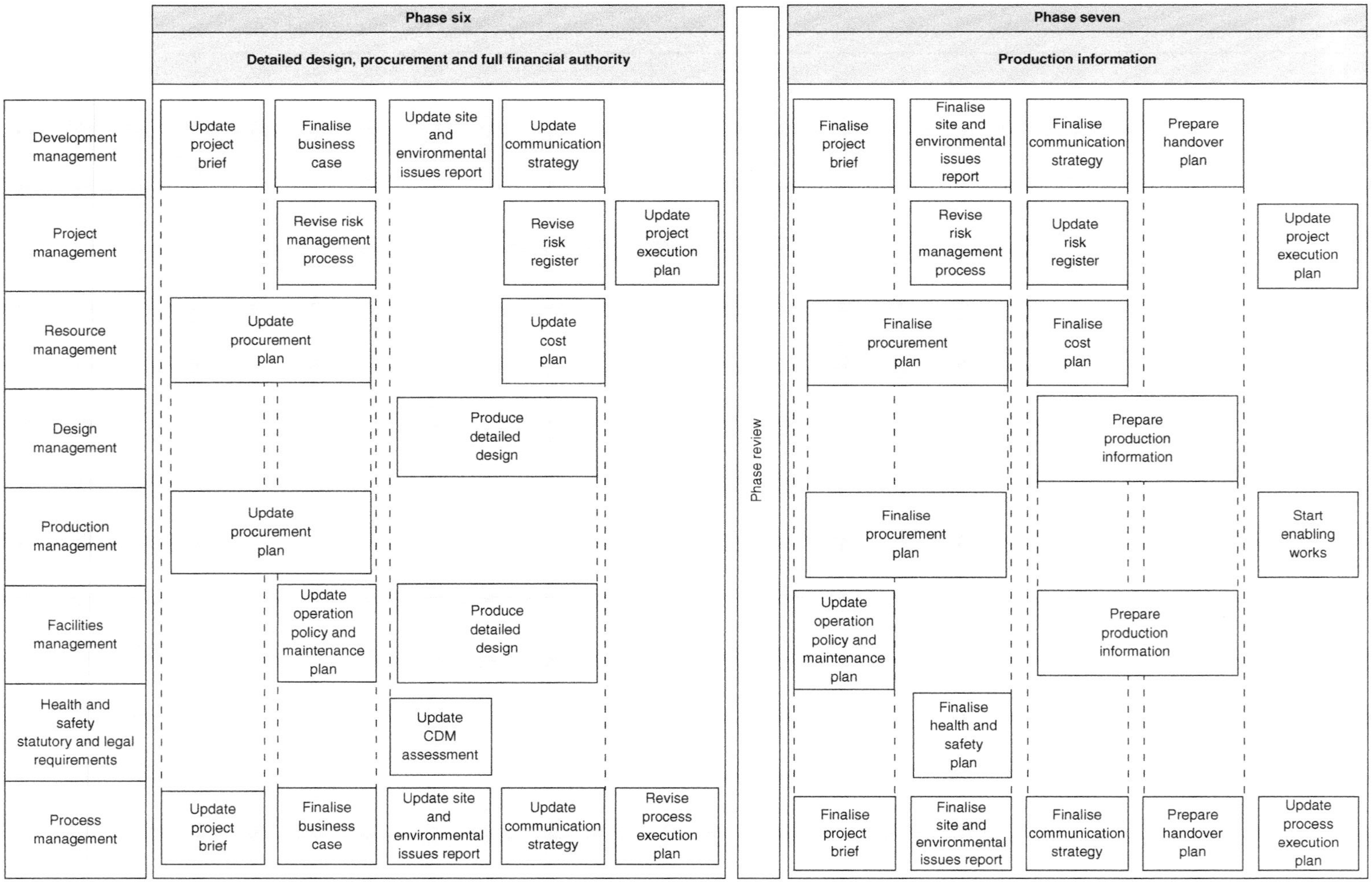

Section I

Phase six

Detailed design, procurement and full financial authority

Phase seven

Production information

Phase review

Row labels:
- Development management
- Project management
- Resource management
- Design management
- Production management
- Facilities management
- Health and safety statutory and legal requirements
- Process management

Phase six boxes:

- Update project brief
- Finalise business case
- Update site and environmental issues report
- Update communication strategy
- Update project execution plan
- Revise risk management process
- Revise risk register
- Update procurement plan
- Update cost plan
- Produce detailed design
- Update procurement plan
- Produce detailed design
- Update operation policy and maintenance plan
- Update CDM assessment
- Update site and environmental issues report
- Finalise business case
- Update communication strategy
- Update project brief
- Revise process execution plan

Phase seven boxes:

- Finalise project brief
- Finalise site and environmental issues report
- Prepare handover plan
- Finalise communication strategy
- Update project execution plan
- Revise risk management process
- Update risk register
- Finalise procurement plan
- Finalise cost plan
- Prepare production information
- Start enabling works
- Finalise procurement plan
- Prepare production information
- Update operation policy and maintenance plan
- Finalise health and safety plan
- Finalise site and environmental issues report
- Finalise project brief
- Finalise communication strategy
- Prepare handover plan
- Update process execution plan

Figure 1.3 Extract from the Process Protocol Map.

Summary

The CIOB Code of Practice, The RIBA Plan of Work and The Process Protocol Map all provide a framework for identifying the planning and scheduling required on a project, a basis for determining and identifying both internal and external requirements and combining these within a project master schedule. PRINCE2 provides a product-based methodology. The importance of these models is to identify different stages of the construction project and the decisions that need to be made to assist in determining how the planning will progress and how the 'diffusion of information' should lead to action and feedback. They highlight that the planning process of the construction client and the construction contractor are different. This difference is shown in Table 1.5 and Table 1.6.

Table 1.5 shows the planning process of the construction client. Table 1.6 the planning process of the construction contractor.

For individual projects the planning and scheduling requirements will also vary depending upon the form of procurement adopted by the client.

Table 1.5   The planning process of the construction client.

Project plan	Action
1. During the design stage	Appraise options
	Confirm business case
	Develop project strategy
	Prepare strategic brief
	Assemble team
	Devise risk management plan
	Choose procurement arrangement
	Risk assessments and pre-construction information
	Health and safety plan
	Commence health and safety file
	Prepare client programme (master schedule)
	Budget and cash flow
	Pre-qualify contractors
	Organise and administer tender stage
	Check tenders
	Choose preferred bid
	Prepare contract documents
	Sign contract
2. During the construction stage	Pre-start meeting
	Check bonds and Insurances
	Check construction health and safety Plan
	Contract administration
	Make contractor payments
	Monitor progress
	Report to the client
	Handover health and safety file
	Administer defects liability period
	Sign-off final account

Adapted from Cooke and Williams (2009).

Section I

Table 1.6   The planning process of the construction contractor.

Project plan	Action
1. Pre-tender planning	Decision to tender
	Pre-tender arrangements
	Site visit report
	Enquiries to subcontractors and suppliers
	Tender Method Statement
	Build-up estimate
	Pre-tender programme
	Build up preliminaries
	Response to pre-construction Health and Safety information
	Tender risk assessment
	Management adjudication
	Analysis of tender performance
2. Pre-contract planning	Pre-contract meeting and arrangements for commencing work
	Place subcontractor orders
	Site layout planning
	Construction method statement
	Master programme
	Requirement schedules
	Contract budget forecasts
	Risk Assessment
	Preparation and approval of the construction health and safety plan
3. Contract planning	Monthly Planning (long term)
	Weekly Planning (short term)
	Progress Reporting
	Cost and Value Reconciliation
	Report to management
	Review/update the health and safety plan

Adapted from Cooke and Williams (2009).

How is the planning process affected by procurement?

Procurement is the name given to the process of identification, selection and commissioning of the contributions required for the construction phase of the project (CIOB, 2011). Different forms of procurement result in different organisational and contractual arrangements. They affect what is done, by whom and when. They do not direct the planning and scheduling for construction but determine the basis for both the agreements between the parties and the process by which the project proceeds. As the project proceeds, information is required by the different parties to the contract in order that they may plan the work needed to meet their contractual requirements.

Procurement methods fall broadly into four categories:

- Traditional
- Design and build

■ Management Contracting and
■ Construction Management

Within each of these forms of procurement there are different variations.

Figure 1.4 shows how to select a procurement route.

The procurement route selected will be determined by the client's perspective on a number of key considerations. The CIOB (2010) recommends that clients examine a number of factors when deciding which route to select. Table 1.7 provides an outline comparison of these different forms of procurement, considering the characteristics of procurement.

Section I

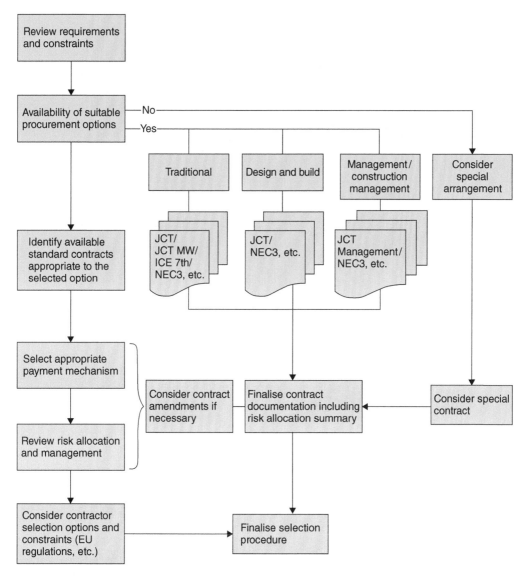

Figure 1.4   How to select a procurement route (CIOB, 2010).

Table 1.7 Characteristics of different procurement options.

Characteristic	Traditional	Design and build	Management contracting	Construction management
1. Diversity of responsibility	Moderate	Limited	Large	Large
2. Size of market from which costs can be tested	Moderate	Limited	Moderate	Large
3. Timing of predicted cost certainty	Moderate	Early	Late	Late
4. Need for the precise definition of client requirements	Yes	Yes	No	No
5. Availability of independent assistance in developing the design brief	Yes	No	Yes	Yes
6. Speed of mobilisation	Slow	Fast	Fast	Fast
7. Flexibility of implementing changes	Reasonable	Limited	Reasonable	Good
8. Availability of recognised standard documentation	Yes	Yes	Yes	Limited
9. Ability to develop proposals progressively with limited and progressive commitment	Reasonable	Limited	Reasonable	Good
10. Cost-monitoring provision	Good	Poor	Reasonable	Good
11. Input of construction expertise to design	Moderate	Good	Good	Good
12. Management of design production programme	Poor	Good	Good	Good
13. Client influence on trade contractors	Limited	None	Good	Good
14. Provision for controlling the quality of construction materials and workmanship	Moderate	Moderate	Moderate	Good
15. Opportunity for the contractor to exploit cash flow	Yes	Yes	Yes	No
16. Financial incentive for the contractor to manage effectively	Strong	Strong	Weak	Minimal
17. Propensity for confrontation	High	Moderate	Moderate	Minimal

Reproduced from Appendix 15 CIOB Code of Practice for Project Management (2010).

The forms of procurement and the range of options available to construction clients do not determine how planning and scheduling are performed but they set the context for the planning and scheduling. This is discussed in the following example. (Further sources of information on different forms of procurement and tendering procedures are included at the website for this book:www.wiley.com/go/baldwin/constructionplansched)

The context of construction project planning

Figure 1.5 is developed from a case study of a major hospital project in West London using a construction management procurement route and shows the overall planning process.

The client's project manager develops the *client's strategic programme* that drives the *tender programme* for the procurement of resource bases such as design consultants and construction manager and also the *architect's design programme*. The contractually binding agreement between the client and the construction manager is the *master programme*. The construction manager's project managers then prepare the *target construction programme* which guides the procurement of trade packages and the *parcel of documents programme* which drives the production of drawings by the architect. This last schedule is non-contractual, but can be used in claims for delays against the client caused by the non-delivery of drawings. Within the target programme trade contractors are given 'windows' for the execution of their responsibilities on site in the *tender restraints programme* and these are formally agreed in the *works contractor's programmes*. Within these programmes the trade contractors then schedule task execution at the level of the WBS that suits them. This is communicated to the construction manager in the

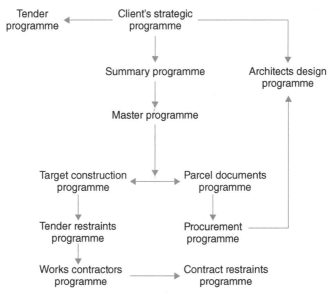

Figure 1.5   The planning hierarchy on a construction project.

contract restraints programme. In order to gain more bargaining power construction managers may or may not reveal the *master programme* to the trade contractors and provide a *target construction programme* that is significantly tighter than the *master programme* so as to buffer the completion date to ensure a satisfied client even if the *work's contractor's programme* slips behind the target completion.

This case study relates to one specific project procured under the construction management procurement process. It leads to a more general question. How does procurement impact overall industry performance?

Procurement and the performance of the UK construction industry

In the United Kingdom there has long been discussion and debate as to how the form of procurement impacts the time taken to complete the work, and the cost and quality of the construction product. This discussion and debate has extended for over 50 years.

The 1964 'Banwell Report' (HMSO, 1964) argued that existing contractual and professional conventions did not allow the flexibility required for a 'modern' construction industry. Banwell considered that change within the industry would be difficult to effect 'until those engaged in the industry themselves think and act together' (HMSO, 1964). He emphasised that 'The industry must think and act as a whole' and argued that the separation of design from construction needed to be overcome if there was going to be overall improvement in the industry and also argued that here was also a need to agree a common form of contract for building and a common form of contract for civil engineering. This it was hoped would lead to a joint form of contract for building and civil engineering works (Lord, 2008).

The recommendations of Banwell failed to introduce major change in the industry. Whilst some were willing to embrace change the majority were not. Despite initiatives by The National Economic Development Office the conservatism of the industry and 'market forces' prevailed.

At the start of the 1990s the UK construction industry was again encouraged to become more productive, more competitive and less adversarial. In 1994 Sir Michael Latham, a former Conservative MP and ex-Director of the UK Housebuilders' Federation presented a report entitled 'Constructing the Team' (Latham, 1994). Latham had been commissioned by Government to end the culture of conflict and inefficiency in the construction industry. The report was well received. It made 30 main observations and recommendations for improving the industry. The main emphasis was teamwork and the creation of working relationships where all parties benefitted. A number of recommendations had implications for the planning scheduling and control of construction works. These included:

- Improved tendering arrangements;
- The evaluation of tenders on the basis of not only price but quality;
- The adoption of partnering arrangements;
- The implementation of proposals for productivity improvements;

- The requirement for trust funds to ensure that construction companies receive payment for the completed work;
- A wider adoption of 'Alternative Dispute Resolution' methods;
- Greater use of procurement methods such as Design and Build.

One result of the Latham Report was the formation of the Construction Industry Board. However, the introduction and adoption of these changes was slower than Latham expected.

The Egan report (1998)

In 1998 Sir John Egan presented his report 'Rethinking Construction' (Egan, 1998). This report had been prepared by a Task Force headed by Egan at the request of the Deputy Prime Minister who had requested Egan to consider the scope for improving the quality and efficiency of UK construction. In his foreword to the report Egan recognised that 'a successful construction industry is essential to all of us' and that substantial improvements in productivity and efficiency were vital 'if the industry is to satisfy all its customers and reap the benefits of becoming a world leader'. The proposals for improving performance included experience gained in other industries where the previous decade had seen considerable improvements in both productivity and the quality of the final product. The report again emphasised the role of teamwork. Through the Task Force the major clients of the construction industry committed themselves to change.

The Report identified five key drivers of change needed to set the agenda for the construction industry at large. These drivers were 'committed leadership, a focus on the customer, integrated processes and teams, a quality driven agenda and commitment to people' (paragraph 17). It set ambitious targets for annual improvements including 'annual reductions of 10% in construction cost and construction time'. It also proposed that 'defects in projects should be reduced by 20% per year' (paragraphs 23–26). To achieve these targets it recognised that 'the industry would need to make radical changes to the processes through which it delivers its projects'. It recommended that 'the industry should create an integrated project process around the four key elements of product development, project implementation, partnering the supply chain and production of components. Sustained improvement should then be delivered through use of techniques for eliminating waste and increasing value for the customer' (Chapter 3). The report stated that 'if the industry were to achieve its full potential, substantial changes in its culture and structure would be required to support improvement' and that 'the industry must replace competitive tendering with long term relationships based on clear measurement of performance and sustained improvements in quality and efficiency' (paragraphs 67–71). The report placed emphasis on the responsibilities of the construction client, stating that 'major clients of the construction industry must give leadership by implementing projects that will demonstrate the approach to construction described in the report and that clients, including those from across the public sector, should join together in sponsoring demonstration projects to make improved performance available to all the clients of construction'. It concluded, 'In sum, we propose to initiate a movement for change in the construction industry, for radical improvement in the process of construction. This movement will be the means of sustaining improvement and sharing learning' (paragraph 84).

In 2002 Eagan indicated how the proposed changes could be accelerated (Strategic Forum for Construction, 2002). Banwell, Latham, Egan. All these have emphasised the need for closer co-operation and a less adversarial way of working. Their thinking led to the emergence of Partnering as a way collaboration.

Partnering

Partnering is 'a contractual arrangement between the two parties for either a specific length of time or for an indefinite period. The parties agree to work together, in a relationship of trust, to achieve specific primary objectives by maximising the effectiveness of each participant's resources and expertise. It is not limited to a particular project' (Latham Report, 1994: 6.43).

Latham recognised that 'Good relationships based on mutual trust benefit clients' and that provided they did not become 'cosy' long term relationships between main contractors and subcontractors could improve construction performance (in terms of not only time but quality) and reduce costs for clients. He anticipated that where major work programmes were under development partnering could prove beneficial by building and maintaining construction teams. Latham proposed that 'specific advice should be given to public authorities so that they can experiment with partnering arrangements [and] where appropriate long-term relationships can be built up' (Latham Report, 1994).

Following Latham the requirement for partnering increased to become a recognised part of the construction industry's ways of working. Several large client organisations fully embraced the new ways of working to increase productivity and produce innovative solutions to the procurement and provision of new facilities and new services. This led to not only project partnering but also strategic partnering and the concept of preferred suppliers and preferred supply chains.

Where partnering is adopted for a single, 'one-off' project, this necessitates new ways of working. It is necessary to establish agreed and understood mutual objectives, determine a methodology for quick and co-operative problem resolution and develop a culture for continuous, measured improvement. Partnering features 'open book' working practices and relationships (Kirkham, 2007).

Client's requirements for contractors to adopt partnering led to new requirements for participating contractors and subcontractors. Partnering meetings facilitated by independent parties in the pre-contract and the contract phase have become a common requirement. At these meetings issues relating to identifying roles and responsibilities are discussed and agreed. Such meetings usually include representatives of all the stakeholders and extend at the pre-contract stage to cover Value Management and Risk Management.

The increased adoption of partnering (and its variants) has led to a number of new forms of contract and legal frameworks including PPC2000 (ACA, 2008) and the ICE Partnering Addendum (ICE Conditions of Contract Partnering Addendum). Where the parties do not wish to enter into a legally binding agreement but wish to create a collaborative working environment charters such as the

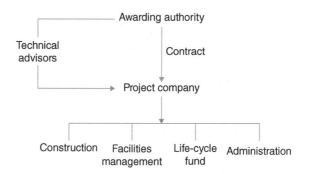

Figure 1.6   The basis of a typical PFI arrangement (Brook, 2008).

PC/N 2011 Partnering Charter (Non-Binding) developed by the Joint Contracts Tribunal may be used.

Public sector construction procurement: The private finance initiative (PFI)

The partnering approach is only one of a number of new ways of working that have emerged over the last 20 years. New forms of procurement have also emerged.

In 1992 the British Government introduced the private finance initiative, PFI. The primary objective of PFI was 'to encourage private investment in major public building projects like schools, prisons, hospitals and roads. PFI is fundamentally different to other methods of procurement in that it is exclusively used for the delivery of public buildings [and infrastructure]; and the procurement involves not only the design and the construction of the building [facility] but also the provision of services within it over a predetermined period known as the concession period. PFI is advocated as a method of risk transfer in capital procurement; the private investment implies the level of government borrowing falls and that risk is transferred from the public to the private sector' (Kirkham, 2007).

PFI (also known as Public Private Partnership, PPP) has emerged as a major form of procurement and attracted a great deal of debate. The procedures for PFI are long and complex and fall outside of the scope of this book. Figure 1.6 shows diagrammatically the basis of a typical PFI arrangement.

(Further sources of information on the different reports made on the UK construction industry are included at the website for this book: www.wiley.com/go/baldwin/constructionplansched).

Having reviewed the context in which construction planners work, this chapter ends by looking at what construction planners actually do and what is current construction planning practice.

What do construction planners do?

To answer this question Winch and Kelsey (2005) interviewed 18 planners from five leading UK construction firms. Most of these planners were currently involved in planning at the pre-tender stage including the assembly and presentation of the tender documentation. About half were involved at the post-tender pre-construction stage. Only a few were further involved during the site works.

Generally, although a number of the interviewed planners had on-site experience the typical pattern is that a planner works either at the pre-tender, pre-execution stages or onsite but not simultaneously. The exceptions tend to be where planners work for some time on a single large project. In such cases it may make sense (from the employer's viewpoint) for their work to carry on to the execution stage.

The form of contract determined the time spent by planners on a single contract. For traditional contracting by single stage tendering the period for the preparation of the construction plan was around 4–6 weeks for larger contracts and 3–4 weeks for smaller ones. The post-tender period to the start on site date showed somewhat greater variation, from 2 to 6 weeks. For two stage tendering, the first stage was similar to the single stage tender period but it was only at the subsequent stage that a price had to be presented. However, for the planners, the time frame was similar to that of the traditional method. Those involved in two-stage planning reported that the plan was a significant factor in progressing beyond the first stage as 'a demonstration of competence' was the paramount selection criterion at this stage.

Under the construction management form of procurement planners tended to be brought into the process earlier. The principal contractors were also involved in partnered contracts and PFI schemes where the tender periods were also longer, typically 3–6 months. On large civil engineering projects the tender periods are longer, averaging 3 months. However, planners often work simultaneously on several tenders, and the actual working time available to them for preparing each tender submission is substantially less than the tender period.

Planners tend to be overwhelmed with information. They typically received large amounts of information that was not relevant to their role and spent considerable amounts of time searching through it. As might be expected, this problem particularly affected the planners working for trade contractors. The quality of the information received was poor.

The methods for dealing with the uncertainty caused by design information deficiencies were:

- Guess the missing information based on experience and past job records.
- Qualify the submitted tender.
- Assess the risk posed by the missing information and adjust the risk premium accordingly.
- Take a strict contractual stance on site with regards to negotiating the cost of variations to the tender drawings/specifications/scope of the works.

Table 1.8 presents a list of the documents produced by the planner as part of the tender team. The list shows the documents produced based upon the number of planners who mentioned their production for external (i.e. client or subcontractor) use. The interviewer specifically mentioned the first four items in the list, the rest were spontaneously mentioned by the informants.

Table 1.9 shows the domain specific knowledge that experienced planners believe enables them to solve planning problems better than inexperienced planners.

Table 1.8   A list of documents commonly produced by the planner as part of the tender team.

Documents produced by the planner as part of the tender team
Documents commonly produced
Method statement
Bar chart schedule
CPA
Design schedule (for design and build projects)
Design information procurement dates
Procurement schedule
Buildability schedule/value engineering options
Quality assurance plans and procedures
Health and safety plans and procedures

The following documents may also be produced
Resource schedules
Phased work location drawings
Environmental protection procedures
Organisation charts, CVs, construction team information
Site mobilisation plans
Site layout drawings
Preliminaries schedules

Adapted from Winch and Kelsey (2005).

Table 1.9   Domain specific knowledge required by planners.

Knowledge required in the following aspects of construction
Better understanding of site processes and works contractors
Better understanding of mechanical and electrical services and co-ordination with other trades
Development of better communication skills including listening
Experience through working on a wide range of projects
Better understanding of contracts and tender processes
Development of the ability to anticipate problems
Better understanding of 3D/spatial aspects of the work
Development of a feel for task outputs and durations
Better understanding supply chain management

Adapted from Winch and Kelsey (2005).

Construction planning practice: a summary

From the research of Winch and Kelsey and others, it is possible to identify the key features of construction planning practice.

Clearly, planning is a key element of construction project management. For the construction contractor planning commences with procuring the work, is a key part of the time spent before work commences on site, and continues throughout the construction process. Planning is a crucial role to play in delivering construction project on time, to cost, safely and at the quality required by the client. Operational managers look to the planner to guide them on future work, to make forecasts on future events: what will happen and when. Senior managers look to the construction planner to measure construction process, to help prevent or at least identify potential construction problems and to ensure that claims for additional costs are supported by appropriate data.

Planning involves not only knowledge and experience of the construction process but also the ability to secure and co-ordinate information from a number of sources, both internally within the contractor's organisation and externally from numerous specialist suppliers. Planning is not a task undertaken in isolation. Planners work in a network of relationships which demand negotiating and facilitating skills and the ability to work collaboratively. Detailed planning needs to be decentralised to those responsible for the execution of the works (Ballard and Howell, 1998a; Barber et al. 1999; Faniran et al. 1999).

Construction planners have to undertake their work under the constraints of time. Often the level of information is limited and there is a need to secure and collate information before making a decision. Where no information is available assumptions need to be made. Planners need to be comfortable working in a world of uncertainty. Planning tasks may be similar from project to project but the form of procurement drives planning practice.

Key points

- Planning is the process of preparing for the commitment of resources in the most effective fashion.
- All those responsible for the management of construction work, client, designer, contractor and subcontractors, need to plan.
- Planning aims to produce a workable schedule that will achieve project goals and serve as a standard against which actual progress can be measured.
- Planning defines what should be done, the activities that should be performed, when they should be performed (the sequence and timing), the methods of operation, who should perform each activity and with what resources and equipment.
- The main objective of planning is to ensure that things happen successfully. This requires objectives to be established, tasks to be identified and progress to be monitored.
- Planning is characterised by the volume of information available to the planner and the uncertainty relating to the information available.
- Planning is a mental activity and as such is subject to the influences of the human psychology.
- Scheduling is the process of quantifying the programme and the production of calculated dates and logic. It covers all aspects of the production of bar charts, networks, method statements and other material relating to the project.

- Different institutions have produced different models for identifying the different phases of a construction project. The importance of these models is to identify within a common framework the different stages of the construction project and the decisions that need to be made to assist in determining how the planning will progress.
- The form of procurement and the range of options available to construction clients set the context for the planning and scheduling and determine the form and provision of information to the client and the other stakeholders.
- New forms of procurement such as PFI and new ways of working such as Partnering have extended the role and tasks of the construction planner.

Section I

Chapter 2
Managing Construction Projects

Li Baiyi[1] and Simon Austin[2]

[1]Doctoral Graduate, Loughborough University, Loughborough, UK
[2]Professor of Structural Engineering, School of Civil and Building Engineering, Loughborough University, Loughborough, UK

This chapter presents an introduction to six construction management schools of thought: the Project Management Book of Knowledge; Simultaneous Management; Lean Construction; A Theory of Construction as Production by Projects; Collaborative Working; and the Perspective of Peter Morris, one of the leading researchers and practitioners of Project Management. Each of these provides a different perspective on the role and execution of planning. All recognise the importance of planning if projects are to be successfully executed. We start with the *Project Management Book of Knowledge* (PMBOK).

Project management body of knowledge (PMBOK)

The Project Management Institute (PMI) first published *A Guide to the Project Management Body of Knowledge* (PMBOK Guide) in 1983. This was a 'white paper', an attempt to document and standardise generally accepted project management information and practices. The first full edition of the guide was published in 1996. The latest English-language PMBOK Guide – the Fifth Edition was released in February 2012 (PMBOK, 2012). The PMBOK represents the dominant management paradigm for construction management and its approach is used extensively in managing construction projects worldwide.

This style of management has a long history, and its evolution has been closely related to the following: (1) the development of systems engineering in the US defence and aerospace industry; (2) the development in modern management theory (e.g. organisation design and team building); and (3) the evolution of computer-based planning techniques (i.e. Critical Path Method (CPM), PERT) (Morris, 1994; Laufer et al., 1996). Because of its long history, some researchers

A Handbook for Construction Planning and Scheduling, First Edition. Andrew Baldwin and David Bordoli.
© 2014 John Wiley & Sons, Ltd. Published 2014 by John Wiley & Sons, Ltd.

refer to this style of management as the 'traditional project management approach'. It is also known as the 'PM theory/approach' (Koskela and Ballard, 2006; Winch, 2006).

The PMBOK approach is based on three theories: planning, execution and control:

1. The PMBOK views management as planning. Detailed planning of tasks is undertaken before the structured implementation of production. Production is monitored and control relates performance back against the original plan to identify any deviations during implementation.
2. The Theory of Execution assumes a 'dispatch' model whereby planned tasks are executed by notification of the start of the task to the executor. Planning is essential to determine what happens and when. Planning, against a predetermined schedule, directs when production commences.
3. The Theory of Control assumes a 'thermostat' model. It is assumed that there is a standard of performance that can be measured at the output, and the process may easily be corrected by the control available so that the standard can be reached.

The concepts, principles and assumptions of each of these theories are summarised in Table 2.1

In regard to the theory of projects, Koskela and Howell (2002a) point out that this traditional project management approach generally adopts a transformation model: production is a transformation of inputs to outputs. Resources – labour, machinery and materials – are changed by the production process into products. Production comprises a number of sub-processes that proceed one after another. The function of production management is to decompose the total transformation into elementary transformations and tasks in order to minimise the cost of inputs and improve the efficiency of the operation.

For the management of projects, the PMBOK approach recognises that 'much of the knowledge needed to manage projects is unique or nearly unique to project management' (PMI, 1996; Choo, 2003). The critical path method (CPM) and the work breakdown structure (WBS) are considered the basis for project management-specific knowledge.

Howell and Koskela (2000) argue that the PM approach is a system for managing *contracts* and that it is based on the assumption that all coordination and operational issues reside within the contract boundaries. They list a number of deficiencies in the assumptions and the theory of this 'traditional' PM approach in the context of modern complex construction projects. These are summarised in Table 2.2.

Koskela and Howell (2002b) point out the traditional project management approach embodies 'the assumption of the independence of sub-processes' and the belief that 'if the transformation process is seemingly not independent from its external environment, or the sub-processes from each other, it can be made independent through physical or organizational buffering'. Laufer et al. (1996) argue that the traditional project management approach 'fit[s] squarely into a world of certainty', and that construction projects are by their very nature uncertain.

Ballard (1999b) argues that the activity centred approach found in the traditional project management approach aims to optimise the project 'activity-by-activity'.

Section I

Table 2.1 The theories of PM approach.

Theory of project		**Conceptualisation**: A project is a transformation of inputs to outputs. **Principles**: 1. The total transformation of a project can be decomposed into manageable and well-understood sub-transformation tasks. 2. A project can be realised in an optimal manner by realising each task in an optimal manner and the tasks in optimal sequence. Corollary: Project performance can be performed by improving the tasks. **Assumptions**: 1. Tasks are independent, except sequential relationships. 2. Tasks are discrete and bounded. 3. Uncertainty as to requirements and tasks is low. 4. All work is captured by top-down decomposition of the total transformation. 5. Requirements exist at the outset and they can be decomposed along with work.
Theory of management	**Theory of planning**	**Conceptualisation**: There is a managerial part and an effector part in the project; the primary function of the managerial part is planning, and the primary function of the effector part is to translate the resultant plan into action. **Principles**: 1. Knowing the current state of the world, the desired goal state, and the allowable transformations of state that can be achieved by actions, a series of actions, the plan, can be deduced. 2. The plan is translated into reality by the effector part of the organisation. **Assumptions**: 1. Translating a plan into action is a simple process, by following directions. 2. The internal planning of a task is a matter of the person to whom the task has been assigned.
	Theory of execution	**Conceptualisation**: Managerially, execution is about dispatching tasks to work stations. **Principle**: When, according to the plan, the time has arrived to begin task execution, it is authorised to start, in speech or in writing. **Assumptions**: 1. The inputs to the task and the resources to execute it are ready at the time of authorisation. 2. The task is fully understood, started and completed according to the plan once authorised.

Table 2.1   (*Continued*)

Theory of control	**Conceptualisation**: There is a process to be controlled, a unit for performance measurement, a standard of performance and a controlling unit (thermostat control). **Principle**: The possible variance between the standard and the measured value is used for correcting the process so that the standard can be reached. **Assumptions**: 1. The process is of continuous flow type, the performance of which is measured at aggregate terms. 2. The process can easily be corrected by the control available.

Adapted from Koskela and Howell (2002b).

Table 2.2   Deficiencies in the assumption and theory of traditional project management theory (Howell and Koskela, 2000).

Category	Assumption and theory	Modern projects
Uncertainty in scope and method	Low	High
Relationships between activities	Simple and sequential	Complex and iterative
Activity boundary	Rigid	Loose
Performance criteria	Activity-based	Need to consider flow between activities
Production management	Not considered	Need to be considered
Model	Transformation	Need to be viewed as a combination of transformation, flow and value generation

Customer value is identified in the design process. Production is then managed throughout a project by first breaking the project into pieces, then putting those pieces into a logical sequence, estimating the time and resources required to complete each activity and therefore the project. Each piece or activity is further decomposed until it is contracted out or assigned to a task leader, foreman or squad boss.

Laufer et al. (1996) state that construction management under this model often '…concentrates on the coordination of sequential and parallel activities and control of performance, backed up by information technology, just as would be the case with flight scheduling problems in major airlines'. It only focuses on the conversion process and ignores the flow and value considerations aspect of production. This leads to uncertain flow processes, expansion of non-value-adding activities, and reduction in the value of output (see also Alarcon, 1997; Ballard, 2000; and Koskela, 2000).

In this management approach, planning is one of five project management processes. The five processes initiating, planning, executing, monitoring and controlling and closing together form a closed loop: plan, do, check, act, which is known as the 'plan–do–check–act cycle'. This is shown in Figure 2.1 and Figure 2.2.

Koskela (2000) argues that this 'management-as-planning' style of project planning and control suggests that management consists primarily of the creation, revision and implementation of plans. Together with Greg Howell in Koskela and Howell (2001) he points out that:

> "This approach to management views a strong causal connection between the actions of management and outcomes of the organization. By assuming that translating a plan into action is a simple process, it takes plan production to be essentially synonymous with action". They go on to state: "However, translating from plans to action is only possible if resources are actually available and interdependence between activities low". (p. 6)

Is the PMBOK approach appropriate for complex construction projects? Laufer et al. (1996) argue that the traditional project management approach is an effective

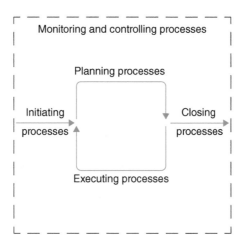

Figure 2.1   The Project Management Processes (adapted from PMI, 2004).

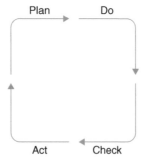

Figure 2.2   The concept of the plan–do–check–act cycle (adapted from PMI, 2004).

tool to manage simpler projects which do not include high levels of uncertainty but may be insufficient for managing more complex projects within today's more dynamic environment.

What are some of the alternative approaches?

Simultaneous management

This approach was first presented by Laufer et al. (1996) and further developed in his book *Simultaneous Management: Managing Projects in a Dynamic Environment* (1997). The starting point was observing that 'skillful managers are able to complete complex and uncertain capital projects in less time without sacrificing cost and quality criteria or leaving customers and users dissatisfied successfully'. After systematic studies, Laufer et al. (1996) concluded: 'Master project managers develop a new style of simultaneous management in which they constantly orchestrate contending demands'.

Laufer argues that simultaneous management means 'planning systematically, making early (but not premature), adequate decisions, involving all parties concerned'. Simultaneous management means leading a team, operating first-rate communications, using simple produces, executing the project's phases early and in parallel (after appropriately splitting the project), monitoring project performance and the environment and judiciously adapting the project's execution to arising contingencies. In this way it is possible to execute challenging projects with excellence and speed and, through simultaneous management, synergy can be achieved. Management of projects is by its very nature dynamic.

This approach highlights the importance of the understanding of uncertainty and the management of uncertainty in project management.

Laufer groups uncertainty into two types: end uncertainty (what has to be done) and means uncertainty (what should be accomplished), and then identifies the different assumptions as to how uncertainties are solved between the current existing approaches and the simultaneous management. He points out that the traditional management paradigms assume that end uncertainty is solved before considering means uncertainty, whilst under simultaneous management both types of uncertainty are resolved gradually and simultaneously. He suggests two steps to cope with uncertainty: (1) de-couple interdependent tasks and isolate tasks plagued by high uncertainty; (2) if decoupling is impossible, manage the interface between interdependent tasks, or employ redundant resources selectively to absorb the uncertainty. (The developers of the ADePT system also take this approach. See Chapter 8.)

As to the position of planning in this approach, Laufer adopts the concept of the traditional 'plan–do–check–act cycle' as shown in Figure 2.2 supported by systematic integrative planning and systematic monitoring. He points out that under traditional project management, performing according to plan with minimal changes is the essence of project management. However, under simultaneous management meeting customer needs whilst coping successfully with unavoidable changes is the essence of project management. (This implies regular, short-term monitoring and control and the need for more resources to be allocated to planning.) Control in the traditional project management style is feedback-based, that is, based on the measurement of project performance. Under simultaneous management, control is anticipation and feedback based, that is, it is necessary not only to monitor project performance but also to monitor and assess the impact of changes in the project environment.

Laufer (1997) summarised nine principles of simultaneous management in three groups:

- **Planning**: Planning includes three principles: systematic and integrative planning, timely decisions adjusted to uncertainty, and systematically managing uncertainty. Decouple interdependent tasks if possible, and isolate tasks with high uncertainty. When interdependent tasks cannot be decoupled, 'absorb the uncertainty by employing redundant resources selectively, or manage the interface between interdependent tasks'.
- **Leadership and integration**: Leadership and integration includes three principles: inward and outward leadership, multi-phase integration and multi-disciplinary teams.
- **Systems**: Simultaneous management is enabled by commitment to three principles: intensive communication, simple procedures and systematic monitoring.

The simultaneous management style of management was generated from practice and has been tested successfully in many real-life settings. These are further explored in Laufer et al. (1994).

Lean construction

In 1990 Womack argued that in the previous decade there had been a significant improvement in performance in the manufacturing and service industries as a result of a new production philosophy 'lean production'. This philosophy advocated combining the advantages of craft and mass production to produce a new way of working. By adopting this approach 'lean production' could efficiently provide volumes of different variety products. This could be achieved at a relatively low cost by using teams of multi-skilled workers at all levels of the organisation and highly flexible, increasingly automated machines (Womack et al, 1990; Womack and Jones, 1996). This approach to manufacturing was extended to construction and has become known as 'Lean Construction'.

Koskela and Huovila (1997), Koskela (1992) and other supporters of the lean construction approach have applied the lean production philosophy to both the design and construction process. Later Koskela (2000) developed a transformation–flow–value model (TFV) and claimed 'the TFV theory largely explains the origins of construction problems...thus the TFV theory of production should be applied to construction'. This perspective, compared to other more traditional approaches is summarised in Table 2.3.

Later, Koskela and Howell (2002a) summarised the theories in lean construction as shown in Table 2.4.

Koskela and Howell divide the theories of lean construction into theories of management and theories of projects. With regard to the theories of management, they point out that the 'lean' approach is based on three theories:

1. Planning theory that is based on the integration of the approach of management-as-planning and the approach of management-as-organisation.

Table 2.3   Integrated TFV view on production (Koskela, 2000).

	Transformation view	Flow view	Value generation view
Conceptualisation of production	As a transformation of inputs into outputs	As a flow of material composed of transformation, inspection, moving and waiting	As a process where value for the customer is created through fulfilment of his requirements
Main principles	Getting production realised efficiently	Elimination of waste (non-value-adding activities)	Elimination of value loss (achieved value in relation to best possible value)
Methods and practices (examples)	WBS, MRP, Organisational Responsibility Chart	Continuous flow, pull production control, continuous improvement	Methods for requirements capture, Quality Function Deployment
Practical contribution	Taking care of what has to be done	Taking care that what is unnecessary is done as little as possible	Taking care that customer requirements are met in the best possible manner
Suggested name for practical application of the view	Task management	Flow management	Value management

Table 2.4   A new theoretical foundation of project management (Koskela and Howell, 2002b).

Subject of theory		Relevant theories
Projects		Transformation
		Flow
		Value generation
Management	**Planning**	Management-as-planning
		Management-as-organising
	Execution	Classical communication theory
		Language/action perspective
	Control	Thermostat model
		Scientific experimentation model

2. Execution theory that is based on the classical communication theory and the language/action perspective that was originated by Winograd and Flores (1986). This conceptualises two-way communication and commitment, instead of the mere one-way communication of the classical communication theory.

3. Control theory that is based on the thermostat model and scientific experimental model that was developed by Shewhart and Deming (1939). This focuses on finding causes of deviations and acting on those causes, instead of only changing the performance level for achieving a predetermined goal in case of a deviation. This adds the aspect of learning to control.

According to the Lean Construction Institute (2005), the aim of lean construction is to extend the objectives of minimising waste and maximising value advocated within lean production management into the development of specific tools and techniques for lean construction project delivery. Lean Construction recognises the dependences and variations that arise along the supply chains of construction projects (Howell, 1999) and identifies the product uncertainties and process uncertainties in the construction process (Tommelein 1997; Tommelein, 2000 – cited in Milberg and Tommelein, 2003). Advocates of the lean construction approach argue that these product and process uncertainties can be actively managed (Howell, 1999).

Lean Construction principles have been adopted world-wide. In the United Kingdom they were strongly endorsed by the influential 'Egan Report' (Egan, 1998) and considered as having an essential part in improving the production record of construction. Lean construction advocates the combination of the approach of management-as-planning and the approach of management-as-organising in construction planning (Koskela and Howell 2002a). However, some have questioned key aspects of the approach. For example, Winch (2006) focused on the root theory of lean construction and pointed out limitations of the TFV theory including the following:

1. An overemphasis on the 'hard' management (i.e. the strong focus on physical processing of materials as an engineering problem);
2. A tendency to ignore the 'soft' management (i.e. the absence of a concept of organisation in the analysis);
3. A lack of consideration of the implications of risk and uncertainty; and
4. Its unitary concept of value.

A theory of construction as production by projects

Winch (2002) argues that 'planning as the creation of a framework for subsequent action and developing plans-as-resources to guide managers as events unfold are compatible with the management as-organizing approach' and proposes a theory of Construction as Production by Projects. This school of thought was first presented in 2002 and subsequently reinforced by several further publications (See also Winch and Carr, 2001 and Winch et al., 2003). Winch views the management of a construction project as 'a holistic discipline', rather than 'a set of fragmented professional domains'. He points out that the management of construction projects is a problem with information – the lack of all the information required to take a decision at a given time.

The root theory root of the Construction as Production by Projects approach states that:

> "Management is essentially an organizational innovation – the identification of a person or small team responsible for ensuring the effective delivery of the project mission for the client." (Winch, 2006, p. 155)

This approach views a project as both an organisation and a transaction between the client and the project participants. It contains four theoretical starting points: transaction cost economics, organisation as information-process systems, the project as uncertainty reduction and the tectonic approach to organisation. It provides a better understanding of value in construction.

Firstly, it points out that 'value in the construction process and the product is inherently contested', and then it identifies three dimensions to the value concept applicable to the construction process:

1. The client's business process;
2. The supplier's business process; and
3. The contribution to society as a whole (Winch, 2002; Winch, 2006).

As for the position of planning in this approach, Winch (2002) states, 'planning is the core competence to the discipline of project management'. However, planning is seen as 'a middle-management, operational discipline, rather than the strategic discipline'. This view is further endorsed with the statement 'planning is the creation of a framework for subsequent action and developing plans-as-resources to guide managers as events unfold are compatible with the management as-organizing approach' (see Winch and Kelsey, 2005).

This approach introduces an important distinction between the requirements of planning as an iterative process, for example, within the design process and planning, and planning as a more linear process, for example, prior to the execution of site tasks. Winch points out that a fundamental assumption of critical path methods and critical chain methods is that task dependencies are sequential. Whilst this assumption is viable for the execution of construction on site, it is not appropriate for design tasks as they may be interdependent. Others endorse this view (Austin et al. (1999) have discussed this issue in detail, their research leading to the production of the ADePT system for planning and managing the design process). The ADePT system is described further in Chapter 8.

In this school of thought, the major concern in improving programming performance is 'the uncertainties associated with task execution duration, which make programming difficult and encourage the padding of durations' (Winch, 2002). Winch reviewed a large number of different tools and techniques before advocating that the problems of uncertainty are best addressed by the use of both the Last

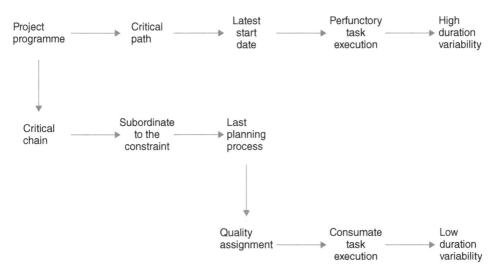

Figure 2.3   Last Planner and Critical Chain combined (Winch, 2002).

Planner and the Critical Chain. He believes that these methods are highly complementary and suggests their combination to deal with both tactical and strategic issues in the managing construction projects. This is shown in Figure 2.3.

Collaborative working

This school of thought first appeared in the 1970s when organisations realised the need to enhance integration through project management in order to better deal with those complex projects that often consist of many dissimilar, highly interdependent components and involve many disciplines (Laufer et al., 1996).

The collaborative working approach is well documented in *Project Management: Designing Effective Organizational Structures in Construction*, by Moore (2002). He focuses on the organisational side of project management and states, 'organisation structure can make or break the project manager'. Johnston and Brennan (1996) called this approach 'management-as-organisation', and point out that this approach enhances the human aspect of management.

Laufer et al. (1996) state that this school of thought views the challenge of project management as the need to ensure integration and teamwork between the different participants and make the team performing as a unified entity. Hence, this approach highlights the importance of the coordination between participants and focuses on process facilitation and the definition of roles.

This approach has been further developed in recent times including the thinking of supply chain management (SCM) introduced during the 1990s. This is widely acknowledged as a major avenue to increase competitiveness and improved business performance. The term was initially proposed to link logistics issues with strategic management (Oliver and Webber, 1992 – cited in Dudek, 2004). A similar understanding is expressed in many contemporary publications. SCM is often regarded as a synonymous term of integrated logistics management (Houlihan, 1985 – cited in Dudek, 2004). However, some experts distinguish between SCM and logistics, arguing that logistics focuses on intra-organisational relationships, while SCM focuses on inter-organisational relationships. Stadtler (2002) states that SCM is 'the task of integrating organisational units along the supply chain and coordinating material, information, and financial flows in order to fulfil (ultimate) customer demands'.

Regardless of whether SCM is understood as integrated logistics management or a broader management discipline on its own, the flows of information and materials and the associated planning and control activities are treated as key elements in this concept (Dudek, 2004).

Collaborative planning was introduced on the basis of these understandings. The earliest contributors focused on technological means available for the exchange of information between independent supply chain partners (Dudek, 2004). In recent times, collaborative planning has gained popularity due to an industry initiative 'Collaborative Planning, Forecasting and Replenishment' (CPFR). However, collaborative planning includes goals, tasks and resources and the development of a conjoint business plan and CPFR may be regarded as primarily a term of business planning.

Kilger and Reuter (2002), viewed collaborative planning as a non-hierarchical, cooperative approach to the coordination of operations, planning tasks across the supply chain. Dudek (2004) has also adopted this understanding and focused on the decision making and negotiation aspects of collaborative planning. He developed a negotiation-based collaborative planning scheme that coordinates master plans of individual supply chain partners each of whom have their own planning domain.

Some researchers in the construction industry have begun to adopt the collaborative planning concept for project planning. For example, Verheij and Augenbroe (2006) argue that project planning is the tactical translation of the strategic objectives into project execution. They have developed a new approach to systematically generate comprehensive project plans by cultivating and enacting the logic and intelligence of incremental and collaborative planning strategies based on a methodology that treats project planning as the execution of a series of structured dialogues between prospective project partners.

Despite the limited applications of the collaborative planning approach the system provides an opportunity for the construction industry to deal with the fragmented nature of industry and improve the collaboration within the supply chain. It is a perspective that demands further consideration, development and monitoring.

Morris' perspective

Peter Morris presents a different perspective. In his book *The Managing of Projects* (1994) and his subsequent paper 'Science, objective knowledge and the theory of project management' (Morris, 2002), he presents two important considerations in project management. First, the management of project is a holistic discipline not a set of fragmented professional domains. Secondly, planning is 'a middle-management, operational discipline, rather than the strategic discipline'.

Morris argues that project management is the total process of 'integrating everything that needs to be done as the project evolves through its life cycle in order to ensure that its objectives are achieved'. He then suggests enlarging the subject to the broader one of 'the management of projects'. Winch, inspired by Morris's work, introduced the important distinction between construction project management and managing construction projects. However, although similar, their perspectives are not identical. With regard to the root theory root of project management, Morris argues, 'there will never be an overall theory of project management'. He contends that the very notion of a theory of project management is mistaken, although admitting that there can be 'some theories', that is, theories of particular aspects of project management. Thus, according to Morris, knowledge of project management will always be personal and experimental and the best that can be done is to 'offer guidance in the form of tools, aids, heuristics, approaches and insights – and some scientifically established and tested theory' (For further details of Morris's perspective see "Reconstructing Project Management", Peter Morris, 2013).

Summary

An overview of these six schools of thought is provided in Table 2.5. This provides a summary of the Theory, Position of Planning, Planning Process and approach to Uncertainty of each school.

Section I

Table 2.5 Overview of schools of thought in construction management.

	Theory	Position of planning	Planning process	Uncertainty
PMBOK	Based on theories of production Adopts the transformation model	Embodies 'management-as-planning' as a core belief	Develops a generic project planning process model	Fits squarely into a world of certainty.
Lean Construction	Based on theories of production Adopts the TFV model, and lean thinking Views construction as one-of-a-kind projects, site production, and temporary multi-organisation	Integrates the approach of management-as-planning and the approach of management-as-organisation Suggests using work structuring and production planning systems	Suggests production planning and focuses on stabilising work flow	Views project management as uncertainty reduction. Believes that in practice some uncertainties are caused by the wrong order of decision making. Believes that uncertainty can be managed, and focuses on reducing the variables prior to production.
Construction as Production by Projects (Winch)	Based on theories of economics Adopts the transaction cost economics perspective Views project as organisations, and projects as a transaction between the client and the project participants	Views planning as 'a middle-management, operational discipline, rather than a strategic discipline' Suggests plans-as-resources to guide managers	Plans are nothing, but planning is everything to point out the importance of planning process	Views project as uncertainty reduction. Views uncertainty management as core purpose in the managing construction projects. Views uncertainty as the difference between the information required for a decision and the information available. Uncertainty arises from complexity and predictability.

	Theory	Position of planning	Planning process	Uncertainty
Simultaneous Management	Manager constantly orchestrates contending demands	Views planning as a core theme of project management	Calls for efforts to be paid for the planning process itself	Suggests timely decisions adjusted to uncertainty. Suggests decoupling interdependent tasks and isolating tasks with high uncertainty. If decoupling is impossible, absorb the uncertainty by employing redundant resources selectively, or manage the interface between interdependent tasks.
Morris	Believes that there will never be an overall theory of project management	Views planning as 'a middle-management, operational discipline, rather than the strategic discipline'	Suggests that better management can be achieved by thinking the planning process	Highlights the importance of the management of uncertainty.
Collaborative Working	Views project management as integration and teamwork between the different project participants. Encourages the team to perform as a unified entity	Embodies 'manage-as-an-organisation'	Implicit	Fits squarely into a world of certainty.

Section I

Key points

- Management may be considered to comprise four functions: Planning, Organising, Control and Leadership
- All the major schools of thought view planning as the 'core competence' to the discipline of construction management.
- The PMBOK represents the dominant management paradigm for construction, the 'traditional' approach.
- In the PMBOK approach planning drives all management action. Management comprises primarily of the creation, revision and implementation of plans.
- Laufer argues that management takes place in a dynamic environment and that managers need to develop a simultaneous style of management in which they 'constantly orchestrate contending demands'.
- Winch argues that the main challenge of construction project management is the problem of uncertainty of information – the lack of information required to take a decision at a given point in time and management must manage this uncertainty.
- Advocates of Lean Construction argue that the product and process uncertainties in both design and construction can be actively managed by a combination of 'management-as-planning' and 'management-by-organisation'.
- Morris argues that the management of project is a holistic discipline not a set of fragmented professional domains and that planning should be viewed as a middle-management, operational discipline, rather than a strategic discipline.

(Further information relating to these different schools of thought is included at the website for this book: www.wiley.com/go/baldwin/constructionplansched).

Section II
Planning and Scheduling Techniques and Practices

Section II

Introduction

There are numerous planning and scheduling techniques available to assist the construction planner. Many of these have been developed over an extended period of time. Most are now assisted by computer software or computer-based systems.

In this section, we include two chapters. The first examines the basic techniques that are available to the planner: to-do lists, bar charts, flow charts, network analysis, linked bar charts, space diagrams, line of balance, ADePT and 4D Planning. These cover a range of approaches to the analysis and presentation of information. Each has advantages and disadvantages; their selection will vary depending on the requirements of the client, the techniques normally adopted by the construction organisation, the project and the time available. The selection of one technique does not preclude the use of another to assist the work of the planner and to communicate the results of an analysis to the management and the workforce.

The second chapter looks at planning and scheduling practice and how these techniques are adopted, adapted and utilised in practical situations. It is necessary to plan, design and structure schedules to link schedule information to the information in other management systems. When preparing schedules, it important to look closely at the activities selected, the sequencing of activities and their duration. The different stages of the project cycle require different information at different stages: pre-tender, precontract and during the contract. We look at

A Handbook for Construction Planning and Scheduling, First Edition. Andrew Baldwin and David Bordoli.
© 2014 John Wiley & Sons, Ltd. Published 2014 by John Wiley & Sons, Ltd.

what information is required both internally and externally when. Having established an appropriate 'time model', it is necessary to monitor progress and manage the model. During construction, therefore, it is essential to maintain comprehensive records of events and to assess the impact of these events upon the construction activities.

Establishing and maintaining the time schedule is not the only task of the construction planner. This chapter also looks at how the cash flow for the contract may be calculated and the method statements that need to be produced. The information used by the planner and the construction team may be uncertain, and the planner needs to be conscious of the risks that may arise due to insufficient information.

Section II

Chapter 3
Planning and Scheduling Techniques

Here we introduce a number of planning and scheduling techniques that are available to assist the planner.

To-do lists

"What is important is seldom urgent and what is urgent is seldom important".[1]

How do we prioritise tasks and our time? Time management is a basic function of project management, and good time management is a requirement for all effective managers. To-do lists are the simplest way of managing time.

There is no shortage of guidance on to-do lists. A simple Google™ search will produce literally millions of sources of information on the value of producing lists that will help you identify priorities and organise your day. Supporters are effusive in their use of the technique. Some even claim, 'It can be fun!' There is a plethora of software tools and templates to assist those who seek more than simply a pad of paper or a block of Post-it Notes™ to organise their time.

Time management is about being organised, setting goals, working in priority order, protecting your time, and exercising conscious control over the time spent on individual activities. There are a multitude of books providing guidance. Typical 'keys' to efficient time management include being self-aware and observant (log how you spend your time), plan every day, eliminate the unimportant, under commit and over deliver (if you over promise you will feel overstressed), 'learn to say no', organise and re-organise, reflect and review to-do lists are an essential part of time management.

---

[1]   A quotation attributed to General Eisenhower.

A Handbook for Construction Planning and Scheduling, First Edition. Andrew Baldwin and David Bordoli.
© 2014 John Wiley & Sons, Ltd. Published 2014 by John Wiley & Sons, Ltd.

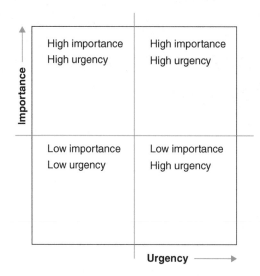

Figure 3.1   The urgent/important matrix.

Some argue that to-do lists are a waste of time and simply lead to failure and frustration: too many items on the list give too much choice. They argue that faced with items of different durations most people will simply choose the shortest. (This is an example of what psychologists call 'heterogeneous complexity'.) It is difficult to look at lists of items in context to decide what to work on. (All tasks look the same on paper.) To-Do lists do not prevent you from choosing the most pleasant tasks over the most important.[2]

It is difficult to argue however that spending some time planning, prioritising, controlling tasks on a daily basis is not worthwhile. Construction planners, by their very nature are likely to adopt structured systems that prioritise tasks (e.g. by ranking them A, B or C) and break down goals into specific actions.

One approach is to differentiate between urgent and non-urgent tasks; important and non-important tasks. Categorising tasks using these criteria enables you to prioritise your workload. Unimportant tasks that are not urgent are dropped. Tasks that are important and urgent are completed immediately and personally. Tasks that are unimportant/urgent are delegated and tasks that are important/not urgent get an end date and are performed personally. (See e.g. the books of Stephen Covey, 2004, 2005.) The matrix is shown in Figure 3.1.

Some supporters of To-Do Lists argue that lists of actions are best linked into calendars and time schedules. If it is important enough to have on the list then it needs to be allocated to a specific time and a specific day. Whether using To-Do Lists for managing your personal time each day/week or simply identifying the priority of work output for your project(s) this approach would appear attractive to planners who by nature and job function are focused on how to complete activities to a schedule of dates.

Most of us adopt lists to organise tasks, even if these are simply held in our memory.

---

[2] To-Do Lists Don't Work, http://blogs.hbr.org/cs/2012/01/to-do_lists_dont_work.html

Above all, find a method time way to manage To-Do Lists that 'work for you' and keep with it.

Bar charts

Bar charts are the most familiar method for presenting plans. But, without sufficient supporting information, these plans are liable to misinterpretation.

Bar charts list activities and identify a time for each activity to start and a time for each activity to finish. They are easy to produce. Bar charts may provide information at different levels of detail. The time scale adopted may be shown in days, weeks, or months. They may be used to present summary information to senior management or detailed information to work gangs. Different bar charts may be used to show activities for the whole project or only a section of the works. The work shown in the chart may relate to a single trade gang or a group of subcontractors. Bar charts can be produced to show a single construction activity broken down into sub-activities. Even when other methods of scheduling are used (e.g. Critical Path Analysis) the output from these methods is normally presented in bar chart format.

The general format for the bar chart is widely recognised. Activities may be linked to show how they relate to each other. Bar charts have become the accepted method for communication of project plans for over 100 years. They are popular and widely used because they are uncomplicated and simple to interpret. (No formal training is required to understand a bar chart.)

But

- Bar charts communicate a schedule of activities but not the plan and how this plan was reached.
- Links between activities on a single bar chart soon become confusing, lines inter-sect, lines overlap, lines cross bars and lines get lost against the background.
- Attempts to shade or colour different types of activities often confuse.
- Links between activities may show inter-dependency but they do not show what needs to be done to complete the task.
- Although the chart may include supplementary information, required dates, handover dates etc., the chart may soon become confused. A key to the information provided on the chart is required.
- Bar charts may be used to monitor progress (see Chapter 4), but they cannot explain the reasons for change. If there is a change you need to re-draw the chart. (And keep a record of why the change in plan has taken place.)
- Most bar charts are drawn assuming the activities start at the earliest start time possible and hence the earliest finish. They do not usually explain float/ contingency.
- The visual presentation of the activities implies a consistent use of resources and consistent productivity over the time period. This is frequently not the situation.

Based on the items scheduled in a bar chart different people may interpret the schedule in different ways. You cannot be certain that all involved with the project will interpret the information in the same way.

However, despite their limitations 'The value of bar charts should not be underes-timated. Their usefulness is not eliminated by their deficiencies' (Hinze, 2012). The

Section II

way to overcome these types of deficiencies is by the adoption of company standards and the provision of additional information whenever a bar chart is used. Details may then be cross-referenced to general activities. Ideally a clear method statement is prepared to the level of detail required by the recipient of the documents.

Establish company 'best practice' and develop standard formats for the presentation of information. Admittedly these formats will change over time but you will be building from a recognised platform for scheduling. You may need to change the format to meet the needs of individual clients, but there is no reason why the presentation of bar chart information cannot become part of your personal or corporate identity.

Always consider what additional information you need to provide with the bar chart that you present. This will depend on for whom you are preparing the schedule. For example, you may well present all or part of a method statement or health and safety plan. For colleagues within your organisation it is always useful to provide details of the resources used and the output rates assumed for each activity. (Although if you do, be prepared for someone to argue with you that these rates are too fast/too slow.) You may wish to cross-reference bar chart programmes to specific drawings or specifications.

Flow diagrams

A flow diagram is a pictorial or graphical model of a process or a project that enables the reader to understand the logic of the process and the alternative paths. Flow diagrams are established tools that are widely used across a number of disciplines ranging from Geography to Engineering, from Education to Medicine, from Computer Science to Business Studies. Put simply, there cannot be many professionals who have not used flow charts or similar graphical techniques to present and explain information or simply to understand information provided to them.

A flow diagram is the generic name for a number of visual diagrammatic methods for displaying data. These include flow charts, control flow diagrams, flow maps, data flow diagrams, process flow diagrams, Sankey diagrams, signal flow graphs, state diagrams, functional block diagrams and IDEF diagrams, all of which have at various points in time contributed to the design and operation of construction and engineering works and systems. Activity-on-arrow diagrams, precedence diagrams and linked bar charts are also examples of flow diagrams commonly used by planners.

Within this section we shall focus on flow charts in general. Details of activity-on-arrow diagrams, precedence diagrams, and linked bar charts will follow later in this chapter together with details of data flow diagrams and IDEF diagrams.

Flow charts

A flow chart is a diagram that represents a process. This representation is achieved by the use of different shaped boxes linked together by arrows. Flow charts are used for analysing problems, designing new systems and managing processes. They have been useful for different industries and different business sectors, including construction.

Flow charts assist in analysing complex processes and identifying faults within systems. The two most common forms of flow chart show an activity or processing step by a rectangular box. A diamond shape is usually used to show a decision.

The first structured methods for documenting processes were developed by Frank Gilbreth in the early 1920s. The engineering community quickly adopted these tools. Since then Gilbreth's methods have been used, extended and developed. Common alternative names include flow chart, process flow chart, functional flow chart, process map, process chart, functional process chart, business process model, process model, process flow diagram, work flow diagram and business flow diagram. Flow charting 'systems' incorporating software to produce, store and edit flow diagrams are now commonly available as software products to cover a whole range of uses.

The range of flow chart symbols used is extensive and has extended to specialist symbols and pictorials for different industries. An internet search quickly reveals a wide range of symbols for use across a range of business activities together with a number of software tools for producing flow charts. Find a suitable set of symbols to meet your needs and appropriate computer software that meets your budget. Alternatively simply draw the chart by hand.

The development of flow charting methods is closely linked with the development of Work Study and emerged from the early work of Taylor, Gilbreth and others.

Work study

Currie (1977) defines work study as:

> "A management service based on those techniques, particularly method study and work measurement, which are used in the examination of human work in all its contexts, and which lead to the systematic investigation of all the resources and factors which affect the efficiency and economy if the situation being reviewed, in order to effect improvement." (p. 48)

Method study is 'the systematic recording and critical examination of the proposed ways of doing work'. Work measurement is 'the application of techniques designed to establish the time for a qualified worker to carry out a specified job at a defined level of performance'.

Whilst the use of formal work study methods within a construction environment may have diminished close to extinction, the basis of these techniques still provides sound guidance as to how to deal with productivity issues.

Method study is a systematic approach that requires the planner to:

- Select the work area or activity to be studied
- Identify the problem
- Record the facts
- Examine the recorded facts
- Develop a new improved method
- Install the new method
- Maintain the new method and check the results

British Standard 3375-2:1993 'Management Services – Part 2-A Guide to Method Study' provides more details of established techniques. Inherent in the adoption

Section II

Table 3.1 A critical examination chart for reviewing construction activities (what, where, when, who, how).

	Primary questions	Primary questions	Secondary questions	Alternatives
Purpose	What is achieved?	Is it necessary? Why?	What is the alternative?	What should be done?
Place	Where is this done?	Why in that place?	Where else is possible?	Where should it be done?
Sequence	When is it done?	Why then?	When else could it be done?	When should it be done?
Person	Who does it?	Why that person?	Who else could do it?	Who should do it?
Means	How is it done?	Why that way?	How else could it be done?	How should it be done?

Adapted from Pilcher (1992) and Currie (1977).

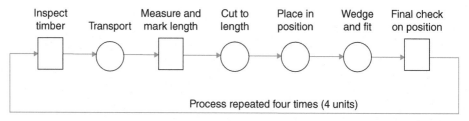

Figure 3.2 A simple flow chart.

of these techniques is a systematic identification of activities and a detailed review of how and why the existing process is being conducted. This may be demonstrated by considering Table 3.1 which provides a critical examination chart for reviewing construction activities.

The charts, diagrams and models used within work study to record, critically examine and identify the best methods for completing work consider the purpose, place, sequence, person and means to develop the optimal method of working. Central to this is the construction of flow process charts, diagrams and models from which activities may be recorded. These are the main legacy of work study.

Process charts may be outline process charts, covering operations and inspections, or full flow process charts covering the activities of men, materials, or equipment. Such charts identify five functions: operation, transport, storage, delay and inspection. Operation includes procedures, accomplishments and anything that relates to further developing the process. Transport includes the movement of men, materials or equipment. Storage holds, keeps or retains material and plant. Delay interferes or delays the process. Inspection verifies either the quantity of goods or the quality of these goods.

Figure 3.2 shows a simple process chart for material through a process. This chart is concerned with *what happens*, the operations that take place and who performs the task, not with *where* the actions take place. The important task is to identify the process.

Process charts are useful for representing a sequence of activities in a production process. By classifying the activities as 'productive' or 'non-productive' it is possible to identify improved methods of working. The use of process charts may not be as common as it was but they remain a useful technique for the construction planner, particularly for analysing the storage of materials and their flow around the site. They are also useful in the development of virtual prototyping sequences within Building Information Management (BIM) (see Chapter 10).

Network analysis

Network analysis requires you to construct a diagram that represents the activities to be completed and to estimate the duration of each activity. This network forms a flow diagram showing the logic behind the sequence of activities. Having checked the logic of the diagram and confirmed the durations, you can then calculate the total time taken to complete the work and the critical path through the network, that is, the sequence of critical activities that form the longest sequence in a project, generally spanning from the start to the finish. The analysis also produces information on each activity, indicating how much time is available before the activity becomes critical to the completion of the project. This is known as float: the period by which a task can be delayed, brought forward or extended without affecting the programme end date. By adding resource requirements to each activity it is possible to calculate the total resource requirements for the project. Scheduling the start and finish of the activities may optimise these resource requirements.

Initially network analysis diagrams used straight lines and circles to show the relationship between events. This form of network analysis was known as 'activity-on-arrow'. Subsequently, the activity-on-node or 'precedence diagram' method as it became known was developed. This became the preferred form of network diagram and is the format most often used today.

Figure 3.3 shows an activity-on-arrow network.
Figure 3.4 shows a precedence diagram.

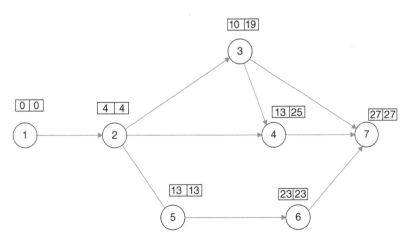

Figure 3.3   An example of an Activity-on-Arrow network diagram.

Section II

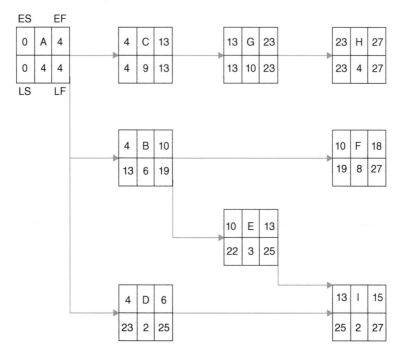

Figure 3.4 An example of a precedence diagram.

For both the activity-on-arrow diagram and the precedence diagram there are rules for drawing and analysing the network. These rules and how to calculate the timings relating to each activity are readily available in all standard textbooks. Details are also included at the website produced for this book. Within this chapter we focus on the fundamentals of how to draw the diagrams and how to incorporate the logic within the links.

Activity-on-arrow networks

Figure 3.3 shows an activity-on-arrow network comprising nine activities. The completed calculations show that the work will be completed in 27 days. In discussing this type of modelling we shall first consider the general rules for drawing the network.

Drawing the network

When drawing an Activity-on-Arrow network the following conventions are adopted:

- Time flows from left to right.
- Numbers within circles represent events.
- The arrowhead determines the direction of flow between events.

- Events are known as Head Events or Tail Events.
- Head events always have a number higher than the tail event.
- The length of the arrow has no significance and is not drawn in proportion to the anticipated duration.
- The orientation of the arrow has no significance. (Most planners prefer to use horizontal and vertical lines.)
- The description of the activity should be written upon the straight portion of the arrow.

Each activity must be given a description. In using the technique each activity is recognised either by this description or by the start and end events. It is imperative to ensure that the logic of the network is correct. Lockyer (1974) encourages the adoption of a rigorous examination of the logic. 'An Activity that depends upon another is shown to emerge from the head event of the activity on which it depends'. This leads to two fundamental properties of events and activities:

> "An event cannot be said to occur (or be reached or achieved) until all activities leading to it are complete
>
> No activity may start until all previous activities in the chain are complete. (p. 36)"

Figure 3.5 shows typical examples of how the network is drawn to maintain the logic of the work to be undertaken.

Figure 3.5a shows the components of each activity which comprise head event (or node), a description, a duration and a tail event (or node) j.

Figure 3.5b shows a single activity with its description and its duration.

Fig 3.5c shows two activities, A and B. Activity B commences after the completion of Activity A.

In Figure 3.5d Activity B commences after the completion of Activity A. Activity C commences after the completion of Activity A. (Activity B is not dependent on Activity C and Activity C is not dependent on Activity B.)

In Figure 3.5e Activity Y can only commence when Activity M is complete, Activity T is complete and Activity V is complete.

In some instances, it is necessary to introduce dummy activities to maintain the logic of the network. Dummy activities are shown as dotted lines. They have no description and no duration.

In Figure 3.5f, a dummy activity has been introduced between Node 151 and Node 161 to maintain the logic: Activity Z can commence after the completion of Activity W and the completion of Activity X. Activity Y can commence after the completion of Activity X. Activity Y is not dependent on Activity Z. Activity Z is not dependent on Activity Y.

Two errors in drawing up the network must be avoided. Figure 3.5g shows a loop in the logic that prevents Activity K commencing. Figure 3.5h shows an activity left 'dangling' with no subsequent activity. A loop in the logic will prevent the analysis of the network. A 'dangling' activity will not prevent the network analysis but may lead to errors in the overall logic. To prevent such errors from arising, it is good practice to commence and end the network with a single activity. This is shown in Figure 3.5i and j.

Section II

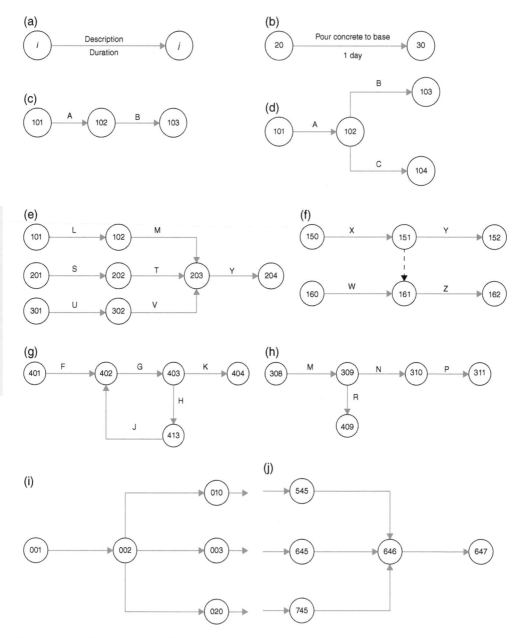

Figure 3.5   Drawing an activity-on-arrow network.

These basic rules apply whatever the type of work and irrespective of the number of activities. When the number of activities runs into several hundreds or even thousands then significant time and resources will be needed to manage the network.

After completing the network and checking the logic the duration of each activity should be estimated and included below the activity line. Activities may be considered in terms of working, days, weeks or months but the chosen unit of time must be used consistently throughout the network. Estimating the duration of activities is considered in Chapter 4. How to calculate the earliest and latest start times of each activity together with the float for each activity and the critical path for the network is shown in the examples provided on the website for the book. See www.wiley.com/go/baldwin/constructionplansched.

Precedence diagrams

Precedence diagrams are networks in which the activities are drawn as nodes. Each node holds all the relevant information for the activity.

Figure 3.6a shows this in diagrammatic form.

The network is drawn in a similar way as an Activity-on-Arrow network, but not all of the same conventions apply.

Drawing the network-precedence diagrams

When drawing a precedence diagram:

- Time flows from left to right.
- The arrowhead determines the direction of flow of production.
- The length of the arrow has no significance.
- The orientation of the arrow has no significance.
- The description of the activity is written on the node.

Each activity must be given a description. (Each activity is recognised either by this description or by the start and end events.) It is imperative to ensure that the logic of the network is correct. The same logic rules apply as for the activity-on-arrow network diagrams.

Figure 3.6 shows the annotation used in the diagrams and typical examples of how the network is drawn to maintain the logic of the work to be undertaken.

In Fig 3.6b Activity B commences after the completion of Activity A.

In Figure 3.6c Activity B commences after the completion of Activity A. Activity C commences after the completion of Activity A. (Activity B is not dependent on Activity C and Activity C is not dependent on Activity B.)

In Figure 3.6d Activity Y can only commence when Activity M is complete, Activity T is complete and Activity V is complete.

In precedence diagrams, there is no need for dummy activities to maintain the logic of the diagram. However, when drawing precedence diagrams the same two errors that need to be considered in activity-on-arrow networks must also be avoided: no loops and no dangles. Therefore a network containing a link as shown in Figures 3.6e is invalid. Figure 3.6f is not strictly invalid but it is inadvisable to draw networks with activities left 'dangling'. It is better to draw the network as shown in Figure 3.9(g) with all activities linked to one final Activity Z.

These basic rules apply whatever the type of work and irrespective of the number of activities as with activity-on-arrow' networks maintaining large precedence networks is time consuming and costly.

Section II

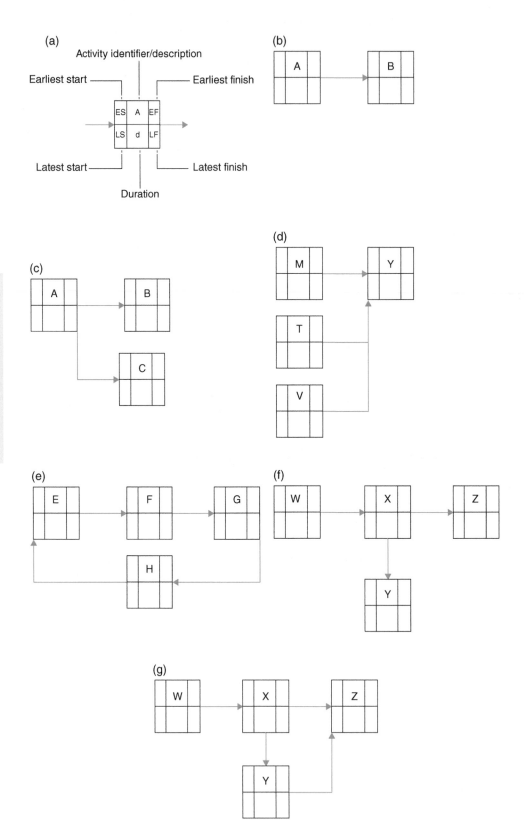

Figure 3.6 Drawing a precedence diagram.

After completing the network and checking the logic, the duration of each activity should be estimated and clearly stated within the node of the activity. As with activity-on-arrow networks the duration of an activity may be considered in working, days, weeks or months. The chosen unit of time must be used consistently throughout the network. Having estimated the activity durations the calculations for the precedence diagram may be performed.

The calculations for the precedence diagram shown in Figure 3.6 are provided at the website for the book, see www.wiley.com/go/baldwin/constructionplansched.

Linked bar charts

Despite the length of time that activity-on-arrow and precedence diagrams have been available for charting and analysing construction work the most popular form of construction analysis and scheduling is by linked bar charts (Cooke and Williams, 2009). One reason for this has been the emergence and widespread use of computer software products that both draw linked bar chart diagrams and also provide planners with the same information as established network analysis software products.

These products allow the user to enter basic activity data (activity identifier, activity name, activity duration, etc.) and then present this list of activities in a screen display format that represents a bar chart diagram. Users may then draw logic links between the various activities to show the construction sequence. The software will then re-schedule the activities displayed on the screen and calculate the completion date for the project and the critical path. Figure 3.7 shows the principles of the linked bar chart.

The basic features and calculations within linked bar chart software are identical with those within network analysis software. With the ability to annotate these linked bar chart diagrams and produce high quality print out copies, this software has become increasingly popular.

Space diagrams

Bar chart diagrams provide a two-dimensional 'picture' of the activities on a project and a schedule of when they will take place. As such they form one kind of chart. Planners show the relationship between construction activities by the use of other forms of diagrams. Mawdesley et al. (1997) call these 'space diagrams'. Space diagrams exist in a variety of formats. We shall highlight two: time chainage charts and multiple activity charts.

Time chainage charts

'Time chainage charts' are diagrams that are prepared for construction projects where the location of the work items is measured on the basis of a linear distance from a specific geographical point. Examples of such projects are the construction of a road, the construction of a railway and the construction of a pipeline. On such

Section II

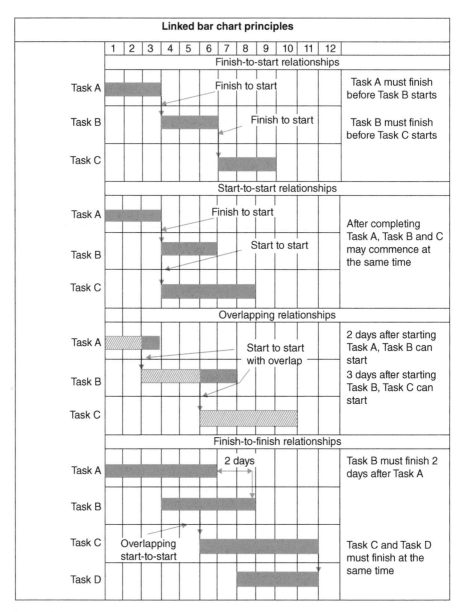

Figure 3.7 Principles of a linked bar chart diagram (From Cooke and Williams, 2009).

projects all construction works (a bridge, a culvert, items of drainage, etc.) are located at a specific chainage (or distance) from the start of the work. This ensures that all constructions including similar items (e.g. culverts, manholes, etc.) have a unique reference.

An example of a time chainage chart is shown in Figure 3.8.

Figure 3.8 shows the main items of construction work for a new section of road. Distance, that is, chainage is shown horizontally. (The diagram shows construction

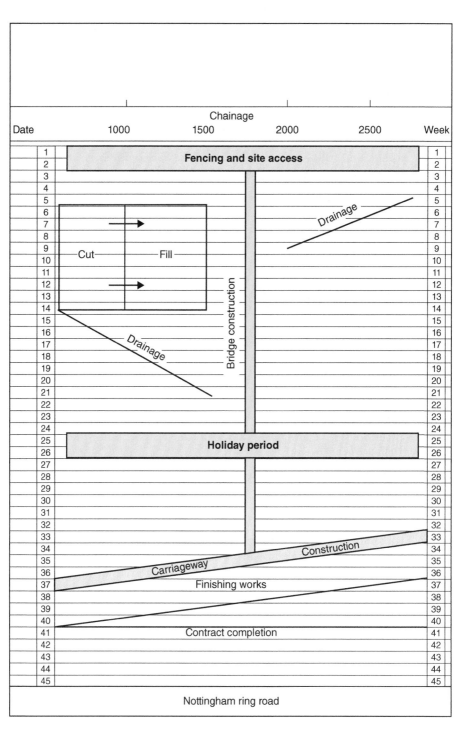

Date	Chainage
	1000 1500 2000 2500 Week

Fencing and site access

Cut ———— Fill

Drainage

Drainage

Bridge construction

Holiday period

Carriageway

Construction

Finishing works

Contract completion

Nottingham ring road

Section II

Figure 3.8   An example from a time chainage chart.

between Chainage 0000 and chainage 2750.) Time is shown vertically. (The work commences in week 1 and is complete in week 40 the following year.)

This is the usual format for representing the work items. For buildings, the convention is to show items with time and distance on the alternative axes: time is plotted horizontally and the level of the building, vertically. The line of balance (LOB) method is an example of a time chainage chart for a building.

Whatever the convention used for the axes, the work represented on the diagram is modelled by a shape drawn between times and locations. The diagram includes the following graphical notation: horizontal lines, vertical lines, diagonal lines, rectangles and parallelograms.

A *horizontal* line represents an activity that takes place across all or part of the project at the same time.

A *vertical* line shows an activity that takes place at a specified chainage and continues for a number of weeks.

A *diagonal* line indicates an activity that commences at one chainage and then moves forwards (or backwards) along line of the works until all the work is completed several weeks later. (An example of this would be Motorway Drainage.) It should be noted that the angle of the diagonal against the time scale is important because it indicates the direction that the work gang will take along the length of the works.

A *rectangle* shows where work will take place along a length of the project over a period of time. A typical example of the use of this form of graphic is when earthworks activity is shown. If the earthworks are represented by more than one rectangle it is possible to indicate areas of both cut and fill.

A *parallelogram* (or sloping box) is used to show activities that take place over a period of time and occupy a significant proportion of the works, for example, the laying of road base.

Limitations of time chainage charts

A time chainage chart provides a graphical summary of the main activities of the construction work and communicates to the reader the overall plan for the works, what work will take place and when.

The limitations of the time chainage chart are similar to those of the bar chart. The graphical devices used to show the construction activities are all approximations. They assume work progresses at the same rate at all locations and there is no difference in the amount of work to be undertaken at each location. Consider again the construction of motorway drainage. The depth of the trench will determine the production rate of the drainage gang. The depth of the manholes along the line of the motorway may vary considerably, which will result in some manholes taking considerably longer than others. Actual production rates will always vary along the line of the works.

Time chainage charts therefore need to be interpreted with additional documentation if they are to provide a full insight into the construction work to be undertaken and the parts of the works that will be critical to the success of the project.

As with bar charts, time chainage charts may be used to monitor the progress of the works. Colouring the different activities on the chart will show progress on individual activities. Problems arise when it becomes necessary to re-draw the time chainage chart. If produced by hand this becomes time consuming. Fortunately there are currently several software products that may be used for this purpose. The availability of these products may well result in a revival in the use of time chainage charts within the industry.

Multiple activity charts

Multiple activity charts are two-dimensional diagrams used for short-term planning. They originate from work study methods originally developed for factory production line and workshop environments. They provide an optimised schedule of the production programme based upon the resources (labour, plant and equipment) available for use.

Multiple activity charts are suitable for the planning of the repetitive elements of building construction work, for example, the construction of a concrete building, the fitting out of high rise residential and commercial buildings. Multiple activity charts are a good communication tool and are especially useful for conveying to work gangs the proposed schedule of construction.

An example of a simple multiple activity chart is shown in Figure 3.9. This shows a worker supporting three machines and the time when each machine is productive. Clearly, action is needed to ensure that the machines are kept productive.

Similar charts may be produced for teams of tradesmen working on different construction activities in specific locations, for example, fitting out hotel rooms and preparing for concrete pours in the basement of a building. There is no single approved format for multiple activity charts except that convention determines that time is shown on the vertical axis of the diagram. Shading may be used to distinguish between different activities and idle time.

Preparation of a multiple activity chart

Uher and Zantis (2011) provide detailed examples of how to prepare a multiple activity chart. Here we summarise their method, you are recommended to refer to their text to view their detailed example. They summarise the steps for preparation of a multiple activity chart as follows:

1. Identify the repetitive cycles of work.
2. Consider alternative methods of undertaking the work.
3. Select the preferred method (Consider dividing the repetitive cycles of work into smaller elements to improve resource usage.)
4. Compile detailed information on the materials and resources needed to complete the work.
5. Analyse the planning strategy for one repetitive area of the project using a precedence schedule.
6. Estimate the duration of the repetitive cycles of work based on previous experience.
7. Determine the size of labour crews for specific key trades and key plant items.
8. Start scheduling by identifying key activities that signify the start and finish of a repetitive cycle of work.

Section II

Time (mins)	Worker	Machine 1	Machine 2	Machine 3
1	Load 1	Idle	Idle	Idle
2	Load 1	Idle	Idle	Idle
3	Load 2	Run	Idle	Idle
4	Load 2	Run	Idle	Idle
5	Load 3	Run	Run	Idle
6	Load 3	Run	Run	Idle
7	Unload 1	Idle	Run	Run
8	Unload 1	Idle	Run	Run
9	Load 1	Idle	Run	Run
10	Load 1	Idle	Run	Run
11	Unload 2	Run	Idle	Run
12	Unload 2	Run	Idle	Run
13	Load 2	Run	Idle	Run
14	Load 2	Run	Idle	Idle
15	Unload 3	Idle	Run	Idle
16	Unload 3	Idle	Run	Idle
17	Load 3	Idle	Run	Idle
18	Load 3	Idle	Run	Idle
19	Unload 1	Idle	Run	Run
20	Unload 1	Idle	Run	Run
21	Load 1	Run	Idle	Run
22	Load 1	Run	Idle	Run
23	Unload 2	Run	Idle	Idle
24	Unload 2	Run	Idle	Idle
25	Load 2	Run	Idle	Idle
26	Load 2	Run	Idle	Idle
27	Unload 3	Idle	Run	Idle
28	Unload 3	Idle	Run	Idle
29	Load 3	Idle	Run	Idle
30	Load 3	Idle	Run	Idle
31	Unload 1	Idle	Idle	Run
32	Unload 1	Idle	Idle	Run
33	Load 1	Idle	Idle	Run
34	Load 1	Idle	Idle	Run

Figure 3.9 An example of a multiple activity chart.

9. Schedule the activities that take place between these two activities.
10. Review and revise the schedule as required, varying the cycle time of the work items until a satisfactory schedule has been obtained.
11. Check that the production logic has been maintained, the contract requirements have been satisfied, the resources are neither over-committed or under used; the crew sizes are appropriate for the tasks; the plant items selected have the required capacity appropriate for the tasks.
12. Check that there are no space limitations that will prevent the work proceeding to the prepared schedule (Uher and Zantis, 2011).

Limitations of the multiple activity chart

Multiple activity charts enable planners to produce detailed work schedules that make optimum use of the resources. Their lack of use in the construction industry rests primarily with the time taken to produce them and the absence of readily available software to assist the planner. However, when produced they provide a comprehensive analysis of the tasks involved and their interaction.

Line of balance

Line of balance (LOB) is an established graphical technique for scheduling repetitive construction tasks. The LOB technique was developed for use in a manufacturing/engineering production environment and then adapted for use on repetitive construction projects. The technique requires the planner to consider carefully the required rate of handover of the construction units to the client and the production rates of work gangs to meet this demand. It is a graphical method that is less complex than traditional critical path methods. It can be used for monitoring and control purposes.

Although LOB has existed for more than 50 years it has failed to become widely adopted in the industry, tending to be overlooked in favour of other techniques. With the arrival of specialist software and the advent of building information modelling its adoption and use may now greatly increase.

Many construction projects incorporate a high degree of repetitive tasks. Such projects include high-rise residential and commercial buildings, hotels, housing estates, and infrastructure projects such as roads, railways, pipelines, and jetties. Line of Balance focuses on the repetitive elements within such projects. Working from a required production output the technique calculates the production rates and resources necessary to meet project deadlines. The resulting schedule is usually displayed in a graphical form showing the handover time for the completed elements and the production schedule for the main sub-elements.

Figure 3.10 shows an LOB diagram for the completion of 30 units of construction. (e.g. 30 houses.)

The diagram shows that after construction commences it takes 64 working days (13 weeks) to complete the first unit. The contractor has agreed to hand over the

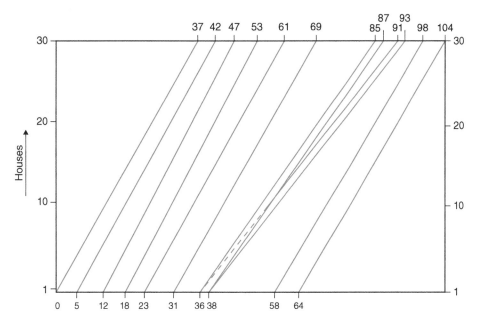

Figure 3.10 An LOB diagram for the completion of 30 houses.

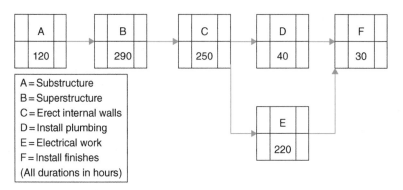

Figure 3.11   A simple network diagram for the construction of each house.

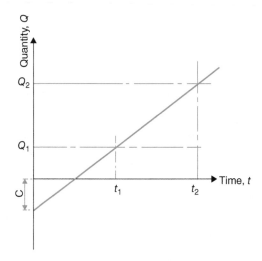

Figure 3.12   An example of an objective diagram.

units at the rate of four per week. Handover of the completed 30 units will take place after 104 days, at the end of 21 weeks.

Each unit comprises six main tasks, identified as A–F. Figure 3.11 shows a simple network diagram for the construction of each house. The durations are based on the original estimate of the time taken by each gang to complete the work in each house. (In practice, the gang size may change to produce the optimum schedule.)

Figure 3.12 is an example of an 'objective diagram' which shows a line-of-balance diagram in general terms (see Pilcher, 1992). This allows you to study the relationship between quantity and time. This relationship is assumed to be linear:

$$Q = mt + C$$

where
   Q = LOB quantity of units produced
   m = required rate of delivery (number per units of time: days, weeks, etc.)
   t = time
   C = a constant

Considering the relationship between Q_1, Q_2, t_1 and t_2

$$t_2 = \frac{Q_2 - Q_1}{m + t_1}$$

From this equation and knowing the anticipated (or actual) time taken to complete any individual unit it is possible to calculate the completion time for any subsequent unit.

For the maximum productivity all the activities within the construction programme must proceed at the same rate. This requires the appropriate number of resources to undertake the work. In practice, for some construction situations it may not be economic to provide this level of resource. Where the output is based on a gang of worker with a fixed number of gang members, for example, a brickwork gang with a fixed number of members, it would clearly be uneconomic to provide a gang of workers who were not fully occupied. This leads to the type of diagram shown in Figure 3.10.

Line of balance – resource scheduling

In practice most LOB schedules are prepared considering the resources required.

Consider again the LOB diagram shown in Figure 3.10. The tasks are shown as six separate sloping lines. From the diagram it is evident that these tasks will have different durations and different production rates. (The gradient of each line is different.) The planner has allowed a 5-day 'buffer' between each task, that is, for each of the units to be produced the planner has allowed 5 days between the end of one task and the commencement of the next. The graph shows when each unit will be completed and when each task within each unit will be completed. It is more than simply a calculation for the overall completion of the project, it is a tool for monitoring and controlling production. For example, if you look at the horizontal line drawn through unit 10 you can see exactly when each of the trade gangs is scheduled to be working on this unit. Similarly a vertical line drawn at 50 days shows which trades should be on site and which housing units on which they should be working.

Key points relating to the LOB technique

- Line of Balance is suitable for medium and to long-term planning. It is best used at the tender or pre-contract planning stage.
- The LOB diagram is drawn on the basis of an assumed handover rate for the units under construction. If you assume that all activities will be carried out to meet the required handover rate then the construction time will be minimised but resource requirements will be maximised and may result in work gangs waiting between construction operations between units.
- An alternative resource based scheduling approach reviews the level of resources required and the balance between different groups of resources. This enables efficient use of committed resources and may be used to optimise the production work for the contractor's work gangs or subcontractors. This approach is the more usual approach.

Section II

- The LOB technique can accommodate single work crews or multiple work crews. (The schedule may be drawn assuming more than one work gang if this is required to achieve the overall production rate.)
- The lines are drawn straight. This assumes that each activity has the same rate of progress throughout the project. (This is an assumption that is rarely achieved in practice but is a close enough approximation in most if not all cases.)
- The 'buffers' between activities may represent specific time allowances (e.g. the time allowed for the curing of concrete) or serve as contingencies, for example, the moving of location and the setting up of equipment.
- Before drawing the schedule it is necessary to consider in detail the work to be completed in each unit and establish a plan for each unit. (This may be drawn up using a bar chart, precedence diagram or multiple activity chart.)
- The LOB chart can be drawn to reflect different construction circumstances, for example, continuity of resource utilisation, varying gang sizes and optimum gang sizes.
- Drawing an LOB requires the planner to consider the relationships between the activities and how they impact each other.
- The delivery (handover of units) may be presented in one of two ways: a weekly schedule stating the unit numbers to be completed by the end of each week, or as the overall diagram.
- Displaying a parallel line to the left of the handover line provides a LOB for each individual trade activity. This is known as parallel scheduling.

An example of an LOB diagram based on resource scheduling together with the calculations and instructions for drawing the chart is provided at the website for the book www.wiley.com/go/baldwin/constructionplansched.

Software for producing LOB diagrams

Universally acknowledged, LOB has not been widely adopted throughout the construction industry. One reason for this is the need to re-work the calculations and re-draw the production graph. In the absence of suitable software, this becomes tedious and time consuming. Such software now exists. Software such as Vico Systems™ software provides flow-line scheduling planning which, based on LOB principles, seeks to align construction operations and minimise waiting time between construction activities. This scheduling system is linked to their BIM software systems. Clearly, with the increased use of BIM there will be new opportunities to incorporate LOB thinking into the preparation of construction schedules.

ADePT

The increasing need for shorter lead times and improved productivity tests the innovative methods of not only construction teams but also engineering design and demands new approaches to design management. Design is an iterative process. One of the limitations of critical path methods is their inability to accommodate logic loops, that is, iteration. The ADePT methodology overcomes this problem and assists designers manage the design process. It utilises data flow diagrams,

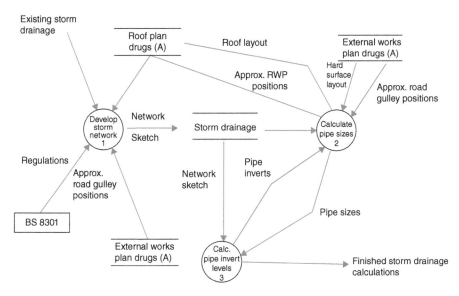

Figure 3.13   An example of a data flow diagram.

information dependency tables, matrix methods of analysis and traditional scheduling techniques to plan design tasks and monitor the production of design work. Each of these aspects of the methodology is reviewed below.

Data flow diagrams

Data flow diagrams were originally designed for use in the development of computer systems (see Yourdon, 1989; Newton, 1996). They identify the tasks to be undertaken and the information required to complete the task. Within ADePT, data flow diagrams show external sources of information and the deliverables of the design process; they do not model the actual design process, only the information required and the outcomes. Data flow diagrams adopt a structured, hierarchical way allowing design elements to be broken down into sub-tasks of increasing granularity. An example of a data flow diagram is shown in Figure 3.13.

A generic model for detailed building design

Using an amended form of the data flow diagrams method known as IDEF0, the ADePT researchers (Austin et al., 1999; Baldwin et al., 1999) produced a generic model for the building design process. Working closely with a number of industry organisations the research developed a generic model for the detailed design process comprising some four hundred-design tasks and several thousand information flows. The model was verified by input from architects and engineers and validated across a number of different types of commercial buildings. Figure 3.14 shows the structure of the model.

Section II

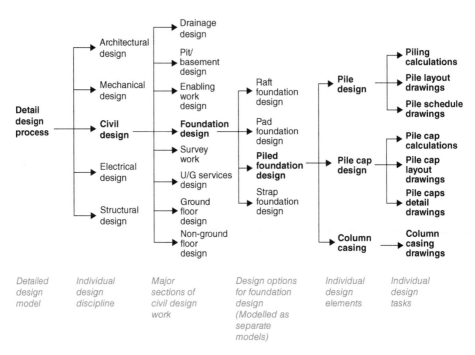

Figure 3.14   The structure of the detailed design process model.

Figure 3.15 shows an extract from the model highlighting the information dependency table.

This generic model provides a basis for analysing the information requirements of designers on any type of building or infrastructure project. The model is independent of the method of procurement adopted by the client and may be extended or amended to meet the requirements of the project under consideration. (E.g., if the particular project requires input from specialist consultants, e.g. fire engineering, then the model may be adapted accordingly.) The project specific model provides a basis for the project team to review the design tasks and the associated information needs. Data from the model and the schedules are then analysed using dependency structure matrix (DSM) analysis methods.

Dependency structure matrix analysis

Don Steward developed the DSM analysis method (Steward, 1981). It was subsequently adopted by a number of researchers and industry practitioners to analyse complex production engineering systems (see Ulrich and Eppinger, 2011). The DSM method is used in the ADePT system to identify interdependent tasks and optimise these tasks to minimise iteration. The method establishes a large matrix showing the design tasks (identified by numbers) on both axes. The squares of the matrix are marked to indicate where there is an information need. (This need is identified in the process model and listed in the information dependency tables.) This matrix may then be analysed and the design process optimised

Activity		Information required		
Number	Name	Name	Type	Source activity
A 3 4 2 1	Lift shaft structure calcs	Ground water levels	Cross-disciplinary	A 2 2 1 3
		Lift plans & elevations	Cross-disciplinary	A 5 7 1 3
		Lift shaft pit/head reqs	Cross-disciplinary	A 5 7 1 2
		Lift type	Cross-disciplinary	A 5 7 1 2
		Precast floor details	Cross-disciplinary	A 3 3 2 3 2 4
		Precast floor details	Cross-disciplinary	A 3 3 2 3 2 7
		Lifting beam details	Intra-disciplinary	A 3 4 1 4
		Plant floor details	Intra-disciplinary	A 3 3 2 3 1 4
		Plant floor details	Intra-disciplinary	A 3 3 2 3 2 4
A 3 4 2 2	Lift shaft structure drawings	AC/vent layouts	Cross-disciplinary	A 4 2 3 5
		Lift control method	Cross-disciplinary	A 5 7 1 1
		Lift door opening details	Cross-disciplinary	A 5 7 1 1
		Lift levelling accuracy	Cross-disciplinary	A 5 7 1 1
		Lift motor room vent reqs	Cross-disciplinary	A 5 7 1 1
		Lift shaft pit/head reqs	Cross-disciplinary	A 5 7 1 2
		Pre-cast floor details	Cross-disciplinary	A 3 3 2 3 2 4
		Pre-cast floor details	Cross-disciplinary	A 3 3 2 3 2 7
		Lift shaft structure calcs	Intra-disciplinary	A 3 4 2 1
		Plant floor details	Intra-disciplinary	A 3 3 2 3 1 4
		Plant floor details	Intra-disciplinary	A 3 3 2 3 2 4
		Subcontractor's info.	External	
		Manufacturer's info.	External	
A 3 4 2 3	Lift shaft structure specs	Lift shaft structure calcs	Intra-disciplinary	A 3 4 2 1
		Lift shaft structure details	Intra-disciplinary	A 3 4 2 2

Figure 3.15   An extract from the detailed design process model.

by re-ordering the matrix and minimising design iteration. The process of re-ordering the matrix is known as 'partitioning'. Reviewing the natural groupings of tasks and changing the order of the tasks taking into account the relative importance of the information requirements enables the design manager to optimise the design process.

Figure 3.16 shows an illustrative example of the information links between 20 design tasks.

Where the information requirement (shown as an X on the matrix) is below the diagonal there is no problem as the information required will have been produced by a previous design task. Where the information requirement is above the diagonal there is a difficulty, the information required has not yet been produced. This will lead to the large iterative loop (the shaded area) shown in the figure.

By assuming values for selected items of information the matrix may be partitioned or re-ordered. The process is re-ordered to minimise the number of dependencies above the diagonal and get marks close to or ideally below the diagonal. Figure 3.17, a revised matrix diagram, shows that by making some estimates of information values the single large iterative loop may be reduced to two smaller iterative loops.

By classifying the information dependencies (e.g. A = Strong, B = Medium, C = Weak) it is easier to optimise the design process. Weak dependencies (e.g. ones where a good estimate can be made) may be ignored and the matrix can be re-partitioned to produce a more suitable design order.

Section II

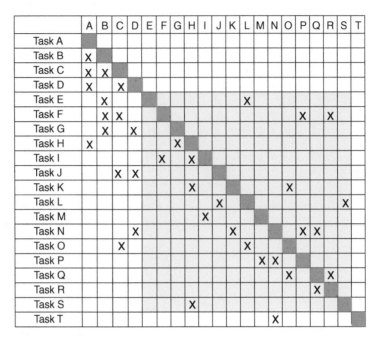

	A	B	C	D	E	F	G	H	I	J	K	L	M	N	O	P	Q	R	S	T
Task A	■																			
Task B	X	■																		
Task C	X	X	■																	
Task D	X		X	■																
Task E		X			■							X								
Task F		X	X			■										X		X		
Task G		X		X			■													
Task H	X						X	■												
Task I						X		X	■											
Task J			X	X						■										
Task K								X			■				X					
Task L											X	■							X	
Task M									X				■							
Task N				X							X			■		X	X			
Task O			X									X			■					
Task P														X	X	■				
Task Q															X		■	X		
Task R																	X	■		
Task S								X											■	
Task T															X					■

Figure 3.16 An example of the information links between 20 design tasks.

	A	B	C	D	G	J	L	E	I	S	O	M	Q	R	T	P	F	H	K	N
Task A	■																			
Task B	X	■																		
Task C	X	X	■																	
Task D	X		X	■																
Task G		X		X	■															
Task J			X	X		■														
Task L							■			X									X	
Task E		X					X	■												
Task I									■								X	X		
Task S										■								X		
Task O			X				X				■									
Task M									X			■								
Task Q											X		■	X						
Task R													X	■						
Task T											X				■					
Task P											X					■				X
Task F		X	X											X		X	■			
Task H	X				X													■		
Task K											X							X	■	
Task N				X									X			X			X	■

Figure 3.17 A revised matrix diagram.

Using this matrix technique tasks can be moved up and down the matrix, which is then re-partitioned, to allow the effects of the change to be analysed. If the contractor requires specific items of information early (e.g. foundation details), the impact on the design process and design solution can be assessed and a cost estimate produced. If the requirements of the client change, the impact on the design process can be assessed.

Producing project and departmental schedules

Output from the matrix analysis may be used to schedule design tasks for the overall project and individual design departments using traditional planning and scheduling software. Start and end dates for each design activity may be established and traditional bar chart activities produced. Data from the design programme may be combined with construction task data to optimise the design and construction process. Chapter 8 describes in more detail how this may be achieved.

Experience in the use of the ADePT technique has shown that having optimised the design process it is necessary to manage the process through to completion using work-flow techniques to monitor production to an agreed work plan. This work plan is best produced adopting the principles of Lean Production (see Chapter 7).

4D CAD

4D CAD (or 4D Planning as it is also known) is the term used to describe the use of computer-based tools to visualise the construction plan in an environment that combines 3D modelling with time. 4D CAD provides the planner with the potential to represent the construction phase of the project visually (as opposed to in bar chart form), and to provide a basis for the evaluation of alternative construction schedules. This assists in marketing and client briefing and promotes collaborative working within the project team. Simulations can be used to ensure the probity of the construction process and assist in the planning of space usage.

4D CAD in construction has been developing since 1987 (see Heesom and Mahdjoubi, 2004) when construction organisations engaged in large complex projects began to use 3D modelling to build manual 4D models with schedule information to provide 'snapshots' of each phase of the project over periods of time. Customized and commercially available tools evolved in the mid-to-late 1990s, facilitating the process by manually creating 4D models with automatic links to 3D geometry, entities and groups of entities for construction activities (Eastman et al., 2011).

The development of new technologies has meant that a new range of products and technologies known as BIM have now emerged as the platform for modelling and simulating all aspects of the building product and the building process. The continued use and development of 4D CAD is now assured. (More details of 4D CAD and BIM are provided in Chapter 10.)

Section II

Key points

- There are numerous planning and scheduling techniques available to assist the construction planner.
- The selection of the technique to be used will depend upon the type of project, the degree of development of the project and the requirements of the client.
- Despite their limitations bar charts are the most familiar method for presenting plans and their value should not be underestimated.
- Flow charts and flow diagrams enable you to produce a pictorial or graphical model of a process.
- The most popular form of construction analysis technique is the linked bar chart.
- Precedence diagrams enable the construction planner to model all aspects of large complex construction projects from inception through construction to completion.
- Space diagrams show the planned construction work over a period of time and at a specific location.
- The ADePT methodology enables the planner to plan iterative processes such as design.
- 4D CAD and BIM have emerged as the new platform for modelling both the construction product and the construction process.

Chapter 4
Planning and Scheduling Practices

Schedule design and structure

With the increasing complexity of large construction projects (both in terms of the construction work and the procurement and organisational structure), the need for some standardisation in schedule design became increasing apparent. It was widely recognised that it was unsatisfactory simply to develop schedules in an *ad hoc* manner and present different schedules with different levels of information. A structure to the levels of information provided was required.

The structure of the schedules presented on a project will depend upon specific factors relating to the project, for example, the project scope and objectives, the project calendar, etc., and decisions relating to the project management and reporting requirements. The Work Breakdown Structure (WBS), activity identification coding, activity content coding and activity cost coding systems will all influence the design and structure of the schedule.

A clear structure for the schedules must be established.

The Chartered Institute of Building, CIOB (2011) recommends five levels of schedule reporting. This number of levels is now widely accepted as the preferred number of reporting levels for a *single* project. These are shown diagrammatically in Figure 4.1.

Level 1 schedule report

The Level 1 schedule report is the highest level of reporting on the project. (It may also be known as the executive summary report or the master schedule.) The report is a summary of all the reporting presented at Levels 2–5. It is best prepared as a single sheet bar chart showing the main activities involved and the timing of

A Handbook for Construction Planning and Scheduling, First Edition. Andrew Baldwin and David Bordoli.
© 2014 John Wiley & Sons, Ltd. Published 2014 by John Wiley & Sons, Ltd.

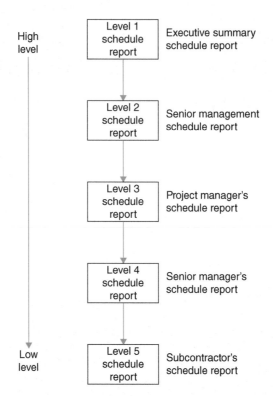

Figure 4.1   Five levels of schedule reporting.

all the key stages on the project. Where possible, milestone dates should be high-lighted. (If this is not possible, then this information should be presented as an addendum.) The critical path through the activities should be clearly shown.

Level 2 schedule report

The Level 2 report is for senior management. It is a summary of the Level 3 sched-ule. The report should be divided into major components on the basis of areas or elements within the project. The report may be a single schedule that combines details from different schedules or a suite of individual schedules. Ideally, one person should produce the schedule. The schedule should enable all concerned within each element of the project to fully understand what needs to be done and how their work interfaces with other elements. The report should enable progress on each element to be monitored. On smaller projects, Level 2 reports may not be required.

Level 3 schedule report

The Level 3 schedule is sometimes known as the project manager's schedule. It shows the detailed timing of all the activities on the project and the critical path through these activities. It may also indicate the level of resources required to undertake the work. Each element of the work must be considered. (Examples of Level 3 schedules include the detailed design schedule listing every drawing to be produced, details of the procurement of all the structural steelwork required, etc.) It is important to clearly identify which activities take place off-site and which occur on-site. Again, ideally, one party should produce the Level 3 schedule. The Level 3 schedule must incorporate details relating to each of the main sub-contractor's activities. These details must be checked and agreed by the sub-contractors concerned. The main contractor may decide not to share details from the Level 3 schedule with the client. This decision will depend upon the contractual arrangements.

Level 4 schedule report

Level 4 schedules comprise information filtered from Level 3 schedules to create more detailed schedules, typically for use by the section manager. This information may be required for a wide range of users relating to specific areas of the project or specific aspects of the work. It will comprise short-term or medium-term schedules, covering the work to be undertaken in the next few weeks or months. The start date and end date of each activity must be clearly shown. If a suitable coding system for construction activities has been adopted and rigorously applied, then Level 4 schedules may be produced directly from Level 3 schedules. Level 4 schedules should be provided for each sub-contractor prior to them commencing their work. Level 4 schedules are sometimes known as 'constraints schedules'.

Level 5 schedule report

Level 5 schedules provide details of the timing of all the activities to be undertaken by the different sub-contractors or trades. A separate schedule should be produced for each sub-contractor/trade. The timings of the activities will align with those in the Level 3, but the schedule is in much more detail showing clearly how the work will be executed. They provide more detail than that shown in the Level 4 schedules. The gang leaders will use these schedules in the execution of the work. They are short-term plans typically covering the work to be undertaken during the next week. The schedule produced for each sub-contractor should show the critical path.

What is required on smaller projects?

The five levels of schedule recommended by the CIOB are typically required on large, complex projects. Smaller projects may use different levels of schedule, and it may be possible to combine Level 3, 4 and 5 schedules into one detailed schedule. On very small projects, the Level 2 schedule may also be omitted. These decisions are dependent upon the size of the project, the type of work, the form of procurement and the number of parties involved.

Section II

What is required when the project concerned is part of a schedule comprising several inter-linked projects?

When the project is a part of a programme of projects or a portfolio of projects undertaken for a client then there is a need to produce an additional level of schedule. The PEO (2005) call this schedule a Level 0 schedule. This schedule is a summary of all the projects within the programme of work. The level of detail provided in this Level 0 schedule will depend upon the number of projects that make up the overall programme. If this is a small number, then each individual project may be represented by say ten activities. If there are a large number of individual projects, then each project may be represented by a single activity.

Creating these schedules

To operate efficiently, the five levels of schedule described above must form a set of coordinated schedules where activities are integrated and where, importantly, the timing of the activities (and sub-activities) is clearly evident and consistent.

Ideally, the schedules should be produced in a 'top-down' way whereby the Level 1 schedule is produced, agreed with all parties and then used to form the basis of Level 2, then Level 3 schedules and so on. Given the time available, the number of parties concerned and often their different locations, this may not always be possible. More often, schedules are created and maintained as stand-alone programmes and then combined to produce schedules at a higher level. This 'real-world' situation demands clear rules for both the timing of schedule production and the content required. Schedule production needs to be planned, agreed and actively managed.

Links with the Work Breakdown Structure

When creating the various levels of a schedule, one often has to consider the impact on the project WBS. From a scheduling viewpoint, there is no real (absolute) requirement to take the WBS into consideration, but experience shows that on complex projects it is helpful to find 'common ground' between the cost/budget and the scheduling.

Work Breakdown Structure

The scope of a project must be clearly defined. It is imperative to know what will be included in the work and what will not. The project scope needs to be managed effectively to control the work and keep track of the inevitable changes that will occur (Association for Project Management, 2012). This demands a comprehensive review of all aspects of the project: the work to be undertaken and how the project will be managed and supervised within the organisation's portfolio of projects. Burke (2003) identifies the WBS as a key tool in managing the scope of the project.

A WBS defines the project's discrete work elements and groups them together in a way that helps organise and define the total work scope of the work in the project. It decomposes a project into smaller elements ad helps focus on the deliverables required. A WBS also provides a framework for detailed cost estimating and control, along with guidance for schedule development and control. This approach to managing projects is widely supported across the industry. Using the WBS methodology, the work within a project can be sub-divided into work packages, the content of which may then be estimated, planned, assigned and controlled.

There are several methods of presenting a WBS. They are all based on developing a hierarchical structure. They all adopt a multi-level numbering system. Most project management software packages facilitate the use of a WBS with network activities and enable activities to be linked to the WBS. Each project will have its differences and this necessitates the design of a new workable template for the assignment of tasks and deliverables according to the selected WBS structure (CIOB, 2011).

An example of a functional WBS is shown in Figure 4.2.

The adoption of WBS means that each element of work is given a unique number. This may be linked to the master programme or to items in the bill of quantities. The referencing system may be used for preparing work packages for sub-contracts, for preparing budgets and for cost control purposes (Cooke and Williams, 2009).

For further details on the WBS approach and its use, see U.S. Department of Defense Handbook 881 B, and/or Australian Standard AS4817:2006 and BS6079-1:2002.

Figure 4.3 shows the link between the five levels of reporting and the WBS.

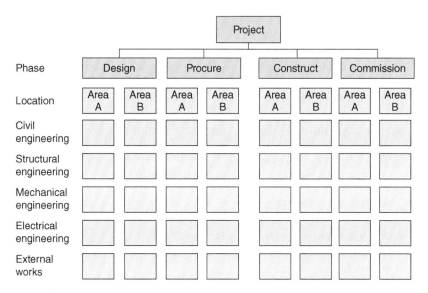

Figure 4.2   An example of a functional WBS.

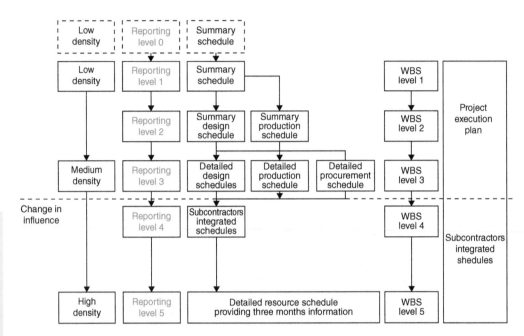

Figure 4.3   The links between the levels of reporting and the WBS (Adapted from CIOB, 2012).

Pre-tender planning, pre-contract planning, contract planning

Pre-tender planning

Construction contractors secure new work through invitations to bid for new projects. The form of contract under which the client proposes to procure the facility determines the basis of such bids. Whichever the form of procurement chosen by the client, the contractor will need to calculate the probable cost of efficiently carrying out the construction work (estimating) and establish a final price and terms for the construction (tendering).

In a 'traditional' form of contract, estimating and tendering is a process that begins with the invitation to tender and ends with the submission to the client (or the client's representative).

This typically comprises the following steps:

- The receipt of the tender documents (including a Bill of Quantities document or an approximate Bill of Quantities)
- Determining the tendering strategy
- Procurement of services and materials from outside the company
- Pricing the works
- Completion of the estimate
- Tender settlement
- Post-tender activities (CIOB, 2009)

The estimate is heavily influenced by the construction method and the construction schedule which determine:

- The sequence and duration of key construction operations;
- The timing and input of specialist sub-contractors;
- Delivery dates for materials; and
- The duration of on-costs such as supervision, transport of materials, welfare and site offices, etc.

Planning and scheduling the construction is therefore a key function within the estimating and tendering process. This is based upon the project study that begins with the receipt of the contract documents and continues until the tender is submitted. It is a process of continual refinement and revision. The preliminary study determines the decision to tender. The main study produces the construction method and the pre-tender programme. The key personnel in the contractor's organisation for this task are the estimator and the planner.

The relationship between planning and estimating depends upon the size of the project and the number of personnel within the tendering team. Sometimes one person undertakes both these tasks. Usually, different people undertake the tasks of the estimator and the planner. This provides an internal 'check mechanism'.

It is essential to gain a full appreciation of the work involved in the project. This is achieved by:

- Assessing the scope of the works and the complexity of the project
- An analysis of the bill of quantities (or approximate quantities) and other contract documents
- Establishing key project dates, events and activities
- A site visit
- Considering alternate methods of construction

Assessing the scope of the works and the complexity of the project

To assess the scope of the works, the planner will seek answers to a number of questions. What is the geographical area covered by the work? How many structures are there? What are the structural elements of each structure? What are the main responsibilities of the contractor?

The geographical area of the construction work will be examined to identify both constraints and opportunities. In a constrained urban environment, lack of space will determine the number and rate of materials deliveries, deliver times, etc. Provision may need to be made within the new building for site offices and accommodation, materials storage, etc. If the work is to be carried out over a wider geographical area, this too will influence the arrangements for the delivery, storage and distribution of materials. In extreme cases, for example the construction of a new road or motorway, a detailed analysis will need to be made of the alternative solutions for materials supply and distribution and the optimum solution selected.

The number of structures will determine the order of construction. Where the structures are similar or even identical, decisions will have to be made on the order for construction and the gang sizes required to complete the work in the optimum time.

The structural elements and the type of structure will determine the construction process, the suppliers and the procurement requirements. They will also determine milestone dates in the construction programme.

The complexity of the project will determine the team required to complete the tender and the resources and methods required to complete the construction. Complex projects demand closer management control and include more risk. They require more attention to detail and a greater planning effort at all stages of the project. Complex projects will require a tender preparation bar chart schedule to be produced, showing who is responsible for the production and collation of all the information needed to produce the tender bid.

The CIOB (2011) lists a number of matters that need to be considered when producing a schedule for construction work. This list is shown in Table 4.1. They comprise a useful checklist when considering *any* new project and a reminder of all the constraints that impact planning and programming decisions when revisiting an existing schedule.

An analysis of the bill of quantities (or approximate quantities) and other contract documents

Within the short time available to produce the tender, there is the need to swiftly gain an understanding of the work to be completed, the constraints and the opportunities.

Table 4.1   A checklist of items to be considered when designing the schedule.

Time for completion	Licenses and permissions
Sectional and key completion dates	Provisional and prime cost sums
Unspecified milestones	Specifications
Access, egress and possessions	Bills of quantities
Information release dates	Environmental considerations
Submittals and approvals	Health and safety
Procurement strategy	Labour and plant resources
Materials' delivery and storage	Method of construction
Temporary works	Sequence of construction
Temporary traffic arrangements	Schedule requirements
Working hours and holidays	Updating requirements
Design responsibility	Notice requirements
Complexity of design	Reporting requirements
Adjoining owners	End-user requirements
Risk allocation	Testing and commissioning
Sub-contractors and suppliers	Furniture and fittings
Separate contractors	Phased occupation
Employer's contractors	Occupation and handover
Employer's goods and materials	Partial possessions
Nominated sub-contractors	Logistics
Utilities and statutory undertakings	Third-party issues

Adapted from CIOB (2011).

At the tender stage, the planner must study the drawings and project documentation to:

- Review the details and identify what is unusual about the project;
- Assess the information available and the quality of the information;
- Identify where the contractual conditions are favourable/unfavourable;
- Assess the complexity of the project;
- Make an estimate of the value of the project; and
- Confirm the basis of payment and the timing of the payments.

The planner must study the drawings and the other project documentation provided and quickly 'assemble' a mental picture of work to be undertaken and the constraints upon the construction team. It is imperative to identify what is unusual about the work in terms of the type of work, the size of the project and/or the physical constraints of the location.

Research has shown that 'novice' planners follow a fixed logical process of examination of the material provided. Experienced planners work differently, making swift assumptions from the material presented and looking for anomalies in the documentation that may either cause problems in understanding the construction or offer commercial advantage.

The quality of tender documentation will vary from client to client and from project to project. Not all the information required to answer all the questions raised in Table 4.2 will be available. The quality of the drawings produced may be variable. Clarification should be sought on discrepancies in the information provided and any additional information required. A list of points to be raised with the client and discussed internally with the tender team should be produced.

Table 4.2   A checklist of items to be included in the site visit report.

Project title (the correct title for the project as shown in the client's documentation)
Project address (address of the site including postal code)
Contact to arrange the visit (name, address, email address, telephone, fax, etc.)
Details of visits to the site (when by whom)
Site position (with relation to public transport, the local office, other construction work in the area, adjacent buildings, fencing and hoarding, demolition, hazards)
General site conditions (topography, trees and vegetation, water courses, site clearance, security problems, weather exposure, space for temporary accommodation, constraints for static plant, e.g. cranes, constraints for moving plant, e.g. excavators, constraints for temporary works, e.g. scaffolding, live services, protection requirements, environmental considerations, etc.)
Ground conditions (borehole details, type of soil, stability of the soil, level of the water table, tidal conditions, pumping requirements, disposal of water, etc.)
Access (local road network, temporary roads, safety, deliveries, traffic restrictions,)
Local facilities (for disposal of soil, telephone services, water services, electricity services, sewerage, garages, refreshments, etc.)
Local contacts (local authorities, statutory authorities, security services, labour agencies, plant hire, local tip charges, etc.)
Additional information (including maps, photographs, videos, etc.)

Adapted from CIOB *Code of Estimating Practice* (2009).

Section II

A review must be made of the documentation provided to identify where the contractual conditions may be favourable/unfavourable to the contractor. Variations from the normal, expected conditions will affect both the risk involved and the potential profit. The specification will reveal the quality of work to be undertaken.

It is important to quickly arrive at an estimate of the approximate cost for the work, the 'rate of spend' and the timing of the payments. This will determine both the level of resources required and the finance. This approximate cost will usually be based on experience of similar projects or simple calculations based on suitable budget estimating methods with adjustments for the ground works and any significant variations in the architect's design.

In addition to the approximate value for the work, it is important to consider the cost of the contractor's work and the rate of expenditure on the project as this may impact both the financing of the work and the resources required. An initial assessment must be made of physical and commercial risks.

Establishing key project dates, events and activities

Before starting to prepare the schedule, the planner must identify the start and finish dates of the project, sectional or phased completion dates, holiday periods and the commissioning or handover dates.

The key project dates will be evident from the client's documentation. On all but the most simple of projects, there will be sectional or phased completion dates where areas of the works are handed over to clients and/or other parties. Similarly, there will be commissioning and handover dates for the completed work. These key project dates should be identified, agreed and shown as milestone dates on the project schedule.

Within the time period for the project (as determined by the start and finish dates), it is important to identify national holiday periods, industry holiday periods and any religious holidays.

The planner will need to identify:

- The important events that must be clearly evident in the programme
- The key construction activities
- The activities that start/finish the different phases of the project

Important events in the schedule of work include start on-site, commissioning, handover, end on-site, etc. Key construction activities will depend upon the type of construction work. Key activities may include the procurement of specialist equipment or services. The schedule should show all the main construction activities and the main logic links between the construction activities. Work to be undertaken off-site should be clearly differentiated from that undertaken on-site.

A careful check must be made to ensure that planning permission has been obtained. If any element of planning permission is still to be secured, this may delay the completion of the works.

A careful check must be made on the activities that occur at the start of the contract, because it is the completion of these activities that normally allow the commencement of a whole range of construction work. An efficient start to construction is essential if high productivity is to be achieved and completion of the works by the required date.

Similarly, full consideration must be given to the activities in the final phase of the schedule. Some activities, for example, Landscaping, are relatively easy to complete, others, particularly those where third-party action (e.g. obtaining a fire safety certificate) is required, are more difficult, and any delay could delay the handover of the works.

Because the schedule produced will be the basis of a contractual obligation, it is important that all elements of the work are identified and described. However, the schedule should not enter into too much detail. It is recommended that the schedule is produced at elemental level, for example, excavation, piling, ground beams, etc.

The site visit

At the tender stage, it is essential that a member of the tender team visits the construction site. A review of the locality using Google Maps™ or similar tools may be of initial benefit but there is no substitute for an actual visit to the location. Ideally, both the estimator and the planner will make a visit to the site.

The timing of the visit is important. If the site visit is made too early, there is a risk that some important questions relating to the construction may not yet have been identified. If the visit is made too late, insufficient time may be left to benefit from the information collected. The visit should be made after the preliminary assessment of the project has been completed and the main construction process identified. It is essential that a comprehensive report is produced. This should be prepared using an appropriate checklist. Table 4.2 gives a suitable checklist of items against which appropriate notes have been made. It is important that the report contains details of not only the construction site but also the locality. The locality, the surrounding services, infrastructure, transport connections, etc. will influence the ability of the contractor to secure the necessary labour and plant and purchase, deliver and store the materials needed for the work.

Considering alternative methods of construction

When submitting a tender bid under the traditional form of procurement, the contractor may submit qualifications to the tender if the performance bond is unacceptable, if the contract duration is considered too short to undertake the work with a reasonable expectation of completion or if the client (or the client's representatives) has imposed extra responsibilities via late amendments during the tender period. The contractor may also make additions to the list of approved sub-contractors if one of the preferred sub-contractors is no longer trading or is unable to submit a price for the works. Such qualifications should be written in general terms and included within the submission. They may also be presented as an alternative contract (Brook, 2008).

Contractors may identify opportunities for alternative designs that provide the opportunity for reducing the cost of the works or reducing the time taken to complete the construction. The client and/or their representatives may be expected to consider an offer for an alternative design that complies with the original bid. If an alternative bid is to be produced, the contractor's tendering team must make an early decision. Such a decision may incur considerable additional work. The contractor should take the opportunity of the scheduled site visit or the visit to the consultant's office to explore the client's reaction to a tender based on an alternative design.

Where an alternative design has been selected, the planner must assess the implications of these changes and their impact on other elements of the construction. The entire estimating team should be contacted to ensure that the ramifications of an alternative bid are fully understood. The contracts manager responsible for submitting the contractor's bid must make a decision on whether the contractor will make a qualified tender bid incorporating the alternative design or present a bid based on the original design plus an alternative bid so providing the client with two alternatives.

Pre-contract planning

Pre-contract planning commences when the client has accepted the contractor's bid.

During the pre-contract period the pressure on the planner is to provide more detailed information than that produced at the time of tender and to help facilitate the start on-site. The planner will need to work closely with the project manager to ensure that their thinking on key aspects of the work is aligned.

At the Pre-contract stage, there will be more information available to the planner. The basis for the contract will be known. If the planner has been involved with the production of the tender, they will be fully familiar with all the information collated and the information that remains to be assembled prior to construction. The main activities at the pre-contract phase will generate more information, and there will now be a continuous stream of information that needs to be assimilated and the implications understood.

By the time the work starts on-site, there will be an abundance of information available, and there is the danger that construction will proceed with information that has been superseded. Effective version control of all incoming documentation and all information produced within the company is therefore imperative. This must be established at the Pre-Contract stage.

Pre-contract planning comprises:

- The pre-contract meeting and arrangements for commencing work
- Placing sub-contractor orders
- Site layout planning
- Construction method statement
- Master schedule
- Requirement schedules
- Contract budget forecasts
- Risk assessment
- Preparation and approval of the construction health and safety plan (Cooke and Williams, 2009)

At the pre-contract meeting, arrangements for commencing work will be discussed together with a review of the tender and adjustments made to the estimate to reflect the decisions taken at the tender settlement meeting.

The tender will include prices based on quotations from sub-contractors and specialist suppliers. These quotations will be reviewed and decisions made on whether to proceed with these companies or seek alternative quotations. When the preferred sub-contractors and specialist suppliers have been selected, the site layout plans will be extended. Method statements made during the tender period will be revisited, revised and extended.

The Pre-tender schedule must be expanded into the master schedule, a detailed schedule that covers the construction operations. Copies of this schedule will be presented to the client's representative who will use it to monitor the contractor's overall progress during construction. The master schedule should show not only the construction sequence but when information is required from the architect and other advisors (Cooke and Williams, 2009).

In addition to producing a master schedule, the contractor may decide to produce a target programme for internal purposes. This target programme will effectively be a compressed version of the master schedule (Cooke and Williams, 2009).

The master schedule is the high-level network showing the whole project and from which more detailed networks are derived. It is the principal schedule for the project produced to comply with contractual requirements. It should be designed and produced in a structured form to meet the requirements of all parties to the project.

The master schedule will be based on the tender programme with amendments agreed prior to the signing of the contract. This schedule will form the baseline schedule, the fixed or 'record' schedule against which current or future activity is referenced. This schedule will become the basis for monitoring production progress.

The Chartered Institute of Building CIOB (2010) recommend that for complex projects the master schedule should be augmented by a planning method statement. The purpose of this statement is to facilitate the understanding and cooperation of all the main participants in the project by providing details of the assumptions made in the preparation of the schedule.

Other tasks undertaken by the planner at the pre-contract Stage include:

- Producing schedules of requirements detailing the resources required so that the buying department can commence with the purchase of the materials and related services;
- Preparing construction budget forecasts based on the original estimate and the decisions made at the tender settlement meeting and prior to the contract signing;
- Undertaking a revised risk assessment for the project and putting appropriate contingencies into place;
- Reviewing and developing the health and safety plan.

In the tender submission, the contractor will have outlined the company's approach to health, safety and the environment and will have made a commitment to ensure that the construction will be undertaken with planning to ensure the safety of not only all concerned with the project but also the community in which the work will take place. At the pre-contract stage, the planning team will formalise the health and safety plans for the site.

Section II

Cooke and Williams (2009) suggest that there are five main issues to be considered:

- Ensuring compliance with the law;
- Making sure that there is an appropriate safety management system in place for the particular site in question;
- The identification of hazards that might give rise to accidents on-site;
- Assessing risk arising from the hazards in question and putting in place appropriate control measures to deal with the risks; and
- Devising safe systems of work.

Contract planning

The main contractor, in order to maintain production control and to ensure that the project is completed on time within the cost limits, will undertake ongoing contract planning throughout the duration of the project.

Contract planning includes:

- Monthly (long-term) planning
- Weekly (short-term) planning
- Cost/value reconciliation
- Reporting to management
- Reviewing and updating health and safety plans (Cooke and Williams 2009)

Monthly (long-term) planning

Monthly (long-term) planning recognises that as construction progresses it is inevitable that work will be disrupted due to unforeseen delays and changes. Such events and their impact must be assessed and addressed.

Updating (and monitoring) progress and forecasting completion requires the planner to review, revise, monitor and update the programme, identifying intervening events, assessing the impact of these events, implementing recovery and revising the planning method statement. It is also necessary to record the changes made and the reasons for the change (CIOB, 2011).

Control based on this approach is re-active. Targets are set, performance is measured, a comparison is made with the anticipated performance and, where performance has failed to meet that expected, remedial action is taken.

This approach to control requires an efficient, effective system for collecting data. This takes time and incurs costs. There is an inevitable time lag between the work carried out and the corrective action that is applied. This traditional approach to manage production is known as the 'control cycle': plan/prepare, initiate action, measure performance, compare the actual with the planned, assess deviation (are we ahead or behind the plan), determine action, re-plan and implement.

The control cycle model applies to the performance of individual plant items, groups of machines and gangs of labour. It applies to production performance at the section, project and company level. Production must be reviewed at each level and remedial action initiated where necessary. At the project level, the actions taken may result in the introduction of different processes, new resources or even a change in management.

In construction, unlike a manufacturing production process, poor performance may only become apparent after the construction activity is completed and there may be no opportunity to make production improvements to the current project unless a similar construction activity takes place at a later stage.

For further details on the traditional control cycle, see the following references: Project Management Institute (2004), Winch (2002) and Koskela (2000).

Weekly (short-term) planning

The Master Schedule will have set the milestone dates for completion of key stages of the work. As the construction proceeds, it will be necessary to undertake the weekly (short-term) planning necessary to provide the teams on-site with information on the work to be undertaken in each location over the next few days. This planning may be based on schedules produced by traditional critical path methods (CPM) at 'high density' or schedules produced by the Last Planner (Lean Construction) approach (see Chapter 7). The planning may adopt the use of virtual construction techniques (see Chapter 9). Whatever approach taken, it is important to work collaboratively with representatives of all the specialist suppliers and subcontractors involved to ensure that the schedules produced are achievable.

Cost/value reconciliation

Contract budgets will normally be prepared to cover the expenditure for labour, plant, materials and the preliminaries. Expenditure will need to be monitored and variances reported to management. This reporting is necessary to assist with cash flow forecasting. An accurate measure of performance can only be achieved by consideration of both the cost of resources and the value of work achieved. This is known as Cost/Value reconciliation. Full details of this approach including the calculations involved and the reporting available is considered in Chapter 6.

Reporting to management

Management will require the regular collection and distribution of performance information. This will usually be undertaken on a monthly basis. Performance reports need to provide information to a number of different stakeholders. Information must be provided at an appropriate level of detail. A simple report might only indicate changes in the scope and schedule of the works together with changes in the budgeted cost. The Project Management Institute (2004) suggests that more detailed reports should include:

- Analysis of past performance
- Current status of risks and issues
- Work completed during the period
- Work to be completed next
- A summary of changes approved in the period
- Other relevant information

The most usual form of presenting such reports is on the basis of variances between baseline and actual performance. The information is best presented

Section II

in tables, charts and graphs that explain not only the current position but forecast performance.

Formal procedures should be initiated to collate, verify and analyse the data and present the information. These procedures should be established and recorded within the project planning method statement.

Review/update of health and safety plans

As the project proceeds, there should be a regular update of health and safety plans for the project. Before the commencement of any new phases of construction, a job safety analysis and/or a review of the risks and hazards should be undertaken (see Cooke and Williams (2009). The ongoing production of health and safety plans and a summary of any health and safety incidents should be included in the monthly report.

Activities: selection, sequencing and duration

Whichever planning technique the planner selects, there will be the need to identify construction and procurement activities, determine their sequence, make an estimate of their duration and understand how they inter-relate. This section considers the first three of these aspects of activities.

Activity selection

The activities shown in the schedule will depend upon when in the project cycle the schedule is produced and for whom the schedule is produced. For the construction contractor, this is usually at the tender stage, at the pre-contract stage and at the contract stage of the work.

Even if the design is complete and the planner has all the information that is required to plan the works in detail (rarely the situation), there is no need to schedule all the activities. However, as the start of construction approaches, there is an obvious need to plan the work in more detail. At the start of construction on-site, it is important that there is a detailed schedule, identifying both the construction process and gangs who will complete the work. The CIOB (2010) recommends that schedules be produced to three different levels of density: 'low', 'medium' and 'high'.

'Low' schedule density 'is appropriate for work which is intended to take place nine months or more after the schedule date'. It may, for example, simply identify a single building or the work of a single trade as one activity.

'Medium' schedule density 'is appropriate for work that is intended to take place between three and nine months after the schedule date. Activities may reasonably be grouped into trade activities of durations not exceeding two months'.

'High' schedule density 'is a pre-requisite of the work that is intended to take place in the short term, say within three months of the schedule date. It is at high density that the work in progress will be recorded, monitored and reported upon'.

The schedule density impacts not only the activities as drawn on the schedule but the activity name, the duration given to the activity and other relevant data that are provided to the recipient.

At the tender stage

At the tender stage, the construction work should be scheduled at a 'low density'. It is important to show clearly the order and sequence of the construction work (different buildings, site works and civil engineering works), the main construction elements, the main trades and the main procurement items. The schedule should identify any items specific to the site location and the contract. The main tender schedule may be augmented by a schedule for procurement.

Whilst the tender schedule need only be produced at a 'low' density, it would be incorrect to leave the impression that none of the planning at this point in time is required to be undertaken at 'high' level density. For certain construction activities, for example, where the construction work that may be affected by access constraints or for construction processes that will utilise new or difficult techniques, it is imperative that detailed planning is undertaken at the tender stage. Such activities are essential to the successful completion of the works.

Here are two examples:

i)   For the construction of a high-rise residential or commercial building, the 'floor cycle' (the time taken to construct each floor of the building) will be critical to completion of the building on time. Any delay in the construction of one floor is likely to be repeated for every floor. For a building of say 80 storeys, the delay could easily accumulate to 3 months. It is obviously imperative that the planner and the tender team are confident in the duration determined for each 'floor cycle' even though the tender schedule may only show the construction as one activity.

ii) For the construction of a new railway station and related infrastructure works, the contractor will only have access to the existing rail track for a short number of hours each night. The contractor must be certain that key construction operations can be successfully achieved within such access periods.

At the pre-contract stage

At the Pre-contract stage, the schedule that is prepared will comprise some activities at high density (the activities that will take place in the first 3 months of the project), some activities at medium density and some activities at low density.

At the contract stage

At the contract stage, the schedule will, for the first part of the project, typically comprise some activities at high density (the activities that will take place in the first 3 months of the project), some activities at medium density and some activities at low density. As the work progresses, a greater proportion of the total number of activities will be at 'high'-density level. Here, it is important to remember the CIOB statement, 'It is at high density that the work in progress will be recorded, monitored and reported upon'. Unless you have scheduled the construction work at a 'high' level of detail, it will be impossible to measure progress accurately. Therefore, it is not a question as to *if* detailed scheduling takes place but *when*.

Section II

Sequencing

Any schedule produced by a planner must show the activities involved in a realistic sequence within the contractually required timings based on the resources available to the contractor. Sequencing is a thought process, whereby the planner considers an appropriate sequence of operations, examining the construction methods, choice of plant, quantities of work and risk; estimates the activity duration and then produces the schedule (Cooke and Williams, 2009). This process is shown diagrammatically in Figure 4.4. This is an iterative process, and it is normal for the planner to undertake this process several times before arriving at the optimum production sequence.

Understanding the proposed construction sequence, its nuances and its implications is imperative for the production of a realistic schedule. For experienced

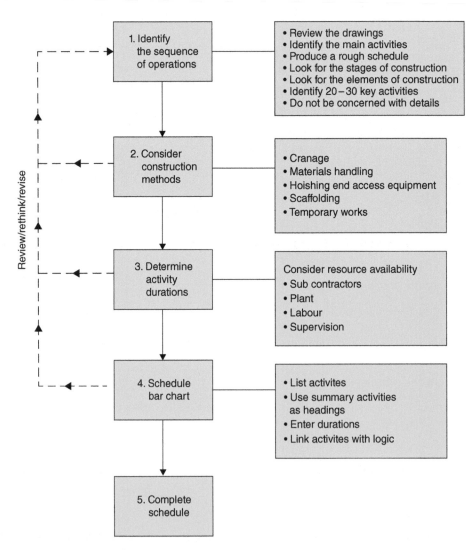

Figure 4.4 The sequencing of construction operations (Adapted from Cooke and Williams, 2009).

planners, this process comes as 'second nature' to their experience, providing them with the basis to complete the task. For 'novice' planners, producing a realistic schedule is dependent upon access to those with the knowledge of the construction process. It is important for both experienced and novice planners to present their schedule and construction sequence to others for review so that any opportunities for improvements may be evaluated.

Traditional CPMs (see Chapter 5) emphasise the discipline of looking at each activity and asking 'What has to be done before this activity can commence?' and 'What can be done after this activity is complete?' By adopting this approach, the planner can ensure both the rigor of the schedule produced and maximise the opportunity to produce a schedule that completes the construction work in the minimum time available.

New design services and new software products now exist to assist the planner in examining the construction sequence. Some design organisations provide design software with access to 4D sequencing software and other guidance as to how construction can proceed (ADAPT, 2013). By combining various software tools, it is possible to explore, understand and communicate construction sequences to other parties. Such tools do not necessarily make the sequencing of construction work easier. They do however make the implications of sequencing decisions more apparent. They also make it easier to communicate the intended sequence to others who may then contribute to decisions on a revised, improved construction operation.

Assessing the duration of each activity

Assessing the duration of activities is one of the most difficult aspects of planning and scheduling. The over-riding concern is to produce a schedule that is realistic and competitive. To do this requires a realistic estimate of the time taken for the activities involved.

This may be done in a number of ways:

- From experience (knowledge gained from previous projects)
- Using specified standards of performance (e.g. the curing time for concrete)
- From calculations based upon the quantity of work (or approximate quantities) and standard production rates
- From specialist sub-contractor and supplier production data
- From historical production records
- Benchmarking against industry or company performance (CIOB, 2009)

For experienced planners, it may be relatively simple to estimate the duration based simply on the overall quantity of work. Typically, planners will use the production figures from previous projects and then amend these to suit the process, detail and quality of the work under consideration. Most contractors' organisations have company or industry standards. These will be made available to all planning staff. Some books provide standard industry performance data. (See, e.g. Brook, 2008.)

Table 4.3 shows an example of typical production data for breaking out concrete from which it is possible to estimate the duration of an activity.

Examples of calculations based on standard production data are provided in the website at www.wiley.com/go/baldwin/constructionplansched.

Table 4.3 Typical production data relating to breaking out concrete.

Breaking out concrete	Average	Units (per excavator)
By machine		
Concrete slabs, not reinforced depth 50–150 mm	32.00	m²/h
Concrete slabs, not reinforced depth 150–300 mm	22.00	m²/h
Concrete slabs, reinforced depth 50–150 mm	11.00	m²/h
Concrete slabs, reinforced depth 150–300 mm	9.00	m²/h
Concrete foundations, not reinforced	2.00	m²/h
Concrete foundations, reinforced	2.50	m²/h
Concrete walls, reinforced	9.00	m²/h
Concrete beams	3.00	m²/h
Concrete columns	3.00	m²/h
By hand		per man
Concrete slabs, not reinforced depth 50–150 mm	6.00	h/m³
Concrete slabs, not reinforced depth 150–300 mm	7.00	h/m³
Concrete slabs, reinforced depth 50–150 mm	7.00	h/m³
Concrete slabs, reinforced depth 150–300 mm	8.00	h/m³
Concrete foundations, not reinforced	6.5	h/m³
Concrete foundations, reinforced	10.0	h/m³
Concrete walls, reinforced	8.00	h/m³
Concrete beams	8.00	h/m³
Concrete columns	8.00	h/m³

Links, dependencies and constraints

In practice, modelling the construction process frequently results in the need for complex relationships between the activities. Whether you are using a linked bar chart or a precedence diagram to model the production process, you may need to introduce different types of logic links between the activities. You may need to introduce specific constraints as to when activities may start or finish. Links and dependencies should be introduced with caution, as they may result in complex or even iterative relationships that are at best difficult to understand and at worst meaningless.

The following example illustrates how these links may be used. Consider four construction activities: excavate trench, install drainage, test drainage and backfill trench. The logic of the construction process dictates that the installation of the drainage commences after the excavation is finished. The trench is backfilled after the drainage has been installed and tested. (Because the trench is shallow and in good ground, there is no need for temporary support to the sides of the trench.) A logic diagram for these four activities would show them as sequential activities.

Whilst this logic may apply to short drainage runs, it is more likely on an extended length of pipework that as soon as a sufficient length of trench drainage has been excavated installation of the drainage will commence. By adding additional activities (i.e. dividing the excavation activity into two activities: commence excavation and complete excavation), it would be possible to construct a model that is more realistic of the actual construction process: the drainage activity commencing after the completion of the activity, commence excavation.

Alternatively, this could be modelled by introducing a start-to-start relationship between the excavation activity and the drainage installation and a finish-to-finish relationship between the end of the excavation and the end of the drainage installation. For example, in Figure 4.5, the logic shows that 1 day after the start of Activity W, Activity X can begin, and 1 day after the start of Activity X, Activity Y can begin. Activity Z commences after Activity Y is complete.

Links and dependencies may be introduced between activities. There are four logic links whereby activities may be tied: finish-to-start, start-to-start, start-to-finish and finish-to-finish. These logic links enable planners to model the wide variety of construction processes that arise.

Also available to assist the planner are constraints.

Constraints are restrictions that affect the sequence or timing of an activity. The client may impose these restrictions, they may result from the construction process or the resources available or they may arise from contractual or business decisions made by the contractor. Constraints that are introduced into the schedule over-ride the timing of events that result from the logic of the model and the durations of the activities.

The Guide to Good Practice in the Management of Time in Complex Projects, (CIOB, 2011), identifies three categories of constraints: flexible constraints, inflexible constraints and what they term 'moderate' constraints.

- Flexible constraints are constraints where 'start and finish dates change according to any changes in logic and the associated resources'.
- Inflexible constraints are constraints where start and finish dates are 'dictated solely by the constraint and will not change as a result of changes in logic or resource availability'.
- Moderate constraints occur when the start and finish dates in the schedule 'respect some changes in logic and associated resources but not all'. (The guide emphasises that moderate constraints should generally be avoided unless they are essential.)

Section II

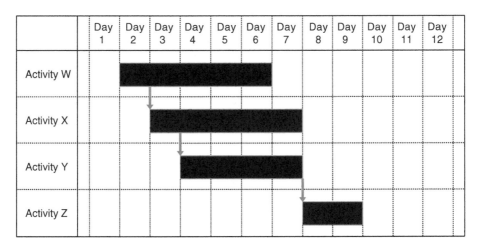

	Day 1	Day 2	Day 3	Day 4	Day 5	Day 6	Day 7	Day 8	Day 9	Day 10	Day 11	Day 12
Activity W												
Activity X												
Activity Y												
Activity Z												

Figure 4.5   An example of linked activities.

Table 4.4 Different types of constraints.

Flexible constraints	As-soon-as possible	Activities start (and hence finish) at the earliest date possible. Float is eliminated from the activity and its predecessors. This is the default setting in some software products.
	As-late-as possible	Activities finish (and hence start) at the latest date possible. Float is eliminated from the activity and its successors. In some software products, this is the default setting for the calculation of the critical path.
Inflexible constraints	Must-finish-on	A date is set for the activity to finish. This date is independent of activity dependencies, lead or lag times or resource levelling.
	Must-start-on	A date is set for the activity to start. This date is independent of activity dependencies, lead or lag times or resource levelling.
	Zero-total-float	This constraint makes the activity 'critical' together with predecessors and successors.
	Expected-finish	A date for the finish of an activity is set. This has the effect of extending and enforcing the activity duration.
	Mandatory-project-finish	The enforced finish date for all the activities within the project.
Moderate constraints	Zero-free-float	An activity is scheduled to finish immediately prior to the commencement of its successor activity.
	Finish-no-earlier-than	The earliest possible date on which the activity can be completed. (The activity cannot finish at any other time.)
	Finish-no-later-than	The latest possible date on which the activity can be completed. (The activity can be finished before this specified date.)
	Start-no-earlier-than	The earliest possible date on which an activity may begin.
	Start-no-later than	The latest possible date on which an activity may begin.

Table 4.4 shows different types of constraints. (This table is based on the details provided in the Guide to Good Practice of Time in Complex projects, CIOB, 2011.)

Float and contingency

The terms 'float' and 'contingency' are often, but wrongly, used interchangeably. Float, also known as 'slack',[1] is spare time in a sequence of events and is a product of the activity durations, sequences and dependencies in a network. Contingency,

[1] Microsoft Project refers to float as 'slack'.

also known as 'buffer',[2] is an allowance specifically added to a schedule or network to take account of unforeseen circumstances.

The extensive use of project management software and the jargon associated with it has tended to limit the vocabulary associated with float. Most software only refers to and identifies 'total float' and 'free float', and these are probably the most important types of float but planners and schedulers must also be aware of 'interfering float', 'negative float' and 'terminal float'.

Generally, but with some specific constraints, float within a schedule provides some flexibility and allows managers some degree of flexibility about the precise timing of activities. Adjustments can be made to the schedule that do not affect the completion of the project to provide a better workflow, to take account of material deliveries or sub-contractor availability and so on.

Knowing which activities have the least amount of float, those with zero float are on the critical path and those which have negative float are causing delay, allows the planner to identify which activities to focus on to reduce the schedule time or to recover delays.

Most experienced construction professionals probably have a good idea where there is float or slack time in their projects if they are simple and straightforward. Large and complex projects require critical path calculations to be carried out to determine where and how much float there is in the schedules. Without such calculations, managers might think they know where the float is but retrospectively find they are mistaken, which can lead to decisions being made on the basis of incorrect assumptions.[3]

Schedules with float are also able to absorb delays more readily. This leads to one of the fundamental questions in planning and scheduling: 'Who owns the float?'

We now explain each type of float. For details of how to calculate the float, see the website for the book www.wiley.com/go/baldwin/constructionplansched.

Total float

Total float is usually defined as the amount of time an activity can be delayed without affecting the completion of the project, and this is generally how it is practically applied. Total float is more precisely the total amount of time available for the activity to be carried out less the amount of time needed for the activity to be carried out.

Figure 4.6 shows two activities, X and Y joined by a finish-to-start dependency at their earliest start dates. Other constraints dictate earliest start of activity Y and the completion date for the project.

In the simplest case, an activity can be scheduled in any position between its earliest and latest dates within the total float period. In the example provided in Figure 4.7, activity X has 5 weeks total float.

The scope for adjusting the timing of activity X, if it is to use its entire total float, is that its start can be delayed until the beginning of week 6. In order to do

---

[2]   Asta Powerproject refers to activities used for contingency as buffers.

[3]   This was a feature of the case *Mirant Asia-Pacific Construction (Hong Kong) Ltd v Ove Arup and Partners International Ltd & Anor* [2007] EWHC 918 (TCC).

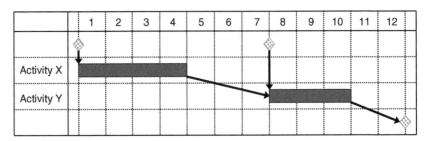

Figure 4.6   Activities scheduled at the earliest start dates.

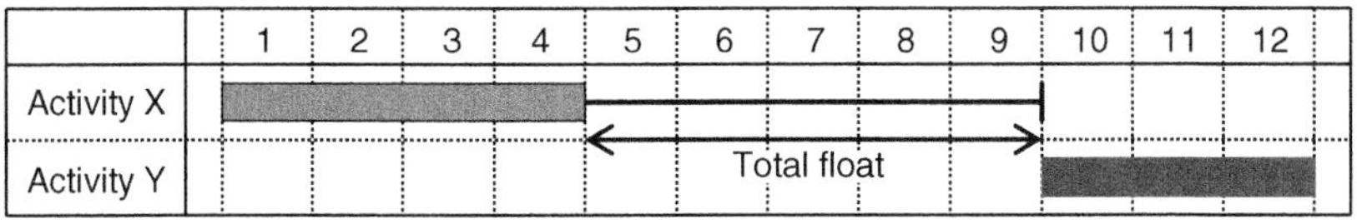

Figure 4.7   Total float.

this, activity Y is pushed to its latest start date. That is the essence of total float: although it gives the greatest potential for rescheduling the timing of operations, it might have a knock-on effect on other activities and might also make previously non-critical activities now critical; see Figure 4.8.

Total float, as its name suggests, is made up of a number of float elements so it can give an inaccurate representation of how much flexibility there is in the schedule.

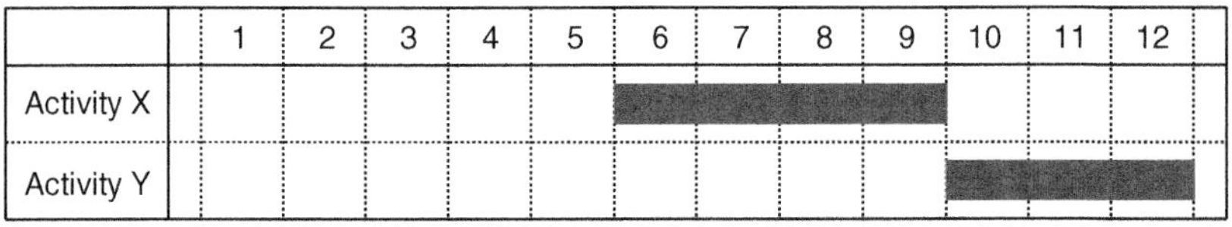

Figure 4.8   Total float fully utilised (activities at latest start dates).

Free float

More valuable to the scheduler is free float. Within the free float period of total float, an activity can move without affecting the completion of the project and without affecting the timing of any other activity, as shown in Figure 4.9. Utilising free float, the start of activity X can be delayed until the start of week 4 without affecting the start of activity Y.

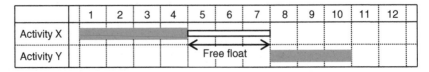

Figure 4.9   Free float.

Interfering float

Interfering float is the second component of total float and is the part that, if used, will not affect the completion date of the project but will affect the start and/or completion of following activities (see Figure 4.10).

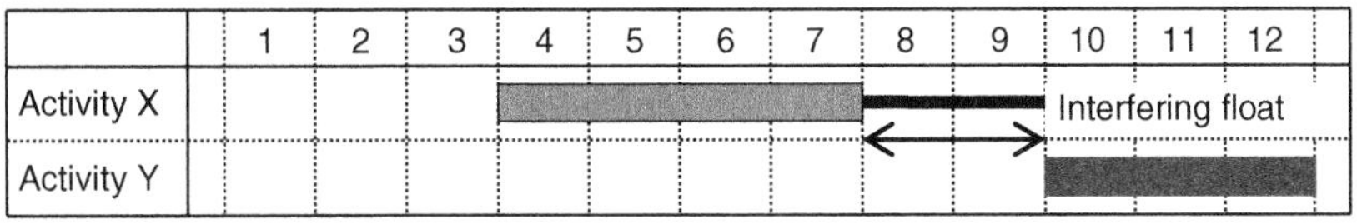

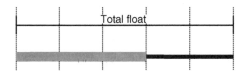

Figure 4.10   Interfering float.

Using the entire interfering float will have the same effect as using all the total float of an activity. This is shown in Figure 4.11.

Figure 4.11   Total float, free float and interfering float.

Independent float

Independent float is rarely quantified in current scheduling as its origins are in critical path networking using the activity-on-line technique. Independent float is less prevalent when earliest and latest activity dates are calculated using activity-on-node techniques, and the authors have not come across any present software that computes it.

Unlike other floats, independent float concerns individual activities and does not flow through the whole of the project. Independent float specifically exists when preceding activities are scheduled at their latest dates and succeeding activities are scheduled at their earliest dates and the time available for the activity to be carried out exceeds its duration.

Consider the programme shown in Figure 4.12; all the activities are shown at their earliest dates.

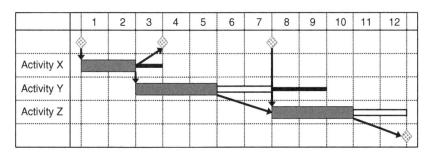

Figure 4.12   Activities at earliest dates: free float and interfering float.

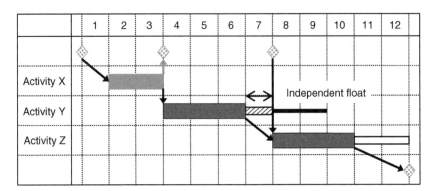

Figure 4.13   Independent float.

Figure 4.13 shows the programme when activity X is scheduled at its latest dates. The independent float on activity Y is not influenced by the timing of any other activity in the network and nor does its use affect the timing of any other activity in the network.

Intermittent float

The predominance in the use of project management software to produce programmes and in particular the linked bar chart has resulted in a new type of float. Again, the authors are not aware of any software that computes intermittent float directly and not all software has the function to designate an activity intermittent or interruptible.

Intermittent float is mainly generated in activities that are connected using start-to-start and finish-to-finish-type dependencies in the precedence network. In Figure 4.14, all the activities appear to be critical (the earliest and latest start dates are the same and the earliest and latest finish dates are the same) and the overall duration of the schedule is 8 weeks; earliest finish of activity Z minus earliest start of activity W, 9 – 1 = 8.

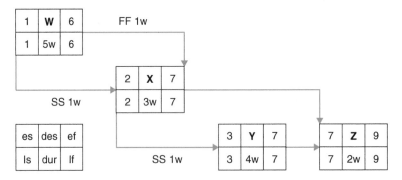

Figure 4.14   Precedence network using start-to-start and finish-to-finish dependencies.

However, when an attempt is made to plot a bar chart of the network, the overall duration increases to 10 weeks (11 – 1), as shown in Figure 4.15.

Most project management software that uses the linked bar chart format adopts this layout and concept. Some have the facility to designate the activity as intermittent or interruptible so that it is displayed as the period from early start to early finish but that its duration remains at the original duration as shown in

Section II

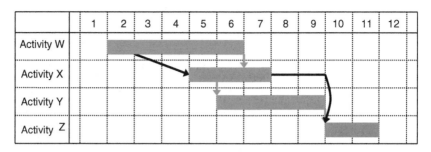

Figure 4.15   Linked bar chart using start-to-start and finish-to-finish dependencies.

Figure 4.16. The nature of the activity will determine whether it is one that can be carried out intermittently.

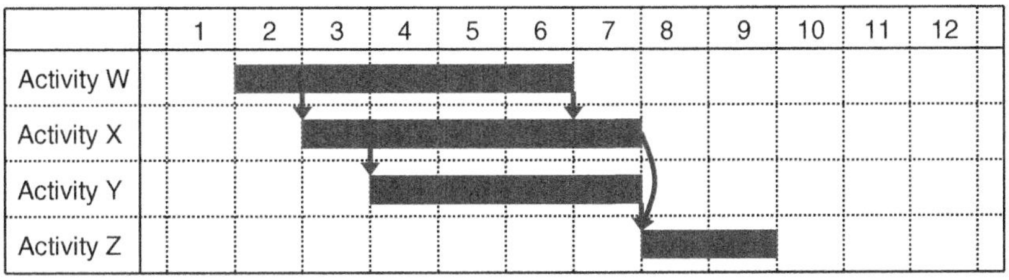

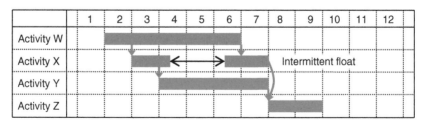

Figure 4.16   Linked bar chart with activity X designated 'intermittent'.

   The problem with displaying the intermittent activity in this way is that the impression is given that the work content is stretched proportionally over the whole of the period. For instance, in activity X the start-to-start link appears to show that 20% of the work, 1 week from 5, needs to be completed before activity Y can start. Whereas in the original network, 33%, 1 week from 3, needs to be complete before activity Y can start.
   A second solution is to split the intermittent activity. There are numerous ways in which the activity can be divided but the split must be some way between the driving dependencies as shown in Figure 4.17.

Figure 4.17   Intermittent float in a split activity.

Negative float

If the end date of a schedule (say, for instance the contract completion date milestone) is restrained with a 'finish on or before' or a 'finish on' constraint and the project falls into delay, then negative float will be generated.

The calculation for negative float is the activity's latest completion date from the original schedule minus the scheduled earliest completion date from the progressed and updated programme.

Negative float is an indication of how much time the completion of the activity must be reduced by to meet the completion date. When a project is severely in delay, many, if not all, activities will have negative float. Those with the highest negative float are those that are most critical and indicate where attention should be focused to recover delays.

Terminal float

Terminal float, or project float, is the spare time in a programme between the planned completion date and the contract completion date – the project is planned to complete early. Terminal float is the free float at the end of the schedule. Terminal float is shown in Figure 4.18.

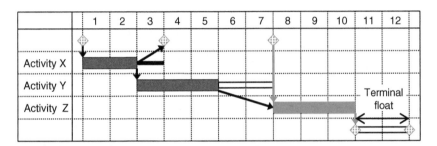

Figure 4.18   Terminal float.

Terminal float can cause the scheduler a few problems as the free float at the end of the schedule filters through the schedule and results in no activities being critical . As all activities have float, there is no critical path evident through the schedule. To overcome this, it is usual to constrain the planned completion milestone such that it is artificially made critical; most software allows the user to 'make critical' an activity.

Planning to complete a project early and generating terminal float can be considered as a contingency or buffer to ensure completion by the contractual date. Because of this, some contracts treat terminal float differently to float elsewhere in the schedule and treat it as a time risk contingency.[4]

Internal float

Even though an activity may appear to have no float, it is possible that there is some float within the duration of the activity itself. Rarely, especially in construction operations, is it possible to precisely calculate the optimum time to carry out a task. (Even if it were, most planners tend to round up their calculated durations

---

[4] For instance, the NEC ECC.

to whole days or even whole weeks.) This built-in float is not evident from the float calculations and can mean that even activities on the critical path have some float!

Building extra time into an activity, especially into activities where there is some uncertainty or where their timings are crucial to the project's success, is more of a contingency and as such is for the contractor's use alone.

Contingency

Contingency is time purposely built into the schedule to allow for specific or unforeseen circumstances that could affect the project. Contingency periods can be built into a schedule by both the Employer and the Contractor. Typically, an employer will state the contractual dates some time in advance of the 'drop-dead' dates to guard against delays to the project completion.

The contractor may likewise may plan to complete the project early so there is some buffer if delays occur (see 'terminal float' mentioned earlier). The contractor may also build in some time at various crucial stages of the project to ensure work-flow is not jeopardised, for instance, between the completion of the foundations and the start of the structural steelwork frame. Similarly, a time contingency could be included to take account of adverse weather during the winter months, affecting weather-sensitive operations.

In the same way that financial contingencies can only be expended by those that make the allowances, time contingencies can generally only be expended by the party that makes them. The employer cannot take advantage of the contractor's contingencies to, say, take additional time to provide information and the contractor cannot take advantage of the employer's contingency to cover for his own delays. In most contracts, sanctions exist if one party does use the other's contingencies; for instance, it may mean paying additional sums for disruption to the contractor's works if the employer uses the contractor's contingency or the payment of damages if the contractor is late in completing and uses the employer's contingency.

There are two primary ways of including time contingencies into a schedule:

- Making a contingency allowance against each activity by either extending the activity duration or, if the planning software allows, adding a separate allowance to each activity; see Figure 4.19.

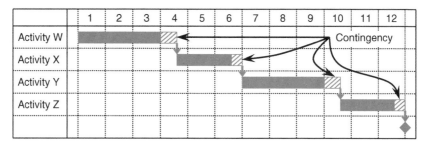

Figure 4.19   Contingency – adding an allowance to each activity.

This has the advantage that each activity knows how much contingency it has and can schedule its work accordingly. The disadvantages are that the schedule never shows activities at their earliest dates – there is a progressive 'delay' to the start of activities so it is difficult to organise work so that it meets the optimum schedule. If an activity finishes without using its contingency, are the following activities able to start earlier or can the work only start at the scheduled dates?

The work can suffer from the effects of Parkinson's law 'Work expands so as to fill the time available for its completion' (Parkinson, 1955). Generally, this will suggest that work is more likely to proceed at the rate shown on the schedule (the time available for its completion) rather than at a faster rate, and hence it is difficult to overcome the effects of activities being planned to take longer than it is required.

Similarly, working to a programme that has over-estimates of activity duration is likely to suffer from 'student syndrome (Goldratt, 1997). This refers to the phenomenon that many people will start to fully apply themselves to a task just at the last possible moment before a deadline. This leads to wasting any buffers built into individual task duration estimates so that getting ahead of programme becomes less likely.

■ Making a contingency allowance at the end of a string of activities or at the end of a section of work; see Figure 4.20.

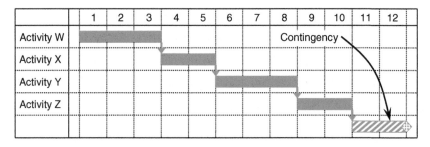

Figure 4.20   Contingency – as an end allowance.

This allows the activities to be scheduled using their normal durations and at their earliest dates. It also gives more scope and flexibility to absorb large delays of a few activities rather than small delays on many activities.

Manipulation of float

Float is a by-product of the network; the activity durations, sequence and logic and, in most cases, the planner does not concern himself with float, or the critical path, when constructing the network. Only when the network is nearing completion, probably after many iterations and rescheduling exercises, does the critical path and the float become apparent. It is at that stage that the planner has the opportunity to 'manipulate the float'.

When examining the schedule, the planner may be able to reconsider resources or methods of working to optimise the programme. For instance, decreasing the resources on an activity and making its duration longer whilst not affecting the overall duration of the project may provide a better solution: less plant;

less supervision or less storage for materials. Similarly, adjusting the duration of activities with no float can increase or decrease the overall duration of the schedule.

Those with more commercial or Machiavellian intent can manipulate the activity durations and float so that the critical path is steered in a particular direction. For instance, if the contract allows for 'nominated' sub-contractors or suppliers, the planner may try to steer the critical path through those activities for which they are responsible in the hope that any delay to those activities will be one which the contract deems is the employer's responsibility. Similarly, engineering the critical path to go through design or information release dates that are the employer's responsibility may give rise to claims against the employer.

Such 'claimsmanship' is rarely fruitful or successful and is not in the spirit of reasonable and moral working practices. The unintended consequences of artificially manipulating float and the critical path generally mean that the schedule is not optimised, and other areas of the schedule will have float that they would not otherwise have. When disingenuous plans are not fulfilled, uneconomic working in other areas is likely to result.

Who owns the float?

The ownership of float has long been debated in the construction industry. It is a valuable asset that allows flexibility and can absorb delays to the programme. If one party uses or has control of the float, it restricts the other's flexibility. It is rare to find provisions in a contract that say to whom the float belongs. There are three views:

The contractor owns the float

It is the contractor who has put the programme together and has planned his work and any changes to the plan are likely to involve additional expense. In most contracts, the contactor merely has to complete by a specified date and how he achieves that is up to him. It would be unfair if the employer used the float and caused the contractor to carry out his works in a different sequence or timing to that he had planned.

The employer owns the float

The employer will say that they are paying the contractor to carry out the works and included in that are the periods within the schedule that the contractor has planned to do nothing.

The project owns the float

Neither party owns the float, and it can be used on a first come first served basis. This seems to be the generally accepted position but both parties must act reasonably. For instance, if the Employer used up all of the float in the programme by continually issuing information late, but not so much as to delay the completion of the project, would it be reasonable for the contractor to subsequently have to pay liquidated damages for some minor delays?

Section II

Monitoring progress and managing the time model

Having commenced construction, it is necessary to monitor progress to check for deviation from the agreed schedule. This may be undertaken for several reasons. First, there will be a contractual obligation to inform the client if progress against the master schedule (contract programme) is not proceeding as anticipated. (Table 4.5 shows typical reporting requirements taken from different standard forms of contract.) Second, there will be the need to advise internal management on progress and any deviation from the anticipated completion date. Third, the impact of any delays and disruption will need to be assessed.

It is imperative that reporting takes place on a regular basis. The method of reporting should be consistent, structured and founded on a recognised methodology. The schedule for progress monitoring and reporting should be set up at the start of the construction work and then adhered to throughout the construction period, unless otherwise agreed by all concerned. It is essential to monitor

Section II

Table 4.5 Typical reporting requirements to the client.

Form of contract	Clause	Requirement
JCT05 standard building contract	2.9.1	Two copies of the schedule for the execution of the works required as soon as possible after the signing of the contract.
	2.28.1	Two copies of the schedule updates following an extension of time.
ICE7	14	Within 21 days of contract award, a schedule showing the order in which the contractor proposes to carry out the works plus a general description of intended arrangements and methods of construction. Revised schedules are required where actual progress varies from the accepted schedule. Further details of methods of construction are required if requested by the engineer, including details of temporary works and the use of plant together with related structural calculations.
NEC	3	Key dates including starting, possession, partial possession and completion. A method statement for each operation showing planned equipment and resources. The order and timing of operations. Provisions for float, time risk allowances, health and safety requirements and contractual procedures. Revised schedule showing the actual progress for each operation and its effect on remaining work. The effects of compensation events and matters giving rise to early warnings and how the contractor plans to deal with these issues.

Adapted from Cooke and Williams (2009).

progress on a regular basis to assess if the project is proceeding as anticipated. Some planners recommend that progress on activities on or close to the critical path is monitored on a weekly basis, with full reporting on all activities on a monthly basis. Others simply monitor the progress of all activities on a monthly basis. In addition to the monitoring of progress required for 'internal' management purposes, the contractor is also required to provide the client with an update of construction progress. The exact nature of this requirement will depend on the type of contract.

On small- and medium-sized projects, it is appropriate to assess and report the progress of all the activities. One person should make this assessment. On large projects, although the reporting of progress will be the responsibility of one individual, data will need to be provided by many. It is particularly important to establish rigorous lines of communication, protocols and timescales for reporting.

The following actions need to be undertaken:

- Review the current working schedule and the assumptions used to produce the schedule.
- Collect and review the production records and progress reports.
- Review work on the activities currently in progress.
- Update the schedule.
- Identify intervening events.
- Assess progress and forecast completion.
- Review contingencies and revise the working schedule to effect a recovery.

These stages taken in sequential order may be considered as what is known as managing the time model. (For an extended discussion on managing the time model see the CIOB, 2011.) Managing the time model includes understanding the assumptions behind the schedule, assimilating new information, measuring the progress made on current activities, understanding events that have impacted the schedule, forecasting the impact of these events and then assuming that there has been (or may be) delay or disruption to the work, reviewing contingencies and revising the working schedule to effect a recovery. The outcome of this process will be updated information on the time schedule for the project that may be provided to the client, to internal management and assist with the production of new short-term work schedules.

Reviewing the assumptions used to produce the schedule

It is always useful to re-visit the assumptions used to produce the schedule. There is little point in monitoring progress against a schedule if you do not understand the basis on which a schedule was produced. Even if you produced the original schedule, you may not remember all the thinking and rationale that went into its production. On large and complex projects, it is necessary to produce a record of the basis of the schedule as part of the quality control plan and then continually update and review this on an ongoing basis. On smaller projects, a simple statement or record of the background to the production of the schedule may suffice. As the project proceeds, any changes in the assumptions relating to the schedule need to be understood.

Section II

Collecting and reviewing production records and progress reports

Mawdesley et al. (1997) highlight that to assess progress you will need not only to collect performance data by measurements from the construction site but also to review records such as:

- Diaries, field books
- Time/bonus sheets
- Quantity surveying (QS) valuations
- Orders/invoices/delivery notes/receipts
- Job card and coding systems
- General correspondence (letters and memoranda)
- Requests for information (RFI), confirmation of verbal instructions (CVIs), architect's instructions (AIs), engineer's instructions (EIs), variation or change orders, measurement memos, daywork sheets, charge notes
- The drawing issue register
- Minutes of meetings
- Progress reports, progress charts, progress (or record) drawings
- Photographs
- Video tapes
- Memory

All of these will help provide details of progress, not only *what* has been done but *when*. It is also essential that all changes are identified and recorded. A comprehensive record system needs to be introduced. The need for good record keeping cannot be over-emphasised. It is frequently necessary to return to these data at a later date, often for claims purposes. When claims for extensions of time or additional payment are being prepared, it will be necessary to determine not only what was done, and why, but *when*. To prepare their legal argument, lawyers focus on the chronological order of events. Accurate records are essential to support such claims.

Reviewing the records for the construction work helps to identify intervening events, events that were not originally anticipated. These events may have adversely affected production or may have had a positive impact on production.

Reviewing the activities currently in progress

Where activities have been completed, it should be straightforward to establish their date of completion and their actual duration. Where activities are currently in progress, assessing progress is not always as simple. Records may exist of when the activity started but not of how the production is proceeding.

To assess progress, it is normal to assess the percentage of work completed for each activity. This is usually calculated by comparing the quantity of completed work with the total amount of work in the activity.

For some types of construction work, this is simple. For example, counting the number of piles driven in a month or the number of columns concreted in a week is easy and non-contentious. Consider though the situation with respect to the bulk excavation of the foundations for a building or the completion of the construction of a roof to a building. In these cases, it may not be as simple to determine the percentage of the activity complete. The tendency is always to over-estimate the amount of work completed when issues relating to payment are involved. (These issues may relate to payment from the client, payments to sub-contractors, or both.)

Some planners develop rules of thumb to assist in measuring progress. It is good practice to assume that the completion of an activity will follow an 'S curve', where output is slower during the initial part of an activity, the learning curve, and slower towards the completion of the activity, due to the requirement to correct 'snags'. This ensures that the planner considers not only the progress achieved but also how recently the activity has commenced and the rate of progress.

Where appropriate some planners will recognise potential difficulties of assessing progress on an activity by dividing single activities into three sub-activities: a start-up activity, the main activity and a 'completion' activity, each with an estimated duration.

Measuring progress can also be assisted by the duration allocated to an activity. Some planners work on the basis that no activity should be more than a specified duration (e.g. 4 weeks), thereby ensuring performance is more accurately identified. Others adopt the approach of basing the progress to date on an estimate of the time taken to complete the remaining work. In this approach, the question is not what has been *done* but what work *remains* and how long will it take to complete this work. This approach is invariably more conservative.

Whichever approach is adopted, it is clear that estimating the percentage of completion of construction activities is not always easy; it is a combination of measurement and experience. To facilitate smooth monitoring and forecasting and to reduce the likelihood of argument over the performance achieved, the experienced planner will avoid problems by agreeing in advance the basis of measurement for potentially contentious construction work.

Any assessment of performance to date assumes that the scope of the works and the quantity of the construction work remain as envisaged when the original programme was developed. Where this is not the situation, a full review is required and any calculations must reflect not only the work completed but the anticipated performance based on the revised quantity.

Updating the schedule

Before comparing actual performance with planned performance, it is necessary to update the schedule.

The CIOB (2011) emphasises that updating the schedule is not the same as progress monitoring:

> "Updating is not progress monitoring, nor is it schedule revision, it is simply the addition of the as-built data to the working schedule and the recalculation of the critical path in the light of the progress actually achieved. The records of progress are used to add as-built start and finish dates to those activities which have achieved either status, and progress data to those started but incomplete at the defined date." (p. 87)

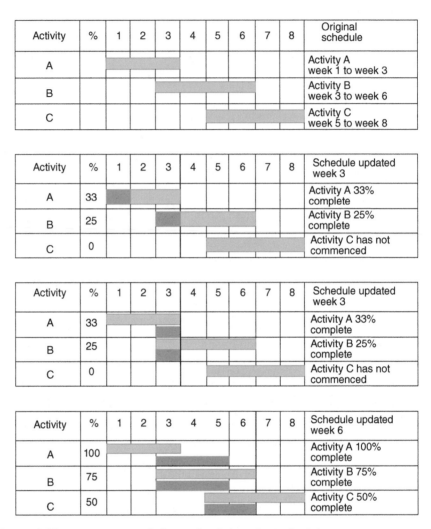

Figure 4.21 Progress recorded on a simple bar chart schedule.

When updating the schedule, it is best to present the relevant information graphically on the bar chart schedule.

One way to do this is to shade the activity bars on the chart to indicate the percentage complete. The time-now line is marked vertically on the chart, commencing at the appropriate week number. Consideration of the activities either side of this line gives an immediate *indication* where construction is ahead or behind schedule.

Figure 4.21 shows an example of how progress may be recorded on a simple bar chart programme. Although it is common to mark up each bar on the chart to show the percentage of the activity that has been completed, a better way to indicate progress is to indicate both what was *intended* to achieve and what has *actually* been achieved by the use of two separate bars.

It is important that wherever possible bar charts are annotated to identify reasons for delay and highlight key areas of the construction that need to be addressed by the management. The addition of the actual start and finish dates for each activity and additional information leads to the production of the 'as-built' programme, a programme that represents the actual sequence, duration and timing of the activities in a project (see Section IV).

Having updated the activities within the schedule, a review of the critical path for the project will indicate whether or not the construction work based on performance to date and the current schedule may be expected to finish on time.

Assessing and monitoring the time model is more than just updating the schedule. The first consideration must be whether or not progress is satisfactory compared with the contractually agreed schedule. (This schedule may be the original Master Schedule agreed at the start of the works or a schedule subsequently agreed with the client.) Construction may have been impacted by delays or disruption. It is therefore necessary to consider the intervening events.

Identifying intervening events

It is important to identify any events that have interfered with the progress of the work and assess their impact on production. The CIOB (2011) refers to such events as Intervening Events. Intervening events are occurrences that were not originally planned for and may adversely affect productivity and/or progress. They are often difficult to detect or at least not immediately apparent. Therefore, they demand close consideration.

Typical Intervening Events include:

- Disruptions to the undertaking of the works
- Delays due to lack of information
- Voluntary variations
- Instructed changes
- Prime costs and provisional sums
- Non-compliance with CDM regulations
- Late information
- Late response
- Failure to grant access or possession
- Failure to obtain consents
- Suspension of the works
- Acts or omissions of third parties
- Neutral events
- Other occurrences

In considering the impact of intervening events, it is important to look at *all* the impacts that arise from an event. Some events can lead to re-work. In some instances, this means the complete removal of the existing work and the re-start of an activity that has already been completed. Some events can generate new activities not previously envisaged. A common outcome of an event is the re-allocation of resources. New resources may be required. These resources may be difficult to obtain and there may be a delay in their procurement. The re-allocation of resources and their impact on other activities may be complex. A single event can result in other events. These new events may occur immediately

Section II

Table 4.6   Information relating to intervening events that should be recorded.

Item	Comment
A unique event identifier	Each event should be given a unique identifier so that in any discussions with the client or the client's advisors reference to the incident is unambiguous.
Description of the event	A clear description of the event should be provided, not just a simple statement.
Originator/authoriser	Where instructions, or a change of requirements, are forthcoming, it should be clear who authorised the changes.
Relevant clauses providing for an extension of time	Where an extension of time is being considered, it is important to identify which clauses in the contract are relevant to the claim.
Relevant contract clauses providing for compensation	Where compensation is being considered, it is important to identify which clauses in the contract are relevant to the claim.
Date on which the event is instructed/occurred	A date, and where possible the exact time, for the instruction should be noted, with confirmation of receipt of instructions (whether verbal or written)
The activities added, changed or omitted	A list should be made of *all* the construction activities impacted by the changes. It is important to identify both new activities and existing activities that are needed to be changed.
The labour and plant resources for each added or changed activity	It is important to consider all changes to the labour and plant resources that resulted from the changes plus any additional materials.
The date and timing of the added or changed activities	The changes and additions will not have occurred on all activities simultaneously. The date and time of the required changes may be important so the date and time of all changes should be recorded.
The location in which any added work was carried out	The location of any added work should be recorded. The event may impact work at other locations and/or work off-site.
The workflow process adopted in carrying out the change	Full details of the processes involved in carrying out the changes need to be recorded. It will be important to detail not only what was done but why and when.

(From 'The Guide to Good Practice in the Management of Complex Projects' CIOB, 2011)

or at a later date. All these aspects need to be considered when revising the schedule.

The impact of such events on current and future construction will need to be fully evaluated. Individually or combined, they may lead to a claim for an extension of time or additional payment. There will be two concerns: the impact on the programme and who is responsible. (This aspect of planning and programming is examined in detail in Section 4.)

Table 4.6 summarises the type of information that needs to be recorded to enable experts to take full consideration of intervening events.

Assessing progress and forecasting completion

In addition to simply updating the current schedule, the contractor should assess progress and forecast completion based on the impact of the intervening events and any other additional information available at the time of the review. This may be achieved by simply reviewing and revising the bar chart schedule, revising/re-drawing the bar chart schedule or re-modelling the project using CPMs. The level of progress modelling and forecasting undertaken will depend on the size and complexity of the project, the contractual situation and the time and resources available.

Critical path methods provide a model of the construction project with which it is possible to produce a range of schedules and reports. These may be used to assess progress and forecast completion for a number of different scenarios. Reviewing the progress of activities on the critical path gives an immediate indication of overall progress on the project. On complex projects, progress on other activities may have changed the critical path. Traditional CPM methods require the model to be updated and the start and finish dates re-calculated. This enables a full review of progress to be made, including changes in the float of each activity. Critical Chain Project Management (CCPM) (see Chapter 5) argues that attention should simply focus on the original critical path and the resources required to complete these activities.

Reviewing contingencies and revising the working schedule to effect a recovery

Assessing progress and forecasting completion provide the contractor with the basis to make decisions as to how the construction should proceed and what (irrespective of the contractual liability) needs to be done to effect a recovery and ensure that the project will be completed by the agreed completion date. Such decisions demand consideration of the uncertainties and risks and resource requirements needed to make the necessary changes.

Other methods of monitoring progress

In addition to managing the time model and monitoring the critical path (described above), there are several other ways of monitoring progress. They include milestone monitoring, cash flow monitoring, activity schedules, planned progress monitoring and earned value analysis.

Milestone monitoring

Milestones are activities of zero duration that are used to identify or highlight key points or events in the project. They are often used to identify the start or completion of sections of the project and are therefore useful for monitoring performance. A milestone schedule (in the United States a milestone plan) may be produced to identify the planned timing and actual timing of the project milestones. Milestones may be adopted to mark specific events such as the completion

Section II

of a section of work, an application for a licence, an application for payment, the start or completion of the work of a key sub-contractor, the building being deemed weather proof, etc.

A milestone must be closely defined so that all involved in the project are in no doubt as to what needs to take place before a milestone may be met. Progress on the project is simply monitored by recording whether or not these milestone dates have been achieved. To be effective, Milestone monitoring needs to be linked to activity monitoring.

Cash flow monitoring

Cash flow monitoring evaluates project progress by the rate of spend – comparing the cumulative cost to date with the cumulative value of work completed. It is therefore simply a financial method of review and takes no account of the construction work undertaken and whether work has deviated from the intended schedule, whether the work on the critical path progresses satisfactorily or what can be done to recover an under spend.

Activity schedules

Activity schedules are lists of construction activities giving details that enable progress to be assessed. Activity schedules identify:

- The activity name
- The quantity of work
- The proposed start date
- The actual start date
- The proposed completion date
- The actual completion date
- The resources required
- The responsibility for completion

Progress may be monitored against these scheduled dates and an updated version of the schedule presented for review by senior management as part of a formal progress report.

Planned progress monitoring

Planned progress monitoring is a method of monitoring performance based on construction activities and their units of time. The method is sometimes known as 'counting the squares'. It requires the production of a bar chart programme for the project. Progress is then assessed against this chart and the cumulative total of activity weeks. To do this you set up and plot the planned progress curve, record and plot the progress achieved, quantify progress and compare the progress achieved with the target and assess the delay or gain against the original schedule.

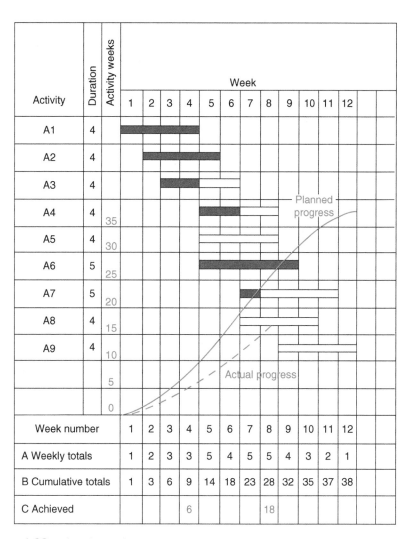

Figure 4.22   The planned progress monitoring method.

Planned progress monitoring may also be used to predict the approximate date for completion based on performance to date.

Figure 4.22 shows the principles of the method.

Earned value analysis

A true assessment of performance needs to take into account not only progress in terms of time but also progress in terms of cost and value. Earned Value Analysis enables you to measure performance in terms of time, cost and scope. It is a quantitative technique that necessities an estimate of both production progress and related cost. The technique may be used to assess progress on an individual construction activity, a group of activities (work package) or the entire project. (Earned value analysis is reviewed in detail in Chapter 6.)

Resources and cost optimisation

Resources

A resource is any goods or services required to complete the work of an activity. Resources include people, equipment, materials, information, safe working space, a safe working environment and money. Resource selection and distribution determine the financial success of a project. Maximum efficiency is obtained by:

- An orderly and even flow of work
- Continuous work without interruption
- An adequate volume of allocated resources
- The employment of appropriately skilled labour resources, technically adequate plant and equipment resources and material resources of the highest quality
- The use of a correct combination of labour and plant/equipment (see Uher and Zantis, 2011)

Some consider that the growth in the use of sub-contractors within the construction industry has reduced the need for extensive resource planning and scheduling by the main contractor. There is however a continued need to review and schedule the key resources required for construction. The prudent contractor checks with the sub-contractor not only when work will commence but how many workers and support staff the sub-contractor will have on-site over the duration of the work. The key resources will vary with the type and size of the project. They will need to be monitored.

In this section, we consider resource aggregation, resource smoothing and resource levelling. These are shown diagrammatically in Figure 4.23.

They are all part of what is known as resource analysis, the analysis and subsequent action required to make best use of the resources available for the work.

Although resource analysis may be undertaken by hand, this is only possible on very simple schedules. Software to calculate requirements and re-schedule activities is required. Such software has been available for over 50 years, and the techniques first developed for CPMs are still applicable. (For a detailed review of these methods, see, e.g. Pilcher, 1992.) Full resource analysis for a project is a time-consuming process and demands data input and review from a number of people closely linked with the project if it is to produce a logical, valid schedule of the resources required.

Resource aggregation

Resource aggregation is the summation of each type of resource throughout the duration of the project. This is usually undertaken only for key resources. Figure 4.23a shows the aggregation of labour resources over a 12-week period. This type of diagram is known as a 'resource profile', a visual representation of the resource requirements.

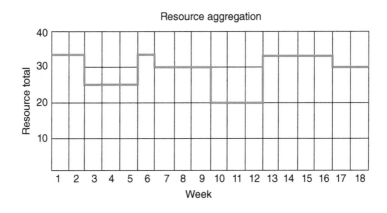

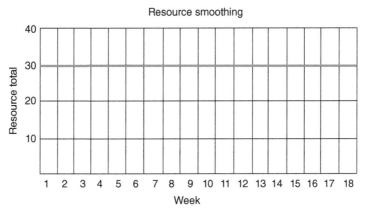

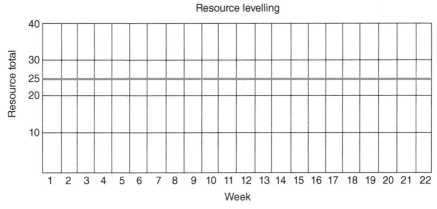

Figure 4.23   Resource analysis: (a) resource aggregation, (b) resource smoothing and (b) resource levelling.

Resource smoothing

Resource smoothing (also known as resource scheduling) is the process of smoothing out the peaks and troughs in resource requirements without extending the project duration. The technique is to delay some activities within their 'float' to remove peaks and troughs in their demands. The scope for smoothing resources is limited by the amount of float available and the constraints of the programme.

This is shown in Figure 4.23b.

Resource levelling

Resource levelling is the process of rescheduling the programme and determining scheduled activity dates to ensure that maximum resource availability is not exceeded. This may cause the duration of the project to be extended. Resource levelling balances the peaks and troughs of resource requirements. It requires you to estimate the amount of resources required for each activity and then schedule the activities using the float available to ensure that the maximum resource level is not exceeded. This then produces a resource-based schedule.

This is shown diagrammatically in Figure 4.23c. Here, the duration of the project is extended because the maximum resource level available is only 25.

This may result in a 'resource-limited project', a project where the availability of resources determines the project duration. That is to say the project could be finished in a shorter time but only if there were more resources available. This typically occurs if there is a shortage of key labour resources, for example, welders, or specialist plant and equipment, such as piling rigs.

The effect of resource levelling extends beyond simply changing the start and end dates for an activity. It changes the float on an activity and the total float on the project. Non-critical activities become critical, and the ability of the schedule to accommodate future delays is reduced. Delaying the start of an activity delays the completion of the activity and may affect the date when an application is made for payment. Resource levelling identifies specific dates for the start and finish of activities. If this does not happen, then the overall work schedule will not be completed by the date(s) anticipated (see Uher and Zantis, 2011).

Resource levelling must be approached in a systematic manner; you need to determine and agree on a basis for re-scheduling the activities within the float available. Priorities must be established and then computations and scheduling completed. Computer software will be needed to examine different 'what-if' scenarios.

Typical computer software will aggregate the resources required in a specified period and compare this requirement with the availability of the resources that has been specified. If the level of resources exceeds this availability, then the program will either:

- Show the excess requirement in tabular form, with the excess highlighted or
- Re-schedule the activities to accommodate the shortfall based on established rules.

Typical rules for re-scheduling adopted by computer software are as follows:

1. Resources are levelled within their normal limits by using up available total float. The project duration is maintained.
2. If resources cannot be levelled within their normal limits, then they are levelled within their total limits whilst maintaining project duration.

3.   If you cannot level resources within their total limits, then extend the project duration.

These rules are the rules used by the Primavera P6 software package™ as identified by Uher and Zantis, (2011).

Different software packages adopt different rules! Therefore, whenever using a software package for resource scheduling, you must ensure that you understand the basis on which decisions are being made. Ensure all in the management team are content to work to these rules.

Software packages work logically to the rules stated. They will not take regard of aspects such as site availability, constraints on working space, the availability of resources may be limited by supplier's ability to deliver, storage of materials may be limited, etc. For this reason, some practitioners argue that unless the schedule is large, resource smoothing is best performed manually using paper, pencil and squared paper. If it is performed manually, then, as with a computer, the rules for making amendments must be established.

A detailed example of resource smoothing is shown on the website for this book: www.wiley.com/go/baldwin/constructionplansched.

Cost optimisation

If the time to complete the project exceeds the agreed date for completion, then penalty charges may be incurred. If you want to reduce the time taken for the duration of the project, you can consider reducing the duration of activities on the critical path by increasing the supply of resources. This will increase the direct cost of the activity but there will be savings in the indirect costs: the site overheads and penalty charges.

Will the savings offset the increased cost of the resources required?

Consideration of this trade-off is known as cost optimisation and may be examined by what is known as 'crashing' the schedule: taking steps to reduce the overall project period by reducing the duration of individual activities. The activities selected for a reduction in duration are selected by consideration of a number of factors: a combination of float (those with Zero Float or Negative Float), the sensitivity of duration to increasing resources, the cost of increasing resources and the availability of additional resources. The adding of resources should not be done indiscriminately or irrationally; not all resources should be crashed, only those that will bring the required benefits (Carrillo, 2010).

Figure 4.24 shows the relationship between the duration of an activity and the direct cost of an activity. When adding resources, the time will be reduced and the direct cost of completing the work will be increased. This increase in cost is rarely linear. Significant cost increases are usually required to reach the minimum duration.

When 'crashing' activities, it is necessary to consider both direct costs and indirect costs. The total cost of an activity is the direct cost of an activity plus the indirect cost. Increasing the resources allocated for the activity reduces the duration of the activity, but a point is reached where the use of additional resources does not result in any overall savings on the project. This, the optimum total project cost, is shown in Figure 4.25.

Section II

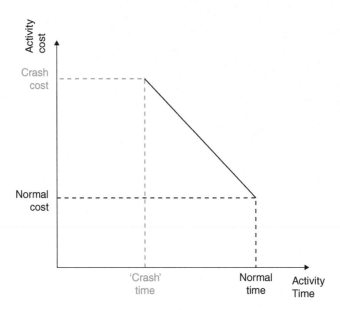

Figure 4.24   The relationship between the duration of an activity and the direct cost of an activity.

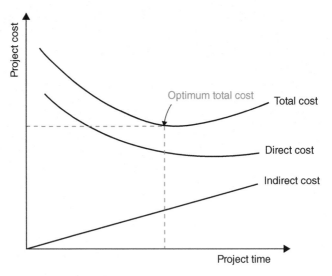

Figure 4.25   Optimum total project cost.

Time cost optimisation became a technique available to planners when the development of CPMs enabled the planner to easily model the project and examine 'what-if' scenarios. You may not use time cost optimisation techniques on a regular basis, but they are a reminder of the fundamental relationship between time and cost on any activity and therefore the trade-off should always be remembered.

Method statements

A method statement is a written description of how the construction will be carried out. The statement may cover all or part of the project. The form, content and level of detail provided in a method statement will vary depending on for whom the document is produced and the stage in the project cycle (Cooke and Williams, 2009). Some method statements will be prepared for internal use, whereas others for external use and distribution to third parties.

The following are examples of different types of method statements:

- The tender method statement (for submission to the client)
- The tender method statement (for internal use)
- The construction or work method statement
- The health and safety method statement
- The planning method statement
- The waste management plan

Table 4.7 summarises the purpose of each of these types of statement.

Format of the method statement

The usual format for a method statement is a written document supplemented by drawings or diagrams. Alternatively, the information may be presented in tabular form. With the increasing use of Building Information Modelling (BIM) and virtual prototyping, new forms of visual presentation are becoming increasingly popular, particularly for large complex projects (See Chapter 9).

The tender method statement (for submission to the client)

At the tender stage, the client (or the client's representative) normally requires a construction method statement to be produced by the contractor and submitted with the tender bid. This will provide confidence that the contractor has considered the technical challenges inherent in the construction and identified viable solutions.

Such a method statement should include details of the proposed construction process and an outline of:

- The site location
- Access details and access restrictions
- Any related design development (where the contractor is responsible for any elements of the design)
- Temporary works requirements
- Safety
- The management and supervision structure for the project
- The quality plan for the project

Where the construction process is straightforward, this information may be presented in a simple table. Where the construction process is more complicated, extended information and additional calculations (e.g. for temporary works items) may be required.

Section II

Table 4.7   Types of method statement and their purpose (Cooke and Williams, 2009).

Tender method statement (external)	To meet the requirements of the tender process.
	To provide the client with confirmation of the contractor's technical ability to successfully undertake the construction work.
	To provide supporting information to the programme of works.
Tender method statement (internal)	To aid the estimator.
	To enable the bid to be based on practical methods.
	To assess alternative proposals at the tender stage.
	To assess plant requirements for inclusion in the tender.
	To provide a basis for the estimator to build up rates.
Construction work method statement	To explain the contractor's proposed methods and sequence of working for checking by the client's representative.
	To calculate activity durations for the programme.
	To decide on gang composition and resource requirements for individual activities.
	To plan activities so that a logical construction sequence is adopted.
	To provide an easily understood document which can be communicated to those who will carry out the work on-site.
Safety method statement	To demonstrate safe systems of working for hazardous operations.
	To be included in the construction health and safety plan together with sub-contractor's method statements.
	To manage any residual risks identified in the risk assessment.
	To provide practical control measures.
	To explain how the work is to be supervised and monitored.
	To show personal protective equipment (PPE) requirements.
	To explain first aid and emergency arrangements in the event of an accident.
	To explain the work method in a 'tool-box' talk or task talk.
Planning method statement	This statement is required for complex projects. The statement records the assumptions have been made in the process of risk management, planning, scheduling. It clarifies how and when the schedule for the project will be updated and by whom.
Waste management	The waste management statement provides details of the arrangements for the collection, holding, removal and recycling and/or disposal of waste materials. Particular attention is given to environmental regulations including the Site Waste Management Regulations.

Section II

The tender method statement (for internal use)

The planner will produce a separate internal construction method statement. The purpose of this statement is to ensure that all in the tender team are aware of the proposed construction method. The method statement will require input from other members of the tender team. Discussion will be needed to produce an agreed method of working. This may require input from external organisations, for example, specialist sub-contractors. It is essential that the thinking of the planner and the estimator is closely aligned to ensure that the cost of all the resources required for construction is included in the cost estimate. This is particularly important for large items of the construction plant (such as tower cranes) and large activities where the pricing of bill items cannot be commenced without consideration of work methods and resources.

An internal construction method statement will be needed to help ensure a safe system of working. If there are dangerous aspects of the construction, for example, demolition, working close to railways or over water, then these activities will need particular attention.

Method statements help ensure high-quality work and may be required as part of the quality management scheme. Details of the construction methods will need to be discussed with suppliers and sub-contractors.

Where the project will require a large configuration of plant and equipment, a method statement will be required to show the mobilisation, utilisation and de-mobilisation of the plant items. Different method statements will be required for alternative designs.

The construction or work method statement

Following the award of the tender and before the commencement of work on-site, the planner further develops the method statements for the key items of construction. The details for each of the items in the tender method statement will need to be extended.

It will be essential to coordinate the interfaces between the specialist sub-contractors. All sub-contractors and all suppliers will need to be provided with arrangements for the delivery and storage of materials and site safety practices. Where there will be extensive temporary works, for example, scaffolding to the building, the designs will require checking and authorisation.

The development of detailed method statements will extend throughout the construction period. Detailed statements and schedules will need to be produced to ensure that key items of the plant, for example, tower crane, operate at optimum efficiency.

The health and safety method statement

Method statements are required to ensure safe methods of working. These should include:

- The time needed to complete the work
- How the work will be undertaken

Section II

- The plant, machinery and temporary works required (e.g. earthwork support, scaffolding, formwork)
- The order and sequence of the work
- The means of access and egress to and from the work place
- The means of preventing falls of people and objects
- The means of preventing unauthorised access to the place of work
- Interference with other trades working in the same location (Cooke and Williams, 2009)

A sample safety method statement is shown in Table 4.8.

Cooke and Williams (2009) indicate that safety method statements are often confused with method statements for construction work and should be an integral part of the overall construction planning method and a legal requirement with respect to the provision of a safe place of work.

Further details of planning for health and safety and the environment are given in Chapter 12.

Everyone managing, controlling and working on construction projects has a responsibility for ensuring the health and safety of the workers. All construction work must proceed under the current laws and legislation relating to safety. In the United Kingdom, the following legislation is particularly important:

- The Health and Safety at Work Act 1974
- The Management of Health and Safety at Work Regulations 1999
- The Construction (Design and Management) Regulations 2007

Under the MM, each notifiable[5] project is required to appoint a CDM coordinator before the initial design work commences or other preparations for construction work have been completed. It is important that the CDM coordinator is appointed early in the project to ensure that health and safety arrangements are considered from the start of the project and at every phase of the project.

Table 4.9 summarises the responsibilities of the principal contractor under the CDM regulations.

For further information, the reader is directed to sources of information on this legislation available at the website for this book.

Planning method statement

On large complex projects, the CIOB stresses the importance of producing a full planning method statement. The purpose of this statement 'is to facilitate the understanding and cooperation of the participants. It should make clear what constraints have been identified, what assumptions have been made in the process of risk management, planning, scheduling, review and update of the schedule and the reasoning underpinning those constraints and choices' p14 (CIOB, 2011).

The content of the method statement will vary over the duration of the project as will those who are required to contribute information and work to the method statement and produce the information required for the overall control of the project.

[5] Projects which last longer than 30 days or involve more than 500 person days of work are known as 'notifiable' projects under the CDM 2007 Regulations.

Table 4.8   A sample method statement.

Item	Details
1. Operation	Construct sheet piled cofferdam
2. Plant items	Mobile (tracked) crane, BSP 900 piling hammer, Komatsu 380 excavator, two 30 tonne wagons, 150mm diesel pump
3. Work sequence(see drawing no – C345 DRW 121)	1. Construct a hardstanding for the piling rig in lock mouth using imported quarry waste. 2. Erect guide frame and install Frodingham 3N piles. 3. Pump out water between hardstanding and piles. 4. Fill behind piles for access. 5. Remove piling frame. 6. Construct top bracing with steel walings and struts. 7. Remove hardstanding and install bottom bracing. 8. Erect secure ladder access to the bottom of the cofferdam.
4. Supervision	Supervision will be provided by the piling foreman. Operation to be monitored daily by the site agent. A banksman will work with a mobile plant. Daily check on crane equipment, load indicators and operation.
5. Controls	The working area will be 'Authorised Personnel Area Only'. When working over water to remove the piling frame, life jackets and safety harnesses must be worn. The working area is to be fenced-off at night with a locking metallic fencing system. Warning notices to be displayed around the cofferdam.
6. Emergency procedures	1. Send for site first aider and/or call for emergency services where necessary. 2. Rescuers must not put themselves in danger. 3. Follow first-aid drill if appropriate. 4. Do not remove evidence. 5. Notify site agent.
7. First Aid	Dinghy to be moored adjacent to cofferdam. Two lifebuoys in wooden locker. First-aid box in site office. Two-way radio system.
8. PPE Schedules	Safety harnesses, hard hats, gloves, welding goggles, high-visibility vests, life jackets, ear defenders.

Adapted from Cooke and Williams (2009).

Section II

Table 4.9 Responsibilities of the principal contractor under CDM 2007.

Those that they appoint, including all contractors and sub-contractors, are
competent.
That the construction phase is properly planned, managed, monitored and
resourced.
That contractors are made aware of the minimum time allowed for planning and
preparation.
Providing relevant information to contractors.
Ensuring safe working coordination and cooperation between contractors.
Ensuring that the construction phase health and safety plan is prepared and
implemented. This plan needs to set out the organisation and arrangements
for managing risk and coordinating work and should be tailored to the particular
project and risks involved.
Making sure that suitable welfare is available from the start of the construction phase.
Ensuring that site rules as required are prepared and enforced.
Providing reasonable direction to contractors including client-appointed contractors.
Controlling access to the site to restrict unauthorised entry.
Making the construction phase plan available to those who need it.
Providing information promptly to the CDM coordinator for the health and
safety file.
Liaising with the CDM coordinator in relation to design and design changes.
Ensuring all workers have been provided with suitable health and safety induction,
information and training.
Ensuring that the workforce is consulted about health and safety matters.

From CIOB (2010).

Table 4.10 shows the main information that should be included in the planning
method statement. This list covers all the information required at 'low' density,
that is, 9 months or more after the date of the document.

At medium density, that is, for work to be undertaken in the period from 3 to 9
months, the information items in Table 4.11 should be added to those in Table 4.10.

At high density, that is, for work activities to be undertaken in the next 12
weeks, the planning method statement should include the items in Table 4.12
which should be added to Table 4.10 and Table 4.11.

The full planning method statement is produced in stages over the duration of
the project and builds from an outline of the key items of work (low density) to a
comprehensive plan for the work (high density). The planning method statement
forms a record of all the production-planning decisions taken and the basis of
these decisions.

Site layout plans

It is essential that careful consideration is given to the layout of the site.

At the tender stage

At the tender stage, the client may request that the contractor submits a site layout
plan. Even if a plan has not been requested, it is important for a site layout plan

Table 4.10   The contents of the planning method statement – *Low density* (CIOB, 2011).

Item	Item description
LD1	A description of the work to be carried out. Design, procurement and development strategy and constraints.
LD2	Third-party and neighbour interests and interfaces.
LD3	A description of the approach to risk management and the risks identified.
LD4	An assessment of the contingencies to be allowed for in these risks.
LD5	A description of the activities contained in the schedule by reference to their activity ID codes.
LD6	The WBS.
LD7	Calendars for working weeks and holiday periods.
LD8	Generic resources anticipated and anticipated resource constraints.
LD9	Permits and licences required and the decision periods relating to each application.
LD10	Material and equipment restrictions and availability.
LD11	The approach to utilities and third-party projects, licences and restrictions.
LD12	The approach to schedule review, revision and updating.
LD13	Activity codes applied.
LD14	Cost codes applied.
LD15	Details of phasing and the zonal relationships of the project.
LD16	Principal methods of construction.
LD17	Details of major plant requirements.
LD18	Site management, logistics assumptions, site welfare and temporary works including scaffolding, access and traffic management.
LD19	Health and safety.
LD20	Environmental considerations.
LD21	Principal methods of procurement and their effects.
LD22	The methods used to estimate durations.
LD23	The assumed sequencing logic and an explanation of any logical constraints.
LD24	A description of the critical and non-critical paths to key dates, sectional completion dates and the completion of the works as a whole.
LD25	Reporting formats, communications strategy and information format.

Table 4.11   Contents of the planning method statement – *Medium density* (these are *additional contents* to those listed for low density) (CIOB, 2011).

Item	Item description
MD1	Identified specialist contractors, sub-contractors and suppliers.
MD2	Key-trade-interface management strategy.
MD3	Design-and-procurement interface strategy.
MD4	Limited possessions.
MD5	Planned overtime.
MD6	Temporary traffic diversions and plant maintenance strategy.
MD7	Resources anticipated and anticipated resource constraints.
MD8	Material and equipment restrictions and availability.
MD9	Utilities and third-party project licences and restrictions.
MD10	Schedule review, revision and updating.
MD11	The methods used to estimate durations.
MD12	Details of plant requirements and their assumed productivity, downtime and maintenance.

Section II

Table 4.12   Contents of the planning method statement – *High density* (these are *additional contents* to those listed for low medium) (CIOB, 2011).

Item	Item description
HD1	Resources to be employed.
HD2	Expected productivity.
HD3	Detailed calculations of the activity durations.
HD4	Detailed methods of construction.
HD5	Details of plant requirements and their productivity, downtime and maintenance.

to be produced for the contractor's internal purposes as the preparation of such a plan will ensure consideration is given to aspects such as:

- The location of offices and site accommodation;
- The location of site services;
- Access to the site, temporary roads;
- The position of plant and equipment (e.g. tower crane locations); and
- Material storage areas.

All of these will have implications on the construction process and productivity.

For some projects, such as those adjacent to existing works and infrastructure works (e.g. railways), site layout plans are particularly important.

At the pre-contract stage

When the contractor has secured the contract, the planner will be required to produce a detailed site layout plan which will meet the requirements of the main contractor and the requirements of all the main sub-contractors. These details must be produced within a short timescale to ensure that work on-site may commence on the allotted date. The proposals must show clearly the order of priority for the establishment of the site.

At the contract stage

At the contract stage, there may be instances where there is a need to re-configure the site layout. This maybe when construction enters a new phase of the work, and this demands changes to the site layout. However, once having established the site layout, the contractor will aim to keep this layout consistent throughout the duration of the project.

Site waste management plans

Increasingly, the Main or Principal Contractor is required to put in place site waste management plans (WSMP) for the removal, recycling and/or disposal of the waste materials that are produced by the construction process. Site waste management plans are discussed in detail in Chapter 11.

Section II

Contractors' cash flow

Contractors need to monitor cash flow carefully. Construction companies that cease to trade do so primarily because of lack of working capital. At the tender stage of the project, a prudent contractor will request the planner (or estimator) to produce a full cash flow forecast to assist the tender adjudication panel assess the cost of capital for the project and make provision for any commercial risks. In this section, we identify the information required to produce a cash flow forecast, outline the calculation process and discuss how contractors may improve their cash flow. The website linked to this book provides an example of a full cash flow calculation(see www.wiley.com/go/baldwin/constructionplansched).

For any business, the monthly cash flow is the difference between the payments received and the payments made to suppliers of materials and services.

A typical contractor organisation will receive payments for:

- Work completed on current construction contracts (work in progress);
- Final account payments for completed work;
- Retentions released on practical completion;
- Retentions released on the issue of final certificate;
- Returns on other investments; and
- Shareholders' funding.

Payments will need to be made to:

- Materials suppliers;
- Plant hire companies;
- Sub-contractors;
- Wages to directly employed labour;
- Staff salaries;
- Head office overheads;
- Equipment and vehicles;
- Miscellaneous expenses.

The company will at any time be involved in a number of construction contracts. This 'portfolio' of projects will be at different stages of completion, and the total value of each project will vary. The projects may be operating under different 'forms' of contract with different payment conditions. The terms for payment for materials and arrangements for payment to sub-contractors may vary, and the overall cash flow for the company will need to take account of all these factors across all the company's projects. Here we look at the cash flow for a single project.

On some projects, contractors may be able to negotiate mobilisation costs, payments made at the start of the contract to assist with the costs of setting up the site. However, for most contracts, contractors are paid in arrears as the work progresses. Applications for payment are made on a regular basis, for example, monthly, following the measurement of the work that has been completed since the last payment application. The work completed is valued and certified. Following review by the client (or the client's representative), payment is received by cheque. The payment terms will vary with the type of contract. Table 4.13 gives details of typical payment terms.

Table 4.13   Typical payment terms.

Contract	Payment terms
JCT 05 standard building contract	Monthly, 14 days from the architect's certificate
JCT 05 design and build contract	Fourteen days from the contractor's application
JCT 05 major project contract	Fourteen days after the issue of payment advice
ICE7 conditions of contract	Twenty-eight days from the contractor's statement
New engineering contract	Three weeks for the project manager's assessment date
JCT standard building sub-contract	Twenty-one days after the appropriate interim certificate for the main contract is payable
Civil engineering contractors association form (blue form)	Thirty-eight days after the main contractor's application to the employer
Non-standard forms	Varies. Payment between 50 and 70 days from receipt of invoice is not unusual

From Cooke and Williams, 2009.

Bank borrowings

Because payment is paid to contractors in arrears, companies require working capital to fund the supply of materials, pay sub-contractors and make payments to direct labour. This working capital is normally secured by a privately arranged loan, a short-term bank loan and/or a bank overdraft.

Head office overheads

The mark-up that the contractor adds to the direct cost of the construction work will include a contribution towards the running of the company's head office costs including head office staff. Companies should be tendering within a defined tendering policy. Depending on the current predicted annual turnover for the company, the tender adjudication team may increase or decrease the contribution to head office overheads.

Working capital

The mark-up added to the direct cost of the construction work should also make allowance for the contractor's cost of finance. This can be calculated by reviewing the cash flow for the project.

Calculating the cash flow for the project

To calculate the cash flow for a project, the following data are needed:

- A graph of contract value against time;
- A graph of contract cost against time;
- The measurement and certification interval;

- The delay in payment between certification and the contractor receiving the cash;
- The retention conditions and repayment arrangements;
- The project costs broken down into labour, plant, materials and sub-contractor categories; and
- The terms of payment for the cost of the resources used.

Each of these items is now discussed.

A graph of contract value against time   A graph (or set of figures) of the contract value against time is required to represent the value of the construction work undertaken in each time period. The contractor will calculate the total tender sum, and this sum may be broken down into various sections such as a section of the Bill of Quantities or a particular trade or section of the works. It cannot however be divided across the time span of the project without additional work.

The planner is required to use the pretender schedule or a simple bar chart to value each activity or bar by the bill rates, including the mark-up for overheads and profits. Each activity needs to be considered as a 'collection' of bill items and the value of the works apportioned accordingly. The cumulative value forecast is only as good as the accuracy of the schedule. By totalling the value over each payment period (e.g. monthly), it is possible to produce the cumulative value of the construction work over the project. This may then be presented in either graphical or tabular form.

A graph of contract cost against time   To produce a similar graph of cost versus time, estimate the cost of each activity on the bar chart schedule using the direct costs from the bill of quantities. This will produce a graph similar to the value versus time graph but representing costs.

The measurement and certification interval   Knowing the week of the first measurement and the time intervals between subsequent measurement periods together with the level of retention, the maximum retention and the maximum retention sum enable the planner to calculate the expected value of the work for which payment will be requested. This information will be stated in the contract documents.

The delay in payment between certification and the contractor receiving the cash and the retention conditions and repayment arrangements   This information, stated in the contract documents, enables the planner to estimate not only the level of payment that will be due after each certification but when the payment will be made to the contractor.

It is usual for the client to withhold part of the payment and not release this sum until the completion of the project. This sum is known as retention. The level of retention is usually a fixed percentage of the work up to a maximum percentage of the total contract sum (e.g. a contract may include the retention of 3% of all payments up to 5% of the total contract sum). At the practical completion of the contract, a proportion of the total retention that has accrued will be released. The remainder of the retention will be paid to the contractor on issue of the final certificate. Retention is 'capital lock-up' directly affecting the level of working capital required by the contractor and sub-contractors. The practice of retention has been

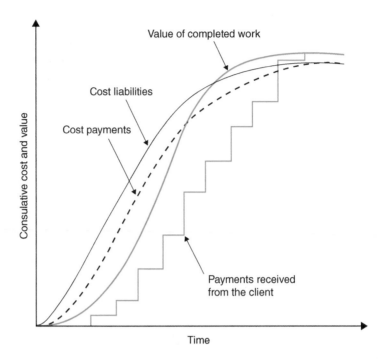

Figure 4.26   The contractor's cash flow for a construction project.

criticised by several industry reports, for example, Latham (1994), but remains an integral part of industry working.

The project costs broken down into labour, plant, materials and sub-contractor categories and the terms of payment for the resources used   The delay between receiving goods and services and paying for them is an important factor in the calculation of the cash flow. The delays between using labour, plant, materials and sub-contractors are determined by current trading conditions.

Labour directly employed by the contractor will be paid weekly. Materials suppliers will typically be paid 30 days from the end of the month of delivery of the materials. Plant hire companies will usually submit invoices for payment when the plant is off-hired. Payment will then be made from 15 to 30 days after receipt of invoice. Sub-contractors will be paid in accordance with the terms of their contract. The terms of these sub-contracts will reflect the terms of the contract between the client and the main contractor including retention.

Figure 4.26 shows a cash flow for a construction project.

Improving cash flow

To improve the cash flow for a project, the contractor has a number of options. These include:

▪ Seeking a review of payment conditions
▪ Completion of the project earlier than scheduled
▪ 'Front-end loading' of the Bill of Quantities

- Over-measurement
- Delays in paying for goods and services

The contract documents will specify the payment conditions including any mobilisation payment, how and when payments may be initiated and any retention of payment. Seeking and securing an advantageous review of payment conditions (increased mobilisation payment, more regular payments, decreased retention) will improve the cash flow by reducing maximum borrowing required.

Completion of the project earlier than originally scheduled will secure final payment and release of retention earlier than expected. Completion of the project early will reduce the time that the contractor needs to pay the costs for operating the site on-costs, hire of temporary accommodation, service costs, etc. Having established a schedule for the project, it may be beneficial (depending on borrowing costs) to employ additional resources to complete the construction work early.

'Front-end loading' is the practice of increasing the rates for items of the construction work that take place early in the project with appropriate reductions for the rates of work items that will take place in the later stages of construction. This keeps the overall cost of the work the same but has the effect of increasing the level of payments early in the project and reducing borrowing costs.

Over-measurement of work in progress, for example, earthworks excavation, will increase the level of payments made to the contractor in the early part of the contract but result in reduced payments as the work item is completed.

Perhaps the easiest way that a contractor can improve cash flow is to delay payment to suppliers and sub-contractors until after payment from the client. This practice, widely criticised, still persists within the industry despite attempts to secure the supply of sub-contractor services on a 'pay-when-paid' basis.

Uncertainty and risk

All the activities of an organisation involve risk. Risk may occur at all levels within an organisation. Organisations should manage risk by identifying their risk principles, developing a framework for managing risk and establishing the processes necessary to ensure that their activities at the organisational level and project level take full consideration of uncertainty (ISO 31000, 2009).

Uncertainty may be beneficial – this is generally known as 'opportunity' – or it may be 'detrimental' – this is often termed as 'risk'. All parties involved in the project must assess their own risk or 'exposure' to adverse events. Risk management (the identification of risk, risk analysis and responding to risk) should be undertaken throughout the project.

For the contractor, this means managing risk from bid preparation, through pre-contract planning, during construction and through to contract completion. The construction planner must be aware of where uncertainty arises, the risks involved and how to contribute to the contractor's management of risk.

In this section, we look at general frameworks and processes for managing risks and some of the techniques available to the planner. Compared with some industries, for example, the financial and insurance industries, the quantitative analysis of uncertainty and risk is seldom undertaken. We explain why.

The Oxford Dictionary of English (Thompson, 1996) defines being uncertain as, 'not knowing or known'. Risk is defined as, 'a chance of possibility, danger, loss, injury etc.' How are uncertainty and risk linked? ISO Standard 31000:2009(E) defines risk as the 'effect of uncertainty on objectives'. The standard adds the following notes: An effect is a deviation from the expected – positive and/or negative. Objectives can have different aspects (such as financial, health and safety, and environmental goals). Risk is often characterised by reference to potential events and consequences and can apply at different levels (such as strategic, organisation wide, project, product and process). It is often expressed in terms of a combination of the consequences of an event (including changes in circumstances) and the associated likelihood of occurrence (ISO standard 31000, 2009).

The construction planner typically views the construction work as a series of inter-related, interdependent activities the result of which is the completion of the work. Planners may be uncertain about many factors relating to construction work: the duration of construction activities, the method of construction to be adopted, the relationship between construction activities, the variability of the ground on which the building is to be sited, the level of productivity of a subcontractor, the variable quality of the materials required for the work, etc.

How can all these events and their potential outcomes be evaluated? How do you cope with uncertain situations: ignore them, seek additional information, make more accurate forecasts, conscientiously adjust or bias, revise the rate of return by adding a risk premium, transfer the risk or seek alternative options?

Uncertainty arises at different stages of the construction process. The planner's response to uncertainty will vary depending on the stage of the project and the risk involved if construction does not proceed as expected. The response should be within the organisations overall risk management procedures.

Risk management

Risk management is a system which aims to identify and quantify all risks to which a business or project is exposed so that a conscious decision can be taken on how to manage risks. This requires the identification of potential risks, risk analysis, determining how the business should respond and initiating the action required.

ISO Standard 31000 (2009) recommends that organisations develop, implement and continuously improve a framework whose purpose is to integrate the process for managing risk into the organisation's overall governance, strategy and planning, management, reporting processes, policies and values. It stresses that the adoption of consistent processes and a comprehensive framework can help to ensure that risk is managed effectively, efficiently and coherently across an organisation.

The management of risk commences with establishing the principles under which the organisation should operate. Examples of these principles include:

- Risk management should be an integral part of decision-making;
- It should be based on the best available information; and
- Should be systematically structured and timely.

These principles form a mandate for the design, implementation, review and continual improvement of the framework for managing risk in the organisation. The risk management process for the project requires that, within the specific context, it is necessary to identify the risks, analyse risks, evaluate risks and decide how the risk should be treated. (Treatment of risk involves the selection of options to modify risks and implementing those options.)

This process is an iterative process. At each stage, it is necessary to monitor and review the work already undertaken and to communicate and consult with all relevant parties.

Risk identification

Risk identification may be undertaken by a number of tools and techniques including a review of project documentation, information gathering, the analysis of checklists, brainstorming, Delphi technique, the use of diagramming techniques, SWOT analysis, assumptions analysis and the input of experts (PMBK, 2004). The risks identified should be recorded in a risk register. The risk register should be structured to collate risks under different headings, for example, commercial, operational, etc.

Risk analysis

Risk analysis is the name given to the techniques available to quantify the impact of uncertainty by analysing the probability of events occurring and the consequences of their occurrence. These methods of analysis may be qualitative, based on assessments against a scale of outcomes, or quantitative, based on sources of data.

Qualitative analysis techniques   Having identified possible risks and produced a risk register, the probability of each risk is assessed together with its impact. Assessment may be made collectively by team members or by input from experts. Pre-defined rules are used to rate each risk. Using look-up lists, tables or probability/impact matrices, an evaluation of each risk is undertaken. (An organisation may produce a risk impact matrix to assess the level of any specific risk against general project objectives, e.g. cost, time, etc.) Alternatively, a general risk assessment matrix may be produced so that given an assessment of the probability of a risk occurring its impact and decisions as to how to manage the risk (or how to take advantage of the opportunity) may be taken.

Quantitative risk analysis   Quantitative risk analysis numerically analyses the effect of the risks identified on the overall project objectives. Similar techniques to those used in qualitative analysis may be used to collect quantitative data on the probabilities and impact of risks identified and listed on the risk register.

Table 4.14 shows a range of cost estimates assembled for a section of a construction project.

Such data may then be used together with probability distributions for forecasting the impact of risks on values such as the duration of the project and the costs of project components. Techniques used for such analysis include sensitivity analysis, expected monetary value, modelling and simulation.

Two frequently quoted, classical quantitative techniques are PERT and Monte Carlo Simulation.

Section II

Table 4.14 A range of cost estimates for a project.

Zone	Element	Low	Most likely	High
A	Design	6	8	10
	Construct	16	20	24
B	Design	3	4	5
	Construct	6	9	11
C	Design	9	12	14
	Construct	18	22	26
Total			75	

PERT

PERT is based on traditional CPM methods but with three estimates: the best, the worst and the most likely for each of the activities in the model. The 'best' duration (*a*) is the shortest duration, a duration that may be expected to be achieved 1 in a 100 times. The worst duration (*b*) is the longest duration, a duration that is also expected to occur only 1 in a 100 times. The most likely duration (*m*) is between these two. (This duration is estimated separately and is not automatically the mean of the best and worst estimates.)

PERT assumes that these three estimates form part of a population conforming to a Beta Distribution so that the best (*a*), worst (*b*) and most likely (*m*) times can be compounded to give a single 'expected time' (te) as follows:

$$te = \frac{(a+b+4m)}{6}$$

This time, te, should be used in all calculations (Lockyer, 1974), and the usual outputs from the CPM model should be produced.

Monte Carlo simulation

In Monte Carlo Simulation, probability analysis is used with the basic CPM network model to produce a risk model as a means to assess project uncertainty. Each activity within the network is considered a variable effecting the overall duration of the project. As with PERT, it is required to provide three estimates for the duration of each activity within the model: a, b and m. Again, the calculations are based on a Beta distribution. Using the Monte Carlo method, the duration of each activity within the model is calculated by the selection of a random number; this number determines the value of the activity duration. For each simulation, the duration of each activity is selected using this random number technique and the time taken to complete the project calculated using the normal critical path calculations. The Monte Carlo approach requires the simulation to be undertaken a minimum of several hundred times (typically between 500 times and 5000 times). This ensures the validity of the Central Limit Theorem, and durations calculated can be assumed to fit a normal distribution described by the mean and standard deviation. The results provide a frequency distribution from which the user can assess the probability of the project being completed within a specific duration. See Figure 4.27.

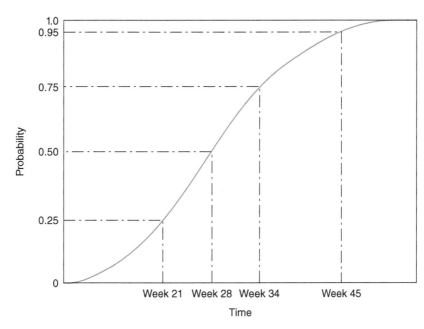

Figure 4.27   Monte Carlo simulation – the probability of project completion.

Further details of PERT and Monte Carlo Simulation are provided in the recommended reading shown on the website www.wiley.com/go/baldwin/constructionplansched.

Risk responses

The outcomes of qualitative and quantitative risk assessment are summarised in Table 4.15.

These outcomes enable you to make decisions as to how to manage the risk or take advantage of the opportunity. To do this, different tools and techniques are available. The PMBOK identifies the strategies. A summary of these is given in Table 4.16.

This section has looked at qualitative and quantitative methods of risk analysis. How does the construction contractor typically use them? As an example, we now look at how contractors price risk in bids.

How do contractors price risk in bids?

The objective when bidding for construction work is to secure the right work at the right price. How do contractors price risks in bids? How does theory align with practice? Latyea and Hughes (2008) reviewed both theory and current practice. This is an overview of their findings.

When asked about risk, most if not all contractors involved in bidding for construction work are able to identify the main areas of risk for a typical project. Major risks relate to conditions relating to the payment for completed work, the design of the work to be undertaken and the ground conditions. Other risks include the weather, contract conditions, the job itself, project location, access,

Table 4.15   The outcomes of qualitative and quantitative risk analysis
(Project Management Institute, 2004).

	Qualitative risk analysis	Quantitative risk analysis
Updating of the risk register	Relative ranking or priority list of project risks	Probabilistic analysis of the project
	Risks grouped by categories	Probability of achieving cost and time objectives
	List of risks requiring a response in the near term	Prioritised list of quantified risks
	List of risks for additional analysis and response	
	Watchlists for low-priority risks	
	Trends in qualitative risk analysis results	Trends in quantitative risk analysis results

Table 4.16   Strategies for managing risk or taking advantage of opportunity
(see PMBOK Guide, 2004).

Category	Response	Further details
1. Strategies for negative risks or threats	Avoid	Change the project management plan to avoid the threat
	Transfer	Transfer the negative impact of the threat along with the responsibility to a third party
	Mitigate	Reduce the threat to an acceptable level
	Exploit	Eliminate uncertainty and ensure that the event takes place
2. Strategies for positive risks or opportunities	Share	Allocate ownership of the risk to a third party ensuring that a share of the benefit is gained
	Enhance	Modify the size of the opportunity and the probability of its occurrence to ensure maximum benefit
3. Strategies for both threats and opportunities	Acceptance	Accept the risk as it is impossible to eliminate all risk from the project
	Contingent response strategy	Retain the original plan but prepare a suitable response strategy to be initiated in the event of pre-defined conditions

project complexity, innovation in design, state of the economy, local government issues and relationship with government councils and agencies.

Some business sectors, for example, the finance and financial services sector, have very advanced systems for managing risk. What is the situation in construction? Latyea and Hughes found that researchers have developed more than 60 different analytical models to assist contractors assess risk within the bidding process. Few, if any, of these were used in practice. Contractors prefer unsophisticated systems where by risk is identified, analysed and an allowance made in the final bid submission.

The short time available for the preparation and submission of the bid and the unsystematic nature of the pricing process prohibit analytical modelling of risk in all but some exceptional circumstances, for example, where the client requests a risk analysis as part of the tender. Contractors do however evaluate and allow for risk. A series of 'gateways' are used to assist in this process.

Tendering is an expensive overhead to any construction organisation. Depending on the type of work and the competition, contractors may only expect to win 1:6 of the bids submitted. Project selection is key. Marketing and promotion is focused on attracting invitations to bid from reliable clients for work that they are confident of executing in the time made available.

The first 'gateway' is project selection. When invited to tender for a new project, contractors select a project where they judge that they can cope with its level of risk. The judgment is generally influenced by the following factors: the client's financial ability, payment regularity, the consultants appointed for the project, the contractors' workload, project location, site, scope of works, contract conditions, clarity of tender documentation and ability to perform the job.

Having been invited to bid, the first decision is whether to accept the invitation. This decision commits members of the estimating team to several weeks of concentrated work and should only be taken if the contractor is committed to winning and undertaking the work.

Having accepted the invitation to bid, the estimating team will exhaustively review the documentation provided to look for risk and opportunity within the project to be undertaken. Typically, a risk register will be established identifying risk items under four main categories: design, commercial, production and financial conditions. A similar register will be drawn up to register the opportunities. Identified risks are analysed and evaluated using a spread-sheet matrix that helps to make allowances for contingencies based on effect or severity and the probability of the event occurring. Risks may be designated a colour code: green, amber or red. Green risks receive no further attention. Amber risks are evaluated as part of the tender risk assessment. Red risks are unacceptable and demand a rethink in how the work will be executed. In extreme cases of red risks, the entire bid may be halted.

The majority of contractors require the estimator to produce the total direct cost for completing the construction work without including allowances for risk, and other mark-ups. These are determined at the tender adjudication meeting.

Tah et al. (1993) indicate that risk may be priced as a percentage in the profit margin, a separate percentage in all the cost, a lump sum in the entire preliminaries bill or a percentage in one bill if the risk is in that bill alone.

Typical percentages range between 2% and 5% of the cost of the project. Where there are concerns about the complexity of the work, the time schedule and/or the ground conditions, this allowance may increase above 10% or may even rise to 20% and above.

These allowances are added in the mark-up to the direct cost. (The mark-up will also include allowances for head office overheads and profit.) The bid manager or the director responsible for adjudicating the bid will make the final decision as to how much to add. Any allowance will be determined by overall consideration of market conditions, the competition and the need for the contractor to win the work.

Key points

- It is widely recognised that it is unsatisfactory simply to develop schedules in an *ad hoc* manner. A clear structure for the schedules to be used on the project must be established. Five levels of schedule reporting are recommended.
- The scope of a project must be clearly defined. A WBS defines the project's discrete work elements and groups them together in a way that helps organise and define the total work scope of the work in the project.
- Construction organisations differentiate between pre-tender planning, pre-contract planning, and contract planning.
- Planning and scheduling the construction is a key function within the estimating and tendering process, because for an accurate estimate of the cost of construction to be calculated, it is essential to know how and when the construction work will take place.
- Pre-contract planning commences when the client has accepted the contractor's bid. During the pre-contract period, the pressure on the planner is to provide more detailed information than that produced at the time of tender and to help facilitate the start on-site.
- The main contractor, in order to maintain production control and to ensure that the project is completed on time within the cost limits, will undertake ongoing contract planning throughout the duration of the project.
- Contract planning includes Monthly (long-term) planning, Weekly(short-term) planning, Cost/Value reconciliation, reporting to management and reviewing and updating health and safety plans.
- Monthly (long-term) planning recognises that as construction progresses it is inevitable that work will be disrupted due to unforeseen delays and changes. Such events and their impact must be assessed and addressed.
- Whichever planning technique the planner selects, there will be the need to identify construction and procurement activities, determine their sequence, make an estimate of their duration and understand how they inter-relate.
- In practice modelling the construction process frequently results in the need for complex relationships between the activities. You may need to introduce different types of logic links between the activities. You may need to introduce specific constraints as to when activities may start or finish.
- Links and dependencies should be introduced with caution, as they may result in complex or even iterative relationships that are at best difficult to understand and at worst meaningless.
- The terms 'float' and 'contingency' are often, but wrongly, used interchangeably. Float is a spare time in a sequence of events and is a product of the activity durations, sequences and dependencies in a network. Contingency is an allowance specifically added to a schedule or network to take account of unforeseen circumstances.
- The ownership of float has long been debated in the construction industry. It is a valuable asset that allows flexibility and can absorb delays to the programme. If one party uses or has control of the float, it restricts the other's flexibility.

- Having commenced construction, it is necessary to monitor progress to check for deviation from the agreed schedule. This may be necessary for contractual reasons, to advise internal management on progress and to assess the impact of delays and disruption.
- In addition to simply updating the current schedule, the contractor should assess progress and forecast completion based on the impact of the intervening events and any other additional information available at the time of the review.
- In addition to managing the time model and monitoring the critical path there are several other ways of monitoring progress. They include milestone monitoring, cash flow monitoring, activity schedules, planned progress monitoring and earned value analysis.
- A method statement is a written description of how the construction will be carried out. The statement may cover all or part of the project.
- The following are different types of method statement: tender method statement (for submission to the client), the tender method statement (for internal use), the construction or work method statement, the health and safety method statement, the planning method statement and the waste management plan.
- On large complex projects, the CIOB stress the importance of producing a full planning method statement to manage the planning and scheduling and facilitate the understanding and cooperation of the project participants.
- Contractors need to monitor cash flow carefully. Construction companies that cease to trade do so primarily because of lack of working capital.
- Over the last 20 years, risk analysis and risk management have become a topic of great interest both within management schools and the industry in general. Some industries, for example, the finance industry, are heavily dependent on the use of risk management systems for their everyday business trading.
- Contractors have failed to embrace quantitative methods of risk analysis to the same extent as some industries relying primarily on qualitative analysis to assess the risk and then determine how to offset or mitigate the risk.
- Contractors' main way of offsetting risk is declining to bid for the construction work or sub-contracting the high-risk element to specialist sub-contractors.

Section II

Section III
Planning and Scheduling Methods

Introduction

Having reviewed the basic techniques of planning and planning procedures and practice, we look at various planning methods: those that have been developed and those that are currently emerging. This in its widest sense extends beyond the 'traditional' concerns of time, cost and quality to look at those other parameters by which successful projects are judged – planning for sustainability, planning for waste management and planning for health, safety and the environment. The section includes eight chapters.

We review two planning systems that have been developed in the last 20 years in an attempt to overcome the 'failings' of traditional critical path methods: Critical Chain Project Management (Chapter 5) and Last Planner® (Chapter 7). Critical Chain Project Management focuses on the uncertainty in schedule activities and identifies the key activities that, based on time and resource constraints, form the 'critical chain' for the construction work. Last Planner is different from traditional planning approaches that simply direct (or 'push') the start of

A Handbook for Construction Planning and Scheduling, First Edition. Andrew Baldwin and David Bordoli.
© 2014 John Wiley & Sons, Ltd. Published 2014 by John Wiley & Sons, Ltd.

Section III

construction activities. Last Planner® looks at what *should* be done, considers what *can* be done and helps you decide what *will* be done (Ballard, 2000). In Chapter 6 we look at Earned Value Analysis (EVA) that enables you to measure progress and then forecast when the project will be completed and at what cost.

Planning is not limited to the construction/production process. It is also necessary to plan, monitor and control design. Design is an iterative process and traditional planning techniques are unable to accommodate iteration. Chapter 8 describes in detail the ADePT approach to planning the design process and the advantages to be achieved by modelling and monitoring the information requirements of the design and construction team.

Chapter 9 introduces Building Information Modelling (BIM). BIM has the potential to radically change the way that commercial buildings and many infra-structure projects are designed, constructed, operated and maintained. BIM is not new but recent advances in information and communication technology have resulted in a surge of interest in its use and the emergence of new techniques and new ways of working. In this chapter, we consider what exactly is BIM, why is it so popular, how BIM can assist the construction planner and how it will change the way that planners work.

Chapter 10 introduces another current topic, the need for sustainability. We consider the drivers for sustainability and industry perspectives. The BREEAM assessment method is outlined and data from three case studies is examined together with data from interviews with practitioners. The successes and short-comings of the current BREEAM system are discussed.

Linked to sustainable construction is the need for waste management. Site waste management plans (SWMPs), now commonplace on UK construction sites, are discussed in Chapter 11. We present the findings of research into the operation of SWMPs as current 'best practice'. We believe that SWMPs are pivotal in reducing the level of construction waste and their widespread adoption is important in the drive for sustainable development.

Chapter 12 considers planning for safety, health and environment. We include details of the SHE Management Model produced by the European Construction Institute (ECI), a major industry research organisation based at Loughborough University. The focus is on the planning element of the model.

Section III

Chapter 5
Critical Chain Project Management

Critical Chain Project Management™ (CCPM) is frequently presented as a revolutionary new project management concept, an important breakthrough in the history of project management. CCPM is based on the concept of a project buffer and argues that managers should manage production by monitoring the buffer time available and allocating resources to critical chain tasks.

Advocates of CCPM claim that the introduction of the CCPM methodology ensures project success, reduces project duration, enables increased project throughput with no resource increases and reduces manager and worker stress, all with minimal investment. Others are more sceptical, questioning whether CCPM is superior to currently accepted project management methodologies.

Background

CCPM has been developed from Goldratt's Theory of Constraints (TOC) and its direct application to project management. In his TOC (Goldratt and Cox, 2012) Goldratt argues that 'any system must have a constraint otherwise its output would increase without bounds or go to zero'. He believes that to succeed in business, organisations should identify their main constraint (the block that is preventing it achieving its target), exploit this constraint, subordinate everything else to the constraint and elevate the constraint to be the primary focus of the organisation. If a new constraint is uncovered, this process should be repeated, otherwise inertia becomes the system constraint and the organisation will eventually fail.

Goldratt applied TOC to project management arguing that traditional project management systems based on critical path methods fail to satisfactorily address

A Handbook for Construction Planning and Scheduling, First Edition. Andrew Baldwin and David Bordoli.
© 2014 John Wiley & Sons, Ltd. Published 2014 by John Wiley & Sons, Ltd.

the issue of uncertainty in activity duration. He argued that this failure, and the tendency of planners to overestimate activity duration, results in projects not being completed to scope, to schedule and to cost. Using other accepted management theories (including the theory of common cause variation and the statistical theory of variances) he developed a new method of working: CCPM. This method was first presented in the book *Critical Chain* (Goldratt, 1997). Since then, CCPM has been used successfully across different industries including the construction industry. It has been used successfully on projects of varying sizes.

How does CCPM differ from accepted best practice in project management?

The CCPM methodology focuses the attention of the project manager on the problem of resource-limited project scheduling for critical activities. This is the major issue of management concern.

CCPM adopts an approach that:

- Specifies the critical chain, rather than the critical path, as the project constraint. This chain is based on the critical path for the project, which includes resource dependencies. Having identified the critical chain, this becomes the point of focus for management. This critical chain does not change throughout the project, and management focus remains on the critical chain.
- Uses 50% probable activity times, and aggregates allowances for uncertainty of estimates and activity performance into 'buffers' at the end of activity chains.
- Uses buffers as an immediate and direct measurement tool to control the project schedule.
- Defines the constraint for multiple projects as the constraining company resource. It links projects through this resource, using buffers to account for activity duration variability.
- Changes project team behaviour by encouraging the reporting of the early completion of activities and the elimination of multitasking (Leach, 1999).

By adopting this approach, CCPM departs from the traditional Project Management Body of Knowledge™ (PMBOK) guide to project management (Project Management Institute, 2004) and other supporting literature. It presents a new way of management based on established project management principles.

Establishing the critical chain

Having researched reasons for project failure, Goldratt identified the main 'constraint' to project success as the 'sequence of dependent events that prevents the project from being completed in a shorter interval'. He argued that resource dependencies determine the critical chain as much as activity dependencies.

Resolving this constraint is the focus of CCPM. A list of activities is produced together with their logical dependencies and an estimate of their duration. The duration of each task is estimated assuming that the likelihood of its completion within the stated time is 95%. A feasible precedence schedule is then calculated based on the logic diagram produced and the duration of each activity. All activities are scheduled on the basis of their *latest* start times.

The total time for the critical path is then reviewed from the perspective of the resource requirements of the activities and the maximum requirements of each resource available. Where resource requirements exceed the maximum available, activities are scheduled with an earlier start.

This schedule is then reviewed. As assumed by CCPM, when asked to estimate the duration of activities, task owners always add a safety margin, that is, all task durations are overestimated. The duration of each task is then revised on the basis of a 50% chance that the activity will be completed in the time stated and with the resources available. (This 50% chance is assumed to be the best estimate for completion, assuming everything within the activity goes according to plan, the resources are available, work is not interrupted and all concerned are able to devote 100% of their time to complete the activity, that is, no 'multi-tasking' or use of resources across different activities.)

Using these revised durations, the total time to complete the construction work on the critical path is re-calculated.

This new duration becomes the allocated time available for the activities. The difference between this new time and the original duration becomes the overall buffer time for the critical chain and is added to the revised chain as an explicit 'activity' (the project buffer) at the end of the construction work.

In this way, the critical chain has been 're-drawn' whilst maintaining the same overall duration. The aim of the project buffer is to ensure that the project date provided to the client can be met whatever the variations that will occur in the duration of activities on the critical chain. The final project buffer and activity durations will need to be agreed with all the parties concerned. There is no one method to finalise the project buffer.

This is shown diagrammatically in Figure 5.1.

In a critical path network, the critical path is 'fed' by other linked activities. If production on these activities is delayed, this may impact the critical path and the critical path may change. The same danger applies to CCPM. To reduce this risk, CCPM protects the critical chain by placing an aggregated 'critical chain feeding buffer' (CCFB) at the end of each path that feeds the critical chain including the paths that merge with the critical path at the end of the project. This, it is argued, provides an adequate measurement and control mechanism to ensure that the critical chain is 'protected' from delays in subordinate parts of the network. The CCPM method also uses resource buffers to protect the critical chain. Resource buffers are advance warnings to the project manager to ensure that the required resources are available to work on the activities within the critical chain. A resource buffer does not have duration; it is a 'flag' to raise awareness.

Monitoring and controlling the critical chain

Within traditional CPM methods the model of the project is monitored and controlled by reviewing the logic of the model and the durations of the activities. The model is updated on the basis of progress to date. This often produces a revised schedule of work and the critical path for the project may change.

Section III

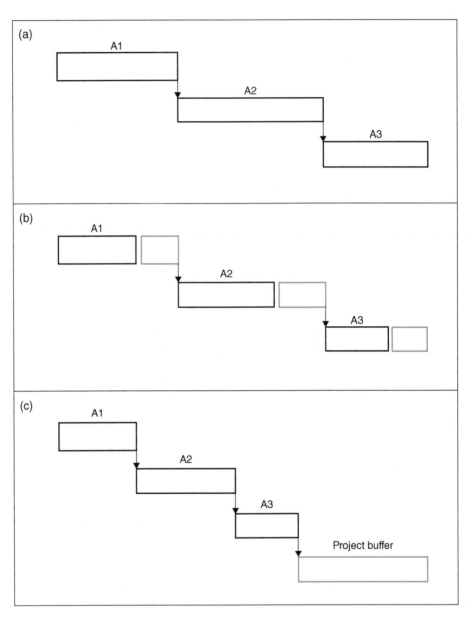

Figure 5.1   The critical chain project buffer. (a) Critical path activities with original duration. (b) Critical path activities with best estimate duration. (c) Revised critical path with the project buffer.

CCPM does not operate in this way. Unless there are major revisions to the scope of the construction work, CCPM simply maintains the original model and monitors activities on the critical chain and other chains leading into the critical chain.

Buffer management is the primary tool to measure and control the project. Monitoring is undertaken by reviewing the buffers created when the schedule was produced. The buffers incorporated within the schedule will inevitably be 'consumed' as the work progresses. Some predetermined consumption of the project buffer is assumed. If the rate of 'consumption' of the project buffer is low, then the project is assumed to be on target. If the rate of consumption is high, then the project is behind schedule. When there is little or no buffer at the end of the project, corrective action needs to be taken. (This usually means securing additional resources.) If the buffer is negative, then a recovery plan must be instigated.

Using CCCPM the construction team is required to adopt a 'dynamic' approach. Management of the work is not driven by predetermined dates for construction activities to commence. Construction work proceeds as soon as the previous task is complete and all the resources are available. If activities exceed their anticipated duration, the construction team is not criticised provided the activities were started as soon as they had the necessary input and the resources were used only on the activity concerned. (Here there is a direct link to the philosophy of approach adopted by the Last Planner System.)

CCPM *expects* some activities to take less time than anticipated, and some to take more. The project buffer accommodates these differences in duration. The CCPM approach manages production by controlling the flow of work into the system. Activities on the critical chain should be started as soon as work becomes available. Other activities should not start before the scheduled start time. Progress reporting is undertaken in the normal way and aligned with normal management reporting requirements (see Chapter 4).

A critical review of CCPM

CCPM has become a popular approach to project management. The methodology has strong supporters. It also has its critics.

Supporters point to the simplicity of the method (compared to traditional CPM, Program Evaluation and Review Technique (PERT), Monte Carlo Simulation and other methods), the ease of collecting data and monitoring performance and the ability of managers to take fast effective action. This has clearly been efficient on some projects and substantial claims have been made for the successful use of CCPM. Leach (1999) for example, states that 'all projects that have diligently applied CCPM have completed the project substantially under the time estimate, fulfilled the original scope, and come in near or under the estimated budget'.

Others point out that the philosophy of approach behind CCPM is completely different to CPM and this results in a 'remarkably different mindset for managers and a different set of management practices'. Managers readily accept CCPM because

- It focuses attention on the primary objective: deliver the project on time.
- Once established, the method does not require regular updates of the project model.
- It focuses the management on resources, something all managers realise is important to the success of a project.

- By starting all activities as late as possible CCPM improves the cash flow of a project.
- CCPM recognises that we operate in a world of uncertainty.

Those who are less supportive of the method argue that

- CCPM leads us to believe that the management of projects can be accomplished through the same rational process that works for production management. They question the validity of such an assumption.
- Whilst focusing on the delivery of the project on time, CCPM ignores other objectives such as maximising project quality, minimizing resource availability costs, maximising the net present value of projects, etc. (Herrolen et al., 1997; Pollack and Johnson, 1999).
- The concept is not new. (For example, Wiest (1964) introduced the concept of a *critical sequence* 'determined not by just the technological ordering and the set of job times, but also by resource constraints; furthermore, it is also a function of a given feasible schedule'.)
- CCPM assumes that it is easy to identify the critical construction sequence. It understates the fact that there may be more than one critical path in the model and consideration of the complexity of sequences of activities in the model is imperative.
- There is no agreed method to determine the size of the buffer.
- CCPM advises against regular updating of the baseline schedule. Some managers consider that regular updating of the project model and review of the baseline are essential.
- Many commercial software packages are unsuited for CCPM because they adopt priority rules and do not include optimal algorithms for resource levelling and resource-constrained scheduling. This may lead to misleading analysis.
- Software packages based on the critical chain scheduling concepts have been developed but the adoption of such software may result in additional costs to the project.

An important aspect of CCPM is resource management. It appears that CCPM is best suited to managing single-unit renewable types of resource such as labour. Examples of its use with other resource types (e.g. non-renewable resources such as money and energy, spatial resources, doubly constrained resources, constrained per period and for the total time horizon) are seldom reported. It adopts 'a crusade' against the use of multitasking, that is, assigning resources to more than one significant task during a particular time window. Goldratt describes this as the 'biggest killer of lead time' (Herrolen et al, 2002). Others argue that some projects, by necessity, require more complex resource scheduling.

There are other matters that need to be taken into account when adopting CCPM. As with any changes within organisations, the adoption of CCPM requires changes in project organisation and processes and the culture of construction organisations. The cost of these changes, including training, may be significant and should not be underestimated.

Key points

- CCPM recognises the importance of considering both the activity durations and the resources required for the completion of the work.
- CCPM makes an important contribution to scheduling and project management because it takes into account the uncertainty of activity durations.
- The CCPM method, based on TOC, uses feeding, resource and project buffers and an underlying buffer management mechanism to provide a simple tool for setting realistic dates for activities and simple project monitoring.
- The concept of producing and protecting a deterministic baseline schedule in order to cope with uncertainties is a realistic approach to ensuring the delivery of the project on time.
- Whilst the concept of determining a project buffer and managing production on the basis of the buffer time currently available is attractive, this may not always be as straightforward and easy as it first appears.
- Is CCPM a new way of managing projects or simply a new presentation of existing methods? CCPM is clearly based on existing methods but brings into existence a new approach based on TOC.
- The CCPM methodology has been successfully implemented on different types of projects in different types of industries.
- However, before they finally decide to adopt CCPM, many construction organisations may seek further empirical evidence of the robustness of the CCPM approach in a construction environment.

Chapter 6
Earned Value Analysis

Earned value analysis (EVA) measures performance in terms of time, cost and scope. It is a quantitative technique that necessities an estimate of both work progress and related cost. The technique may be used to assess progress on an individual construction activity, a group of activities (work package) or the entire project. It can be used on any type of construction project and is best used with graphical tools and supporting software to highlight current and future trends. As a technique, EVA has led to earned value management (EVM), a management methodology for controlling a project. This methodology has strong support from industry practitioners because it facilitates decision-making and performance improvement. It is widely accepted by project managers and the clients of construction. Some major clients, such as The Department of Defense of the United States of America, have produced detailed implementation protocols. We explain the basis of the technique and the EVA methodology. We outline what is needed for an EVA system and how to use this system successfully, identifying common pitfalls and how to overcome them.

Terminology and definitions

The use of EVA requires familiarisation with new terms. Different users of EVA adopt similar but different terms and definitions, and several produce their own glossary of terms. This text uses the following terms and definitions. To understand the technique more quickly, it is helpful to familiarise yourself with these terms before reading the chapter.

A Handbook for Construction Planning and Scheduling, First Edition. Andrew Baldwin and David Bordoli.
© 2014 John Wiley & Sons, Ltd. Published 2014 by John Wiley & Sons, Ltd.

Cost Performance Index (CPI)

The ratio of earned value (EV) divided by the actual cost (usually expressed as a percentage).

Cost variance

A metric for showing cost performance derived from EV data. It is the algebraic difference between EV and actual cost (cost variance = earned value – actual cost). A positive value indicates a favourable condition. A negative value indicates an unfavourable condition. The cost variance may be expressed as a value for a specific period or as a cumulative value to date.

Earned value analysis (EVA)

A technique that compares the budgeted project costs, actual project costs and value of the work achieved to determine, *inter alia*, the status of the project, the likely completion of the project and out-turn cost of the project.

Earned value management (EVM)

A management methodology for controlling a project based on measuring the performance of work and the EVA technique.

Earned value management system (EVMS)

The process, procedures, tools and templates used by an organisation to undertake EVM (Lukas, 2008).

Budgeted cost of work scheduled

The budgeted cost for the work scheduled to be completed at any point in time. (Budgeted cost of work scheduled is sometimes known as planned value (PV).)

Budget at completion (BAC)

The total budget for the activity, work package or project.

Section III

Actual cost of work performed (ACWP)

> The cost of the work accomplished.

Budgeted cost of work performed (BCWP)

> The measure of achievement or the value of the work done. It is also known as earned value (EV).

Earned value (EV)

> The percentage of the total budget actually completed at a point in time. This is also known as the BCWP (Lukas, 2008). Note: EV is not the same for the contractor as the value of the work completed.

Performance measurement baseline

> A time-phased budget plan for accomplishing the work. It is against this budget that the contract performance is measured. The budget includes the sums assigned to scheduled control accounts and the applicable indirect budgets.

Schedule Performance Index (SPI)

> The ratio of the EV divided by the PV (usually expressed as a percentage).

Schedule variance (SV)

> A metric for the schedule performance derived from EV metrics. This is the algebraic difference between EV and the budget (schedule variance = earned value – planned value). A positive value is a favourable condition; a negative value, unfavourable. SV may be expressed as a value for a specific period of time or as a cumulative value to date.
>
> For a detailed explanation of the terminology of EVA see also BS6079-1:2002; AS4817:2006; ANSI748B and CIOB (2011).

The basis of the EVA

> EVA considers the scope of the project, the schedule of work and the cost of the work undertaken. It is normal to complete all the budgetary estimates in monetary terms although the technique may be used on the basis of man hours or man

days. The technique requires the production of a full work breakdown structure (WBS) for the project together with associated cost estimates (producing a WBS is described in Chapter 4).

When using EVA, the WBS is completed in a hierarchical structure to the level of detail required. At the lowest level of detail are the elements to be monitored. These elements should be mutually exclusive to ensure that no problems will arise when allocating costs. The PV of each construction activity must then be assessed. This may be determined by simply allocating a proportion of the total PV for the project to each activity or, where a Bill of Quantities document has been used to measure the work, assigning specific bill items, or groups of items, to the construction activities.

Each of the construction activities must then be allocated to a schedule that clearly shows the start and end of each activity. From this schedule it is then possible to calculate the budgeted cost of work scheduled (BCWS), also known as the PV for each week/month of the project. This is normally expressed on a cumulative basis. It is this cumulative PV that is used in the calculations and compared with the cumulative figures for the BCWP and the ACWP to measure the performance on the project.

Figure 6.1 shows a typical PV curve for a project and the BAC. The exact form of this curve will vary for each project depending on the activities, costs and timing of the construction work.

Figure 6.2 shows the PV and actual cost at the time of review.

This gives rise for concern; the actual cost for the planned work is more than the budgeted cost.

Figure 6.3 shows the PV, actual cost and EV at the time of review. Examining the data there is clearly a problem with this project. Not only does the actual cost exceed the budget, the construction is also behind schedule (the EV is less than the PV).

Figure 6.4 shows another project. In this scenario, the contractor has completed more work than planned and at an actual cost less than the anticipated EV. This is clearly a considerably better situation.

Section III

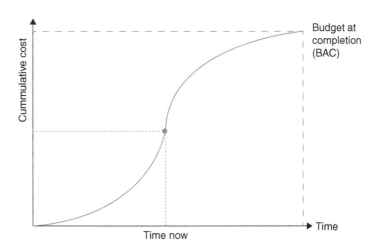

Figure 6.1   Planned value curve and budget at completion.

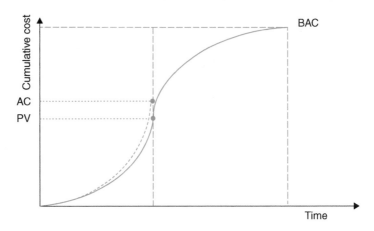

Figure 6.2 Planned value and actual cost at the time of review.

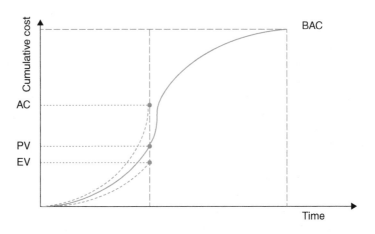

Figure 6.3 Planned value, actual cost and earned value at the time of review.

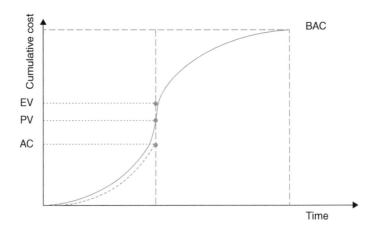

Figure 6.4 Planned value, actual cost and earned value curves for a different project.

Earned value analysis calculations and their interpretation

Having calculated the PV, BAC and EV for the project, EVA requires you to monitor construction progress on a regular basis and to keep accurate cost records that enable you to assess the cost of the construction work completed to date.

After each progress review, EVA enables you to assess performance by calculating the cost variances (CVs) and schedule variances (SVs) using a series of simple calculations. This allows you to review how the planned performance compares to the actual performance and to make forecasts for future performance.

The formulae for these calculations are:

Cost variance:

$$CV = EV - AC$$

Schedule variance:

$$SV = EV - PV$$

Cost Performance Index:

$$CPI = \frac{EV}{AC}$$

Schedule Performance Index:

$$SPI = \frac{EV}{PV}$$

$$\% \text{ complete} = \frac{EV}{BAC}$$

$$\% \text{ spent} = \frac{AC}{BAC}$$

$$CPI = \frac{\% \text{ complete}}{\% \text{ spent}}$$

These calculations enable you to show the overall status of the project. The two most useful key performance indicators and the basis for calculations of future performance are the SPI ('How are we doing against plan?') and CPI ('Are we efficient?') (CIOB, 2011).

It is imperative to remember that the overall status of the project is shown in *monetary* terms. Put simply, EVA shows the costs that have been incurred for your project and whether these costs have achieved the value of production that you expected at a given point in time.

The figures calculated show the overall situation for the project. To understand the status of individual work packages or activities, it is necessary to review the data in more detail. Provided the cost data are available, similar calculations may be made at these different levels to review exactly which activities are going according to plan and which are not. When these performance indicators are calculated on a regular basis, one is able to plot the values on a graph and gain a picture of how performance is varying over time.

Forecasting

From the calculations made, when progress is reviewed it is possible to make forecasts of the anticipated project completion and the cost at completion.

To do this you will need to calculate the cost estimate at completion and the time estimate at completion based on performance to date.

- Time estimate at completion (TEAC)
- Time variance at completion (TVAC)
- Cost variance at completion (CVAC)
- Cost estimate at completion (CEAC)

Figure 6.5 shows how the situation at the time of update in the example under review may impact the completion date and cost. On the basis of performance to date it may be expected that the project will be delivered later than schedule and over cost. This forecast assumes that there will be no further problems with the construction and no variations within the scope of the work, the anticipated construction activities or the construction costs.

An example of EVA calculations

The following example shows the calculations involved in EVA.

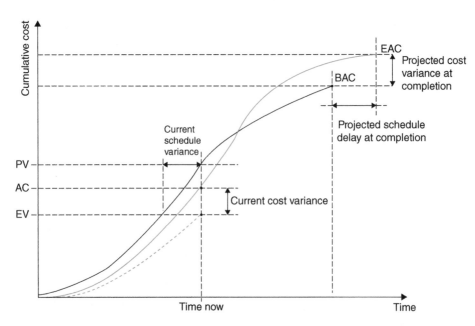

Figure 6.5   Forecast completion date and budget at completion.

In this (illustrative) project renovations to a building require structural, electrical and plumbing work. The budget for the work is £1050000. The work comprises eight work packages that will be completed over a 12-month period. A work breakdown structure for the construction to be undertaken is produced to identify the budgets for each item of work. These budgets are then allocated across the construction activities. Figure 6.6 shows the time schedule for the work with added budget data.

Figure 6.7 shows the updated schedule to show progress after 6 months of the project. After 6 months, Activity 1, Activity 2 and Activity 3 are complete. Activity 4 is only 50% complete and will require an extra month to complete. Activity 5 is 33% complete. Both these activities are behind schedule. Activity 6 is 33% complete. This activity is ahead of schedule. The ACWP at the end of this 6-month period is £430000.

For the overall project:

The BCWS at the end of month 6 was £460 000.
The BCWP by the end of month 6 was £410 000.
The ACWP at the end of month 6 was £430 000.

The cost variance	$= EV - AC$	
	$= £410000 - £430000$	$= -£20000$
The SV	$= EV - PV$	
	$= £410000 - £460000$	$= -£50000$
The CPI	$= £430000/£410000$	$= 1.05$
The SPI	$= £410000/£460000$	$= 0.89$

	Month												Total BCWS
Activity	01	02	03	04	05	06	07	08	09	10	11	12	
1	40	40											
2			40	40									
3			50	50									
4					100								
5					50	50	50						
6							50	50	50				
7									80	80	80		
8											75	75	
Monthly BCWS	40	40	90	90	150	50	100	50	130	80	155	75	1050
Cumulative BCWS	40	80	170	260	410	460	560	610	740	820	975	1050	1050

Figure 6.6   EVA example – the time schedule for the work with added budget data.

Section III

	Month									%	Total BCWS
Activity	01	02	03	04	05	06	07				
1	40	40								100	80
2			40	40						100	80
3			50	50						100	100
4					50					50	50
5					50					33	50
6						50				33	50
Monthly BCWS	40	40	90	90	100	50					410
Cumulative BCWS	40	80	170	260	360	410					410

Figure 6.7   EVA example – the schedule updated to show progress after 6 months.

The project is behind schedule and over cost (the CPI is greater than 1.0 and the SPI is less than 1.0).

The project is 410000/1050000 × 100%   = 39% complete
The percentage spent is £430000/1050000   = 41% of the total budget

From the updated time schedule is appears that the problem activities are Activity 4 and Activity 5 but only a full analysis of each activity can allow the planner to reach this conclusion. To do so, it is required to establish the actual cost data relating to the work on each activity.

Table 6.1 shows the analysis of the items of work at the end of month 6. This reveals that Activities 4 and 5 are in fact the 'problem activities'. Both are behind schedule and over cost. It is also noted from the table that Activity 1 was completed under the original budgeted cost but Activity 2 was over budget. According to the data collected, Activity 6 is ahead of schedule and on cost.

On the basis of the performance to date, it is possible to estimate the actual cost to completion and the schedule time. Whilst this may simply be calculated on the basis of the figures to date, it is more appropriate to look carefully at the work to be completed before simply extrapolating these data through to the end of the construction work. This also requires a more detailed knowledge of the reasons why Activity 4 and 5 are overspent and behind schedule.

These are the basic calculations used when undertaking EVA. Provided that you are able to make an accurate assessment of progress to date and have accurate cost data performance to date at hand, you can assess progress and make forecasts for

Table 6.1   EVA example – The analysis of the items of work at the end of month 6.

Activity	BAC	BCWS (PV)	Percentage complete	BCWP (EV)	ACWP	SV	CV
1	80	80	100	80	60	0	+20
2	80	80	100	80	100	0	−20
3	100	100	100	100	90	0	+10
4	100	100	50	50	60	−50	−10
5	150	100	50	50	70	−50	−20
6	150	0	33	50	50	+50	0
Total	660	460		410	430		

completion of the work. To manage performance effectively, a systematic approach to data collection and analysis is required. A tabular approach using spreadsheet software is recommended.

Earned value management systems

Both the Association for Project Management (PMBOK) and the Association of Project Managers (APM) refer to EVMS. They emphasise that the adoption of EVA in an organisation requires a systematic approach.

In his review of 'Implementing Best Practice in Hospital Project Management Utilising EVPM methodology', Raf Dua (1999) emphasises that 'for cost/performance measurement to work effectively certain systems must exist'. He identifies the following systems as being necessary for the implementation:

- A time-based schedule
- A work breakdown structure
- A costs collection system
- An objective method of assessing progress
- A responsibility/authority matrix

You must have an agreed schedule of work that can form the basis of each progress update. The scope of the work must be known and be analysed in a detailed work breakdown structure. There must be a system in place to collect and allocate cost data. A consistent, objective, quantitative method must be used to assess construction progress. All these requirements must be completed and the responsibility for undertaking each task allocated. Systems must be set up to cover data collection, data collation, data processing, the review of output and the reporting of the results.

The monitoring systems used must be appropriate to the organisation and agreed by all those involved with the project. The constraints of scope, schedule and cost must be carefully controlled. Any changes to the scope of the work, the timing of the work or the anticipated cost of the work will change the anticipated values. All changes must therefore be monitored, approved and controlled against the performance baseline. The EVA system (and hence the project) is certain to fail if all changes to expenditure are not authorised and recorded against appropriate cost

Section III

headings and included in regular monthly cost reports. All costs must be directly linked to a recognised deliverable and a documented activity. All cost changes must be communicated to the relevant stakeholders. It is essential to have systems in place to monitor and report changes and to update the data within the system.

The Department of Defense of the United States of America, as the client for a range of systems and facilities, recognises that the correct implementation of EVM ensures that the project manager is provided with contractor performance data that

- Relate time-phased budgets to specific contract tasks and/or statements of work
- Objectively measure work progress
- Properly relate cost, schedule and technical accomplishment
- Allow for informed decision-making and corrective action
- Are valid, timely and able to be audited
- Allow for statistical estimation of future costs
- Supply managers at all levels with status information at the appropriate level (Department of Defense, 2006)

With the requisite that EVM be used for all projects above US$20 million, this means that the government and major clients can neither impose single-solution approaches for the use of EVA nor specify integrated systems to be used by all contractors. Hence, the 'Guidelines for Implementation' do not identify a specific system but establish a framework for an integrated management system. They provide a comprehensive basis (108 pages) for implementing an EVMS. These guidelines have been developed over the last 50 years and are now published as an American Institute/Electronic Industries Alliance Standard ANSI/EIA-748 (2007). They provide comprehensive information on content, format and process and identify the key process areas and cross process areas for each of the five main categories of the system: organisation; planning, scheduling and budgeting; accounting considerations; analysis and management reports; revisions and maintenance. This emphasises that implementing an EVMS impacts all parts of the organisation.

Problems and pitfalls of EVA and how to overcome them

EVA can provide successful performance management for the project provided that basic mistakes are avoided. Problems and pitfalls include:

- Requirements for the system are not clearly identified and documented.
- Progress is not monitored against updated schedules.
- Progress of the construction work is incorrectly assessed.
- Cost analysis is performed with inaccurately allocated cost data.
- There is no agreed understanding of what comprises the PV of the work.
- Failure to prepare a fully completed WBS integrated with the cost estimate and time schedule.

Provided that the work content and the sequence of the schedule used as the baseline are the same as the current schedule, the main benefits of EVA will be twofold: the identification of what work has been achieved against the plan; and what it has cost to reach that level of achievement.

However, '[a]s with all methods for monitoring progress, if a different work content or sequence of working from that planned is adopted, a meaningful comparison between planned and actual will be difficult, if not impossible' (CIOB 4.6.10.6, 2011). This means that when undertaking performance update, it is essential that the costs and value calculations are compared against a schedule where the performance achieved is measured against the performance that was expected. This becomes increasingly important as the project proceeds and the client demands variations in the work content. It is essential to have an effective change management system to identify all changes to the proposed work and ensure that the impact of these changes are fully reflected in a revised cost schedule and revised value estimates.

The system of cost control adopted must produce accurately allocated cost information. A common mistake is only to include costs that have been invoiced to the project. However, it is important to include all costs including those that have been accrued but for which no invoice has been received. The cost of all the resources that have been used to undertake work must therefore be included in the analysis. (Consider, for example, a situation where an expensive excavator has been hired for the project and has been working for 2 weeks but the hire company has not yet invoiced the contractor. This 'accrued' cost should be included in the calculations.) Similarly, careful allocation of costs must be included for items of plant that require a cost to bring them to the site, an operational cost and a cost to remove them from the site, for example, a tower crane.

It is essential that all concerned have the same understanding and agreement as to what is meant by the PV of the work. Different organisations may, depending on their estimating and cost control system, define budgeted costs and the constituent parts of these sums in different ways. One must ensure that one knows exactly what comprises the PV for one's project and calculations.

EVA requires a fully completed WBS to be integrated with the cost estimate and time schedule. If the estimate has not been prepared to align with the WBS, it will at best be inaccurate or, at worst, unrepresentative. Similarly, the time schedule and the schedule of activities on the project will need to be appropriately drawn and all WBS elements correctly allocated. If this is not achieved and updated as work proceeds, then the performance analysis will invariably be incorrect.

The principles of EVM may be used on any type of construction project irrespective of size or complexity. Traditionally, EVM has been adopted on large construction projects. Many organisations base their decision on whether or not to use EVA on a simple threshold project value. For projects above a specified figure, EVA must be adopted; below this figure, other control measures are used. Alternatively, the decision as to whether or not to use EVA may be taken following careful consideration of the work to be completed by the project team. If the contract demands that EVA be used, then the decision is easy.

EVA requires that the progress of the construction work be correctly assessed. Assessing the progress of construction work is not easy (see Chapter 4). For EVA it is important that there are established rules for determining when the value within each item is recognised as being 'earned'. You may decide, for example, that as soon as 50% of the item is complete you can claim 50% of the value.

Section III

Alternatively, you may decide that the value of the item can only be claimed when the activity is 100% complete. Whatever the decision rules established, it is important that the progress data collected from the site are representative of the true progress achieved. Measuring progress in EVA should be based on quantitative metrics of performance to ensure that within the system implemented there is a clear focus on where potential problems exist.

EVA identifies variances. EVA cannot identify what has caused a departure from required progress, and the effect upon the critical path for the project. It cannot identify what can reasonably be done to recover from a predicted delay to completion (CIOB 4.6.10.7, 2011). Where a significant variance has occurred, further investigation of both cost records and site records is needed to identify the cause.

Whilst the implementation of EVA may appear straightforward, it is not always easy and demands considerable time and energy to ensure that the commitment of the staff produces accurate reports on progress so that accurate forecasts to completion are available for management use.

(For more discussion on the problems of using EVA, see 'Earned Value Analysis – Why It Doesn't Work' by Joseph Lukas, 2008 and other recommended reading on the website for the book: www.wiley.com/go/baldwin/constructionplansched).

Key points

- The EVA technique and its management control process, EVM, have the support of government, professional and business organisations.
- Many consider that EVA is the most appropriate tool for assessing progress because it covers time, cost and scope.
- EVA is flexible and capable of producing timely information at all levels.
- The U.S. Ministry of Defense, the Project Management Institute, the Association of Project Managers and other professional institutions all support EVA.
- Some companies are reluctant to use EVA because they consider it too complex.
- In practice, EVA is as complex or as easy as you want to make it. It may be used at project level or at activity level, or any level in between.
- EVA requires accurate cost and performance records and the commitment of all involved to ensure realistic forecasts. An important contributor to the successful implementation of EVM is strong and visible leadership support.

Chapter 7
Last Planner®

Background

In the last 10 years, many construction organisations have turned towards collaborative planning to meet target schedules and increase production. Some companies have adopted a system called Last Planner®.

Traditional planning and control consist in formulating and declaring what should be done, then monitoring progress to determine what was achieved and comparing this with what should have happened. Last Planner® changes the way 'should' is determined. It is based on collaborative planning and adds a methodical process for making ready tasks that should be done so that they can be done. (This is known as Constraints Removal.) Last Planner® also secures explicit promises from front-line supervisors and crews regarding what will be done (this is called 'reliable promising'). Recognizing that we are unlikely to be perfect planners (after all, all plans are forecasts, and we know for a fact that all forecasts are wrong), the goal is to never make the same mistake twice. Hence, Last Planner® analyses the root cause of problems and identifies corrective action.

'Last Planners' are the people who are directly in charge of the work teams, that is, those responsible for organising the work groups and supervising production on a day-to-day basis. 'Last Planners' work collaboratively with their counterparts from other companies to prepare the work schedules for the next work period. They use planning workshops to analyse where savings in time may be gained. They agree and initiate work assignments and monitor subsequent performance daily. Where anticipated production is not achieved, they consider why this is so and the changes that need to be made to the schedule. This leads to continuous production improvement. The system uses techniques such as the 5 Whys to analyse why production targets are not met. Advocates of the Last Planner

A Handbook for Construction Planning and Scheduling, First Edition. Andrew Baldwin and David Bordoli.
© 2014 John Wiley & Sons, Ltd. Published 2014 by John Wiley & Sons, Ltd.

System® LPS, argue that the system increases the chances that projects will be completed on time and that when using the system productivity is improved.

The development of Last Planner®

Traditional planning techniques such as critical path methods (CPM) are excellent for analysing the logic of construction activities, identifying critical activities and producing a model from which it is possible to undertake resource analysis, produce schedules for activities and identify milestone/completion dates. CPM helps to clearly identify key contractual dates for the start of activities on site and completion of construction work. CPM confirms production deadlines and contractual obligations. However, CPM is not the best tool to direct production on site.

It has long been recognised that CPM is ill suited to short-term production planning. Experience has shown that when scheduling at high levels of detail, there is less chance that the activities will be completed as envisaged. Whilst CPM schedules can be expressed at any level of detail, the practical limit arises from the work required to keep highly detailed schedules updated. It is a question of granularity. There is clearly an optimum level of detail for CPM planning past which it is expensive and impractical to use the technique on a weekly/daily basis. There is research evidence to support this view. Ballard (2000) found that the planned percentage complete (PPC) of activities on sites managed with traditional CPM systems was typically 50%. At that level, trade supervisors have only a 50:50 chance of actually being able to do the work 'promised'. Many give up intensive planning and preparation.

The Last Planner® system of production control was first used in 1992. Since then it has been developed and adopted on construction projects in a number of countries and its use has increased considerably over the past 10 years. On some projects, the process has resulted in considerable improvements in production performance. (Using Last Planner®, Grana and Montero in Peru improved their project profit margin from 7% to 17%.)

Glenn Ballard and Greg Howell first used the term Last Planner® to describe the person or group that produces assignments. They used Last Planner® to name the system because they wanted to emphasize that everyone has managerial responsibility for project success, especially the front-line supervisor. They also intended to convey the message that the planning decisions of the front-line supervisor are qualitatively different from planning decisions made at higher organizational levels because they are not inputs to other planning processes, but directly drive action.

Ballard and Howell introduced and formalised short-term planning on the basis of a 'pull-type' system approach. The Last Planner® System (LPS) is now fully established as a Lean Production–based management system. It is a cornerstone of the Lean Construction methodology. Applying LPS to construction projects shows that improvement in all the four principle dimensions of project performance – cost, schedule, quality and safety – is possible (Nova Award Nomination, 2003). LPS has been extended to all stages of construction planning: pre-contract planning, tender planning and production planning. It has also been used in a design office environment.

The system comprises:

- A philosophy,
- A set of rules and procedures, and
- A set of tools for implementation.

Collaborative planning between site management and all the trade teams involved in the construction work is integral to the success of the LPS. In collaborative planning all parties discuss and agree the work schedule. They consider and review the critical dependencies between their operations to ensure that interactions between them are understood. They use backward scheduling, the process of planning a project by starting with the deadline and working backwards through the component steps to reveal the latest possible start date. This method can be used at any level of detail: from the entire project to multi-craft tasks spanning only one day.

Last Planner® rests on a belief that the world is inherently stochastic – all plans are forecasts and all forecasts are wrong. The further into the future we attempt to forecast, the more wrong we will be. The greater the level of detail we attempt to forecast, the more wrong we will be. If this is true, and it seems to be true, why plan in too great detail too far ahead of the event?

CPM indicates when activities should start but does not ensure that you will be able to finish them within the agreed schedule. Strict adherence to the dates for starting work identified in the contract programme ensures that contractual obligations are met but it can lead to one trade being out of sequence with other trades, disruption, abortive work and contractual disputes. Traditional planning systems are 'push' systems. They look at the project objectives, collate information and plan the work on the basis of what should be done (Ballard, 2000). This is shown in Figure 7.1.

Strict adherence to CPM target dates pushes you towards starting production even when you know you will not be able to complete the task to the agreed schedule. Last Planner® works on the basis that there is no point in starting construction activities if the construction team does not have all the resources required to complete the work.

The Last Planner System® (LPS) is a 'pull'-based system that overcomes these criticisms. This is shown in Figure 7.2.

The LPS schedules the construction activities to start when everything required for completion of the task is available. All the trades involved with the construction activities agree production schedules collaboratively – helping to ensure work will be completed to anticipated timescales.

Section III

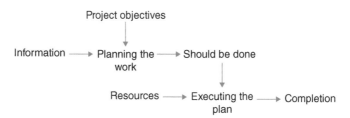

Figure 7.1   A traditional 'push' planning system.

Figure 7.2   Last Planner® is a 'pull'-based system.

Last Planner® incorporates five main principles:

1. Plan in greater detail as you get closer to doing the work
2. Plan collaboratively with those who will do the work
3. Undertake a constraints removal process on planned tasks publicly with those who can remove constraints
4. Make reliable promises
5. Learn from past failures; identify and act on root causes to prevent reoccurrence (Ballard et al., 2009)

Principles of the Last Planner System® (LPS)

Last Planner® started as a system for short-term production planning immediately before construction. This recognises that you only need to plan in detail prior to the commencement of work. To improve the planned percentage complete, PPC to construction activities it was decided to only allow into weekly work plans tasks that were (1) well defined, (2) sound (all constraints removed that could be removed prior to the plan period), (3) sequenced and (4) sized to the capacity of those who were to do the work. Process constraints are identified, analysed and removed.

Site superintendents and representatives of the trade organisations directly involved with the production (the 'Last Planners') discuss and agree the production tasks that are to be completed within the next 'lookahead' period. (This period is normally between 3 and 12 weeks – typically 6 weeks. It depends upon the type of work, the type of project and the lead times required for information, labour, plant and materials.) This 'lookahead' review process includes:

- Breaking down the master schedule activities into work packages and operations;
- Shaping the work flow sequence and rate of production;
- Matching the work flow with the production capacity;
- Developing detailed methods for executing the work; and
- Maintaining a backlog of ready work (Mossman, 2013).

The process involves consultation, the sharing of information and the use of wall-charts and 'stickies' to schedule and agree tasks. The Last Planners work together to agree and 'make ready' the construction tasks for completion. They consider what needs to be done and whether all the resources and information needed are available to complete the tasks successfully. To complete the work successfully it is necessary to consider seven 'construction flows': people, information, equipment and materials, completion of prior work and maintaining a safe space and working environment. All those who have been involved in producing the

Section III

schedule make a commitment to achieving it via the 'promise conversation cycle'. They consider the 'Conditions of Satisfaction' (conditions imposed by the person/organisation initiating a process that specify how success of the process will be measured) and then seek commitment from everyone involved and their promise to produce.

The Promise Conversation Cycle is *not* a question and answer session followed by authorisation and instruction. It is a negotiation process. When the work programme is established, the final decision to go forward to production is made on a weekly basis via a weekly work plan produced collaboratively by the site manager and representatives of each trade. Work is only started when all the resources and the information required are available.

The success of the weekly work plan is measured against PPC (also expressed by some as planned promises completed), and this is used to monitor the performance of the individual work teams and the overall production performance. When targets are not met, you need to investigate the reasons production fell short of that anticipated. You can use the '5 Whys' technique to identify why production did not go ahead as planned.

The 5 Why Analysis is a question-asking method used to explore the cause–effect relationships underlying a particular problem. The goal is to determine the root cause of a problem. You identify the causes of the problem by asking 'Why did this happen?'. For each of the causes identified, you ask again 'Why did this happen?'. This is repeated five times. This is considered enough to identify logical root causes. The solutions to the root causes can then be identified and implemented. It is recommended that a tabular diagram or a 'fishbone diagram' be used to analyse the problem and present the solution.

When you continuously monitor performance achieved and reasons why some activities did not meet their handover targets, you can make sure that production is improved. In the case of repetitive tasks, lessons learned can be applied to the project in which they were learned. You can use the lessons learned for future construction projects. This is particularly important when you are planning construction projects with repeat activities.

Implementing the Last Planner System® (LPS)

Using the LPS involves six main stages: initial programming; collaborative programming; programme compression; make ready, lookahead, production planning, and production.

Figure 7.3 shows how LPS can be applied in an actual construction company.

Initial programming takes place before the contract is awarded. The construction company produce a feasible programme acceptable to the client. This is based on the project objectives and the information available. Initial Programming may be considered in the same perspective as Tender Programming.

Collaborative programming takes place before construction on site, or after construction has started and before a new phase of production begins. Collaborative programming works from the initial programming to plan how to achieve each milestone, usually expressed as the completion of each phase of work; for example, site preparation, excavation and shoring, underground utilities and substructure,

Section III

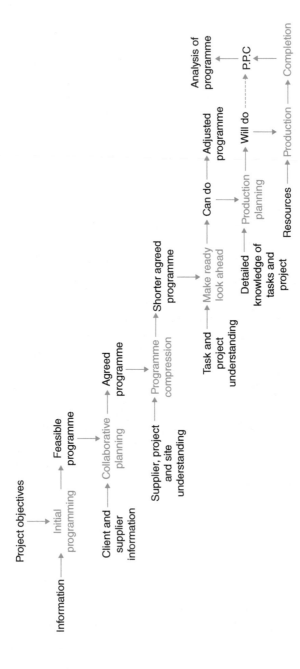

Figure 7.3 The LPS overview flow chart.

superstructure, etc. Phases of construction may be broken into sections to facilitate more detailed planning. Phases may be collapsed when project durations are short and representatives of the involved trades can be assembled. (You know you're moving from one phase to another when the players change.)

Representatives from all stakeholders (including the client, if needed), meet together at a *programming workshop* to collaboratively produce an agreed programme. They examine the feasible programme in detail to identify and agree a way of working that will produce a feasible programme acceptable to all. In some cases, it may be required to compress the original programme to meet client deadlines.

Representatives at the Programming Workshop look at ways to achieve overall *programme compression* and eliminate waste. Backward scheduling is used, working back from the expected completion date to see how time can be saved. They agree the operational process and each trade estimates the time their activities will take. They reorganise tasks to reduce the time allocated. The programme is then reviewed and compressed. Representatives have to work collaboratively and be prepared to help minimise the work schedule via trade-offs.

It should be noted that compression of the initial program is not always desirable. Some clients do not want handover of the constructed asset until a certain date when they are prepared to accept and operate it. However, it frequently happens that an initial backward planning process reveals a logic network that is too long. In order to meet a predetermined milestone date, the team must compress that initial logic network. It is also necessary to produce a logic network with sufficient slack (float) to absorb expected variation in the duration of critical tasks.

It is important that the time estimates are based on the amount of work to be done and the resources to be used. It will not work if representatives include extra time for possible problems. Team members are asked to provide average durations for their tasks, as opposed to padding them against 'bad luck' – a practice that results in excessive buffering because not every task will have durations beyond the average. This practice is similar to that recommended by Goldratt in his Critical Chain (see Critical Chain Project Management, Chapter 5).

Unlike the Critical Chain Method, Last Planner® allocates buffer time to critical tasks based on team judgement, as opposed to keeping it in a general buffer at the end of the phase schedule, to be drawn down as needed to protect the milestone date. The goal is to keep start dates reliable to the extent possible, recognizing that specialist subcontractors work on multiple projects simultaneously and in close succession. They try to make reliable promises to others as well.

This programme then goes on to the *lookahead/make ready* stage. The latest task and project information is examined to assess what needs to be done before the next stage of construction. This ensures that when the final production plan is produced, it will be based on what *can* be achieved given the information and resources available including site space.

The lookahead process shapes the workflow sequence and the rate of production. Trade gangs and work gangs are given work packages to undertake. Work methods are developed in detail. The question of how to retain a backlog of ready work but not include 'waste' is important. Workable backlog consists of tasks that are sound, but not critical. The key issue is to release only workable backlog tasks that can be done now without paying a penalty later. This is an issue of proper task sequencing. Commercial pressures often compel work being done out of sequence in order to avoid capacity loss; but this too can often increase the complexity of later tasks; for example, when small piping is installed ahead of large piping, making it difficult to put the large pipe into place at a later date.

The 'lookahead' process may require higher-level schedules to be revised and updated to align with the production programme. This happens when scheduled tasks cannot be 'made ready' in time to hold their scheduled start dates. At this point, commitments to complete are made by everyone involved. These commitments are conditional upon the performance of the other parties.

Production planning typically takes place weekly. 'The aim is to make "quality assignments" and shield production units from workflow uncertainty, enabling these units to improve their own productivity and to improve the productivity of production units downstream" (Ballard and Howell, 1998). At the production planning meeting the Last Planners (trades foreman, gangers, and subcontractors' representatives) meet to plan the work in the next production period. They confirm the work that is currently being undertaken and the work that has been made ready. Each last planner/team leader prepares a production schedule for their team. These production plans are collated, agreed and the Last Planners commit to them. These plans comprise the agreed production plan for the next production period. At the end of the production period, the activities are reviewed.

The percentage of the activities completed on time and to the required standard (PPC) is a measure used to review the performance of the trade gang. In the LPS system, representatives of each trade/work gang commit to achieving the agreed programme of work. Promises are recorded and become a basis for evaluating performance. Last Planners only 'promise' to meet a commitment when the *conditions of satisfaction* for their task have been identified and they are confident that the task can be done. They consider how and when the preceding trade will *make ready* the work and what their own trade/work gang needs to do to *make ready* the construction for the following trade. This process is termed by some as the 'promise cycle'. This is shown in Figure 7.4

This figure shows that for each trade a request for work to be completed should be met with a promise to deliver or a decline to promise. The promise is only made after clarification and negotiation of the conditions of satisfaction. After completing each construction task, Last Planners 'declare completion' of the task to site management. This allows the site management to check that the task is on

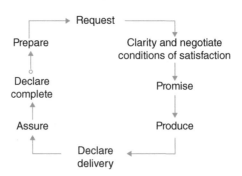

Figure 7.4   The Promise Cycle.

time and to an appropriate standard for the next trade gang to undertake their task. Both self-inspection and successive inspection are used in lean construction, so it is not only site management that checks, both the performer and their immediate customer also check to see that work has been done correctly.

Improving production performance

An important part of the LPS is the analysis of production performance. Examining production on the basis of the PPC enables all involved with the construction work to learn how to improve productivity.

The 5 Whys tool can be used to analyse activities not completed to the agreed schedule. Analysing the reasons for incomplete promises on a Pareto Approach across all the trades is the basis for continual improvement, by identifying and solving the most important problems such as unclear information or insufficient operatives. Teams that underperform and affect the performance of others are identified and brought to the attention of the main contractor. Lessons learned can be carried forward to the next phase of the project and the next project. These lessons need to be used to change existing work practices and develop a process of continuous improvement and collaborative working.

Benefits of the Last Planner® System

Last Planner® can significantly improve production performance. Ballard (2000) says:

> "With adherence to the Last Planner® rules, with extensive education and involvement of participants, and the use of techniques such as task explosion, constraints analysis and make ready actions shielding production from uncertainty through the selection of quality assignments planned performance completion levels of 76% or higher can be consistently achieved, a significant improvement on performance using traditional planning methods." (Section 10, p. 3)

Mossman (2013) emphasises that each of the key parts of the Last Planner® process brings its own benefits: For example, collaborative planning:

- Prepares team members for action together
- Enables team members to discuss details much sooner
- Sorts out sequencing and other issues that would be difficult to change at the production stage
- Enables team members to test options to improve workflow, buildability and program reduction
- Identifies unclear design details
- Builds commitment to programme and reduces overall program period

The benefits of the 'Make Ready' process include:

- Tasks are ready for production when required
- Safer working – planning involves hazard analysis and method statements
- Greater certainty of time, materials and equipment – less waste

Section III

Experience has shown that when using the LPS, subcontractors work to achieve the targets set. Professional pride develops and not letting the team down becomes important, no one wants to admit failure to deliver on time at the next meeting. LPS provides evidence that subcontractors have acted within the terms of the contract. Records show when work was undertaken in which areas and the time taken for production.

One specialist contractor who implemented the system found that using LPS improved the managerial skills of their foremen on site and led to a freeing up of senior management time. The managing director commented:

> "The Last Planner® System [enables] our site supervisors to plan their workload on a weekly basis and assess their team's performance on a daily basis [and] to make an accurate prediction of the labour required on a weekly/daily basis. This plan is based on facts, not a site manager's wish list. Once supervisors understand Last Planner and are confident in using the documentation, it can reduce the frequency of senior management visits to site. The foremen are capable of handling situations as they arise as their decisions are based on facts that are documented weekly." (Mossman, 2012, p. 4)

Barriers to the adoption of Last Planner®

What are the barriers to the adoption of Last Planner® and how do you overcome these problems?

Last Planner® requires a new approach to production planning. Collaborative working requires time and attention, extensive education and facilitation. Adopting the LPS means that managers and contractors have to accept a decentralised control system. This may be unacceptable to some. The idea of centralised control still dominates the construction industry. Many managers want to control information so that they can control staff and subcontractors.

Adopting the LPS requires a change of thinking, changes at operational level and a change of management philosophy. Project managers need to accept responsibility for effective planning and learning taking place, and review and respond to the measured results. They should routinely participate in Last Planner® meetings to emphasize the importance of this way of working, their commitment to its use and to provide help in removing constraints when needed.

Some find it difficult to link the weekly work plans to the activities listed on the master schedule. (This has always been a problem of short-term production planning. Working collaboratively simply emphasises the problem.) Last Planners need the confidence to ignore activities and activity start dates on the master schedule and focus on achieving the milestone dates for the next stage of the construction work. This is overcome if the master schedules are expressed in milestones between phases and the phases down into processes in pull planning sessions with those who will do the work in the phase. The processes within these phases should then be broken down into operations during the 'Lookahead' process. Operations are then broken down into steps. Using virtual prototyping, physical prototyping or first-run studies enables you to test the proposed operations.

The LPS relies on your ability to collect and collate information from all the trade organisations involved, to decide upon changes to the schedule. It requires you to analyse performance and formulate the reasons for failure. It requires you to initiate change. Everyone involved must be committed to the system and must benefit from the amount of work they need to put in to successfully use and maintain the system.

A description of how collaborative planning based on the Last Planner™ approach was introduced into a major UK construction company is shown in Appendix 2.

Key points

- Traditional planning systems are 'push' systems, Last Planner® is a 'pull' system.
- Traditional planning systems are widely recognised as being ill suited to short-term construction planning.
- Last Planner® was first developed for short-term planning; the technique has now been extended into a full methodology that is applicable in both the design and the construction stage of the project. It is applicable at all stages of the construction process and is increasingly being adapted to the design process.
- Last Planner® requires the construction team to plan collaboratively with those who will be undertaking the work.
- Each construction task is only allowed to proceed when all the constraints to production have been removed.
- Those undertaking the work are formally required to commit to undertaking the work they propose.
- The LPS methodology requires the construction team to investigate why any activities are not completed as anticipated and then to learn from any failures to prevent future delays. A range of techniques may be used to analyse why activities are not completed as scheduled.

Chapter 8
ADePT–Planning, Managing and Controlling the Design Process

Background

Planning, managing and controlling the design process has always been difficult. All schedules are established on the basis of assumptions, but too often, design schedules quickly become out of date, output falls behind expectations and staff lose confidence in the programme. The design process is notoriously difficult to manage. Research shows that the cost of design exceeds the budget in a third of all construction projects and design deliverables are delivered late over 40% of the time. The result of such delays is cumulative. Late delivery of design information invariably means increased design costs. If information fails to reach the construction team in time, there will be delays in production. Design changes necessitate expensive rework.

This chapter provides details of the approach taken by the Analytical Design Planning Technique (ADePT). This technique models the flow of information required by designers and then uses 'matrix modelling' techniques to optimise the design activities and produce a schedule of design tasks. Monitoring design production ensures efficient delivery of design information. The approach is based on collaborative working between designers and the construction team that produces innovative solutions to design and construction problems and makes significant savings in cost and time.

The need for improved management of the design process and better integration between design and the other processes in the delivery of a building structure has long been recognised. The design process is iterative. Designers make assumptions to enable them to further develop their design ideas and there is inevitably a need to refine or even fundamentally change such design decisions. Design coordination tends to be focused on individual design disciplines. When design changes are made, it is often difficult to assess the full impact of the changes, and unless these are 'tracked' problems may easily arise later in the design process or, even worse, during construction. Clients' lack of understanding of the design process may lead to problems when their needs are ill-defined or not fully understood. When clients introduce changes in their requirements, this necessitates a design review and frequently a redesign.

The Rethinking Construction Report (Egan, 1998) highlighted the problem of the lack of integration between design and construction as a major weakness in the construction industry. Since then, attempts have been made to improve the interface between design, procurement and production. However, figures published by the Department for Trade and Industry (2004) show that little has changed. Design goes over budget in a third of all construction projects. Design outputs are delivered late over 40% of the time. Late information contributes to production problems onsite. Only 44% of construction processes are delivered within budget and only 60% are on time. Clearly there is more to do.

Design management is based on the production design deliverables: drawings, specifications, etc. These outputs of the design process need to be released at required times to meet the needs of other designers and, ultimately, the construction team. Timely release of design information is imperative to avoid claims for delays.

The traditional form of building procurement assumes that design is complete at the time of tender. This requires significant upfront investment by the client. In recent years, there has been a considerable increase in the use of the Design and Build form of procurement. In Design and Build, it is common for contractors to manage projects from the schematic design stage onwards. Other forms of procurement, such as Private Finance Initiative (PFI) and Public Private Partnership (PPP) have placed different demands on the construction contractor and this has resulted in many contractors establishing design management departments with the aim of better integration between design, construction and operation of the facility. Placing the construction contractor at the centre of the design and production process has not, however, solved all the problems. Whilst it may have reduced issues of the 'buildability' of designs, it has not always improved the management of the design process. Often the design team is treated as simply another member of the contractor's supply chain and managed in the same robust way.

A new way of working

There is a need to focus on the delivery of the design information required by the recipient (be it client, designer, supplier, subcontractor or contractor) at the optimum time. Design must be integrated with procurement. Design must be integrated with production.

The new forms of procurement have led to new roles and new staff positions. Terms such as design manager, design team leader, lead designer, lead consultant and design coordinator have emerged and are often, incorrectly, used interchangeably.

Gray and Hughes (2012) identify two key requirements relating to effective design delivery: the provision of accurate fully coordinated and complete information and the timely provision of that information. They argue that the first requirement is that of the Lead Consultant, usually the Architect. (This role may alternatively be termed the design team leader or lead designer.) The second requirement is one of management and this the responsibility of the design manager who must establish the necessary 'framework' to enable the design team to produce fully coordinated and complete information at the time required by the rest of the project team. This is an active, not a passive, monitoring role. The design manager must perform as a facilitator to ensure the needs of all parties are understood and (as far as possible) accommodated.

However, experience shows that simply appointing a design manager for the project is not enough. There need to be changes in how the design team works.

All schedules are produced on the basis of assumptions. In the design process, these assumptions are often liable to change. Experience has shown that the number of unpredictable variables in a design and construction project means that it is usually impossible to rigidly follow the original schedule. In order to ensure the consistent delivery of design deliverables, it is necessary to understand the designer's information requirements and plan production, taking account of the iteration inherent within the design process. To manage the production of design information it is necessary to manage iteration.

The ADePT methodology, introduced into the industry in 2000, is based on the development of a schedule of activities produced with an understanding of the complexity of the design process. It enables schedules to be produced that recognise where iteration exists. It enables assumptions to be made about the importance of each element of information required in the design and accommodates these assumptions in decision-making processes. ADePT utilizes data flow diagrams, matrix methods of analysis, logic networks and traditional network techniques. A description of the ADePT methodology was included in Chapter 4. It comprises four stages. These are shown in Figure 8.1.

In the first stage, the scope of the design process and dependencies between activities are defined. In the second, the optimum sequence of the design process is determined based on the dependencies between activities and the necessary iteration within the overall design process. The third stage entails the representation of the design process in the form of a schedule, enabling the integration of the design process with procurement and construction. In the fourth stage, the design process is monitored and the flow of work is controlled (Austin et al., 1999). Each of these stages is now described in more detail.

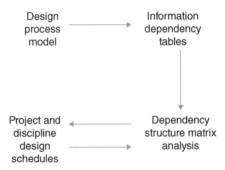

Figure 8.1 The ADePT methodology.

Defining the scope of the design process

The full scope of a design project must be identified in terms of the design activities required, the dependencies between them and the information required to complete the design activity. Steele (2010) describes how this may easily be achieved for a new project through a series of facilitated workshops where representatives of the architect, other design disciplines and construction advisers meet to identify and prioritise the design tasks and the information requirements. To collate these data and produce such a model is time-consuming but has the advantage of building close collaboration between the members of the design team. However, the ADePT methodology includes a generic detailed design model for all the main design disciplines. This model has been verified and validated across a number of building and infrastructure projects, and adopting this generic model and adapting it to meet the needs of a new project greatly reduces the time required, as in most cases more than 90% of the information required already exists. Alternatively, the design model may be directly adapted from previous projects.

Process sequencing

Having established the design activities and the information between them, it is necessary to consider the sequence of activities within the design process. Using specially developed software (called the Design Structure Matrix (DSM)), the ADePT technique enables the sequence of design activities to be optimised and iteration within the design process to be minimised.

This optimisation process takes into account any assumptions that the design team feels it needs to make with respect to the order and priority of the work. The calculation of a design sequence, including clusters of interrelated design tasks, prioritises the availability of outputs associated with the most critical dependencies. In order to eliminate some of the main iterative loops, it is necessary to make assumptions about the importance of the design information. Dependencies between design activities are normally weighted on a three-point scale: high, medium and low. Within the software there are facilities to take account of different levels of importance when ordering the design tasks. Inevitably, following sequencing, there will still be some interdependent iterative groups of activities that remain within the process. These activities are typically multidisciplinary. The start and end points of the iteration are key milestone dates in the design process and represent 'hard' gates where design decisions have to be coordinated and agreed by all members of the design team. This requires the design team members to work concurrently to finalise design decisions.

Scheduling

The sequenced design process will need to be transferred to a time schedule. Data from the ADePT software may be easily transferred to proprietary project management software. This enables the information required for managing design production to be presented in a standard bar chart format. The schedule for the

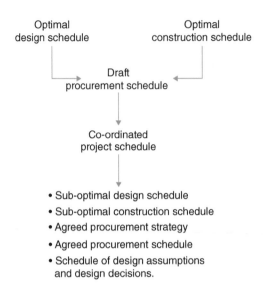

Figure 8.2   The production of an integrated design and construction schedule.

design production needs to be integrated with the construction schedule. The production of a coordinated project programme where design production is combined with construction must be based on an agreed procurement strategy and schedule. An integrated schedule usually results in a suboptimal design programme and a suboptimal construction programme. This is shown diagrammatically in Figure 8.2.

Controlling the design workflow

Because design is an iterative process, progress with the design of the building often quickly deviates from the agreed design programme. If appropriate action is not taken the design programme may quickly become meaningless. The design process needs to be monitored and controlled. Table 8.1 shows details of typical information that may be used for management purposes. Where work has already commenced, the production of design information is linked to the construction process by construction milestone dates. Where construction work has not commenced but information is required for tendering purposes, the production of design information is linked to tender milestone dates. Information is normally supplied in an annotated tabular/graphical format.

Practical implementation

A facilitated approach to planning

Experience from projects that have adopted the ADePT methodology has shown the importance of collaborative planning of the design process. It is essential to incorporate the knowledge and experience of *all* members of the design team. It is

Table 8.1   Information for managing the design process – Key construction information milestones achieved and milestones at risk.

Key construction milestones achieved	*This is a record of specific construction milestones achieved within the period.* This table records when the design team delivered information within a given reporting period together with the target milestone date. If the milestone was due to be delivered before the reporting period in question, it will identify the lateness of the milestone.
Key construction milestones at risk	*This chart identifies the construction milestones at risk.* This table records and predicts which key construction information release milestones *may* be at risk based on current progress, that is, those milestones that the design team may struggle to achieve. If the predicted date is later than the required date, the milestone is at risk and is highlighted in red. If the predicted date is less than 4 weeks earlier than the required date, the milestone could potentially be at risk and is highlighted in orange. Where the information to be produced is on schedule, the item is highlighted in green. The key aspect of these reports is that the milestones are not yet late but are predicted to become critical unless the project team takes mitigating measures to address the reasons due to which this element is delayed and rectify the situation.
Reasons for failure	*This chart records the given reasons for failure in any given reporting period.* This chart enables the manager to review a particular period in detail and look at the proportion of 'reasons for failure' assigned to each design discipline. 'Reasons for failure' are key to overcoming delays and minimising their impact on successive activities and the design process as a whole. The process of the disciplines reporting 'reasons for failure' on activities that they have been unable to complete is an important way of encouraging them to assess and question why an activity has not been completed. Once the information is collated, it is vital that the management and design team work together to review them, agree solutions to prevent further delays to the activity and ensure that the reasons being reported are accurate prior to the issue of performance-monitoring data outside of the design team.

Section III

particularly important to include not only the in-house members of the design team but also designers from the contractor, key suppliers and main subcontractors. A facilitated approach to planning the design will help ensure success. Projects that have adopted the ADePT technique have demonstrated that the use of a facilitator, such as the design manager – to define the high-level structure of a design plan, involve the design team members at appropriate times in defining the design scope and identify issues around the interfaces between design, procurement and construction – enables a consistent and meaningful schedule to be produced. The facilitator's role is a skilled one and requires technical knowledge, good communication skills and a balanced understanding of the project itself, the design process generally and planning in general, as well as the ability to foster positive and collaborative contributions from team members (Steele, 2010; Waskett et al., 2010).

Ideally, a number of workshops will be held to review the scope of the work, identify key activities, identify risks and produce an outline of the design process. Such workshops give the design team members the opportunity to share their perspective of the design to be undertaken, identify priorities and deliver a shared understanding. It is important to focus on individual disciplines' information requirements, not their outputs.

Dealing with iteration

The ADePT approach identifies iteration within the design process. By using the DSM software and considering the importance of individual information elements to the overall design, it is possible to reduce, but not eliminate, iteration. Typically, the number of activities within the iterative loops is greatly reduced. Activities at the start and end of iteration become important in managing the design process. The activities within the iteration cycle must be shown within the schedule. Usually, grouping activities together, and running them concurrently over a period of time, enables the planner to show this graphically.

It is necessary to define tactics to manage the design team as they work concurrently on an interdependent design problem. There is no single solution as the number of activities and deliverables, the number of team members involved and the time required to develop the design will dictate the most suitable approach. What is important is that each of these issues is considered in turn and that an appropriate procedure is implemented. By establishing this procedure the management remains focused on important iterative coordination problems and provides a basis for concurrent working (ADePT Management Ltd, 2010).

Integrating design with procurement and construction

When integrating the design programme with a construction schedule, design information and document release dates must be linked to the dates when these deliverables are required for procurement or construction. Design information is typically presented and distributed on the basis of its location within the building. Procurement and construction proceed on the basis of work packages. This requires a mapping between the design process and the work packages. This link usually determines the structure of the construction schedule.

The anticipated release of design information will seldom align totally with the dates on the construction/procurement schedule. Where information is due to be delivered before it is required, there is clearly no problem and it may be possible to reschedule work within the design team and make design savings. Where design information is due to be released after the required date on site, it may be necessary to review the design process and introduce additional design assumptions and fixity. Alternatively, incomplete design information may be released early on the understanding that this may have cost implications. In such circumstances, it is imperative to understand which information is missing and the impact of this lack of information upon the cost of the project (Steele, 2010; Waskett et al., 2010).

Managing constraints and measuring progress

It is important to know when and where production is deviating from that anticipated so that, where necessary, corrective action can be taken. Having produced a schedule for the design work the design process still needs to be actively monitored and controlled. Reports on each design team should be produced on a regular basis, typically fortnightly with a full review made monthly.

Progress may be measured with a number of tools and techniques. The following techniques have proved useful in managing production: planned percentage complete (PPC); key milestone prediction tracking; a traffic light system for monitoring production performance (green = improving production, likely to meet target; amber = holding performance; red = deteriorating performance); overall activity completion rate monitoring (elapsed days completed and activities completed) (ADePT Management Ltd, 2010; Steele, 2010).

Figure 8.3 shows extracts from typical reports.

Performance should be monitored for all the main participants: client team; main contractor; architect; design disciplines, for example, mechanical and electrical; etc. A regular summary of performance should include details of key tender milestones achieved, key tender milestones missed, milestones at risk and construction information milestones at risk.

Where there is a problem, it is necessary to identify and log key mitigation strategies and agreed actions. Reasons for failure should be logged and listed under suitable headings, for example, resource not available; awaiting design information; schedule priorities changed or incorrect; awaiting third-party information; etc.

ADePT and lean construction

The ADePT technique has been used successfully in conjunction with Lean Construction methods.[1] Lean Construction pre-empts deviation from the target programme by working towards key milestone dates, analysing constraints to the completion of the design tasks, working to resolve these constraints and then only proceeding with the work when all the required information is at hand. This assists in maintaining a meaningful programme and reducing potential risks of redesign and/or late supply of information.

The Lean Construction approach was originally produced for short-term construction planning. (Details of Last Planner are discussed in Chapter 7.) The adoption of the Last Planner approach to scheduling design work has proved extremely successful. It is imperative for the design manager to carefully review designers' 'work plans' (activities to be undertaken in the next period) and 'look-ahead schedules' (activities due immediately following the next period). The intended output is best presented in the form of simple to-do lists that are much easier for the design team members to understand than detailed schedules where changes in priorities may not be immediately apparent.

Following the Last Planner approach, when considering the work of the next period, designers are only asked to undertake activities which are free of constraints and that they feel they are able to complete without delay. They are asked to formally

[1]   See http://www.leanconstruction.org/

Section III

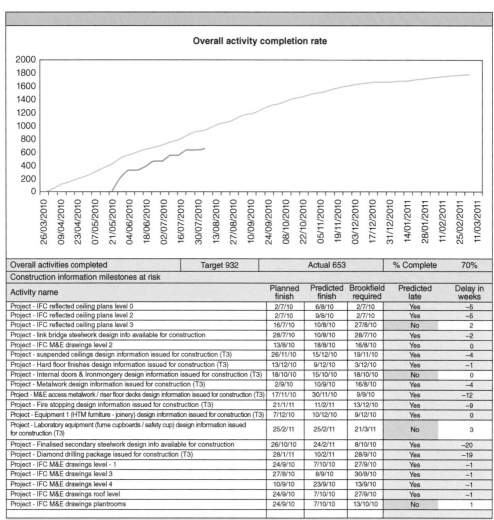

Overall activity completion rate

Overall activities completed		Target 932		Actual 653		% Complete	70%

Construction information milestones at risk

Activity name	Planned finish	Predicted finish	Brookfield required	Predicted late	Delay in weeks
Project - IFC reflected ceiling plans level 0	2/7/10	6/8/10	2/7/10	Yes	-5
Project - IFC reflected ceiling plans level 2	2/7/10	9/8/10	2/7/10	Yes	-5
Project - IFC reflected ceiling plans level 3	16/7/10	10/8/10	27/8/10	No	2
Project - link bridge steelwork design info available for construction	28/7/10	10/8/10	28/7/10	Yes	-2
Project - IFC M&E drawings level 2	13/8/10	18/8/10	16/8/10	Yes	0
Project - suspended ceilings design information issued for construction (T3)	26/11/10	15/12/10	19/11/10	Yes	-4
Project - Hard floor finishes design information issued for construction (T3)	13/12/10	9/12/10	3/12/10	Yes	-1
Project - Internal doors & Ironmongery design information issued for construction (T3)	18/10/10	15/10/10	18/10/10	No	0
Project - Metalwork design information issued for construction (T3)	2/9/10	10/9/10	16/8/10	Yes	-4
Project - M&E access metalwork / riser floor decks design information issued for construction (T3)	17/11/10	30/11/10	9/9/10	Yes	-12
Project - Fire stopping design information issued for construction (T3)	21/1/11	11/2/11	13/12/10	Yes	-9
Project - Equipment 1 (HTM furniture - joinery) design information issued for construction (T3)	7/12/10	10/12/10	9/12/10	Yes	0
Project - Laboratory equipment (fume cupboards / safety cup) design information issued for construction (T3)	25/2/11	25/2/11	21/3/11	No	3
Project - Finalised secondary steelwork design info available for construction	26/10/10	24/2/11	8/10/10	Yes	-20
Project - Diamond drilling package issued for construction (T3)	28/1/11	10/2/11	28/9/10	Yes	-19
Project - IFC M&E drawings level - 1	24/9/10	7/10/10	27/9/10	Yes	-1
Project - IFC M&E drawings level 3	27/8/10	8/9/10	30/8/10	Yes	-1
Project - IFC M&E drawings level 4	10/9/10	23/9/10	13/9/10	Yes	-1
Project - IFC M&E drawings roof level	24/9/10	7/10/10	27/9/10	Yes	-1
Project - IFC M&E drawings plantrooms	24/9/10	7/10/10	13/10/10	No	1

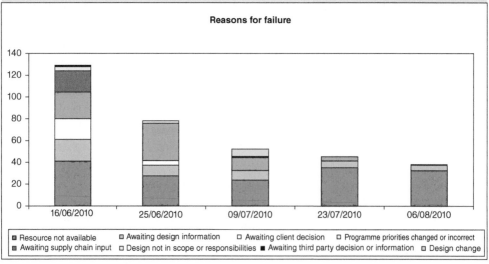

Reasons for failure

■ Resource not available   ▨ Awaiting design information   ▢ Awaiting client decision   ▢ Programme priorities changed or incorrect
▨ Awaiting supply chain input   ▢ Design not in scope or responsibilities   ■ Awaiting third party decision or information   ▢ Design change

Figure 8.3   Typical reports for reporting design production.

commit to meeting this delivery of design information. Where designers believe that they are unable to complete design tasks due to constraints such as lack of resources, incomplete information, etc., these are not included in the work schedule. The design manager will then work with team members to remove any constraints in order that those activities can be completed in the next work plan period. As with the Last Planner® system, progress is reported on the basis of the PPC. The proportion of design activities completed within the specified period is calculated and used to monitor the performance of both the team and individual team members.

Summary

Despite increased awareness of the importance of an integrated design process, and some tangible steps towards achieving this goal by the industry, projects generally continue to be delivered late and go over budget. The analytical design planning technique offers an approach to planning and controlling design processes that is more effective than is typical in current practice. Practical implementation involves a structured, facilitated approach. This provides opportunities to establish the optimal sequence of the design process and to understand the interface between design and construction. A mounting body of evidence from projects that have implemented this approach has revealed a range of benefits with significant impact. Many of these can help build the 'integrated team' proposed by the Strategic Forum for Construction and endorsed by Constructing Excellence in the Built Environment, through robust modelling and optimisation of the design process and coordination of information flow.

Key points

- Conventionally planned design programmes soon deviate from schedule and staff lose confidence in the programme.
- An approach that takes account of iteration in the design process should be introduced.
- The ADePT methodology allows you to manage iteration. It includes defining the scope of the design process; process sequencing; optimising and programming (scheduling) the design process; and controlling the design workflow.
- Planning the design process is best undertaken collaboratively with representatives of all the main parties and a facilitator acting as the design manager.
- Design production must be integrated with procurement and construction.
- Rarely does all the design information meet the target tender dates. When this happens, it is necessary to introduce assumptions to ensure that all design information is available or to release incomplete information and assumptions made in the pricing that can then be firmed up later.
- In managing a design project it is important to understand where the project is deviating from programme. Methods such as PPC adapted from the Last Planner® system have proved useful to monitor and control design work.

Section III

Chapter 9
Building Information Modelling (BIM)

What is building information modelling (BIM)?

Building information modelling (BIM) may be defined as

> "the digital representation of physical and functional characteristics of a facility creating a shared knowledge resource for information about it forming a reliable basis for decisions during its life-cycle, from earliest conception to demolition." (Sinclair, 2012, p. 3)

Such modelling provides a platform whereby all aspects of the building product, process and operation may be simulated and the design optimised. This may be achieved through the generation of models representing building components (or elements) by the use 'intelligent objects', that is, objects that comprise not only geometric definitions but also associated data and rules which will allow automated modifications to the overall building model whenever changes are made to associated objects.

It is the 'intelligent objects' within the model that ensure that changes to the model are fully examined and enable all stakeholders to understand the implications of the changes. Because users are working with one model (not different versions of 2D drawings), they may have the additional confidence that they are always working with the latest information.

BIM extends beyond 3D modelling to provide the basis for understanding all aspects of the building's performance:

> "BIM is an activity, ...not just a technology change, but also a process change. It is our belief that BIM is not a thing or a type of software but a human activity that ultimately involves broad process changes in design, construction and facility management." (Eastman et al., 2011, p. xi)

A Handbook for Construction Planning and Scheduling, First Edition. Andrew Baldwin and David Bordoli.
© 2014 John Wiley & Sons, Ltd. Published 2014 by John Wiley & Sons, Ltd.

BIM is not simply an extension to 3D drafting. It is the production and use of models that allow changes in one view to be automatically reflected in other views. It is the production of models where objects are parametrically defined and where the effects of changes to specific elements within the model automatically amend other parts of the model.

BIM is not new

In some application areas (e.g. piping, factory-made houses, prefab concrete components, constructional steelwork), BIM has been the relatively common practice since the early to mid- to late 1980s. These specialist construction sectors soon realised that extending the use of 3D CAD modelling to incorporate all aspects of their design and construction process enabled benefits in productivity and the potential to change their business processes. This acceptance and the desire to develop and adopt advanced modelling techniques were not however universal throughout the construction industry.

The development of BIM has been limited by problems in interoperability – the ability of all parties to the construction project to easily transfer data between the different systems adopted for modelling the building product. Initially, there was a dependency on 'translation software' to link between individual products. This was clearly not viable in the longer term and led to the development of neutral data exchange formats. Now, international standards such as iFCs are increasingly available, but these are not, as yet, all encompassing. When they are, increased interoperability will facilitate the use of BIM and the benefits will become fully apparent.

BIM is changing and will change the way of working for many construction organisations. This is the experience from other industries. As the use of modelling software and the ability to transfer data from one product model to another grew in other industries, advances in product and process modelling revolutionised the way that products were designed and produced. The adoption of these techniques in industries such as the aeronautical and automotive industry enabled consistent productivity improvements to be gained. This is the same expectation in construction.

In building, architects such as Frank Gehry recognised early the potential for BIM tools and adapted them to design and construct the signature buildings they visualised but construction companies were unable or unwilling to produce. Using the software produced by Dassault Systèmes™ and extending its use to construction, Gehry was able to complete a series of buildings including the Guggenheim Museum in Bilbao, the Disney Building in Los Angeles and the Stata Building at the Massachusetts Institute of Technology in Boston. (The software developed and used by Gehry is now available under the product name Digital Project™.)

Similar advances in software modelling tools and related technologies now provide a range of products and services that have been increasingly adopted by all sectors of the construction industry. These products and how they have been used in the construction sector are well documented in the BIM Handbook (Eastman et al., 2011); a guide to information modelling for owners, managers, designers, engineers and contractors which provides a comprehensive guide to BIM tools, parametric modelling and its adoption in the construction industry. Details of other publications are also provided at the website for the book.

Why now?

BIM is currently attracting considerable attention in the construction industry. Why now?

Put simply, BIM is, rightly or wrongly, being seen as a panacea to continuing problems or lack of productivity and the continuing problem of delivering infrastructure projects on time and at cost. In the United Kingdom, the government has mandated the use of BIM as part of the government's drive to shave 20% from construction costs (CIOB, 2012). More importantly, experiences from projects that have adopted BIM from all over the world have shown the benefits that can be achieved. BIM is viewed as a benefit to all. The adoption of BIM does not benefit one party in the project process to the exclusion others. BIM brings benefits to all the stakeholders within the construction process. Table 9.1 outlines some of the advantages available from BIM.

Given the advantages of BIM, many major construction industry clients are increasingly requiring engineering and construction organisations to adopt BIM

Table 9.1 Some of the potential advantages of BIM to the construction client, architect/designer and contractor.

Stakeholder	Potential advantages
Client	Improved building performance and quality Improved project delivery times Reduced project costs Better management and operation of the facilities Integration of facility operation with other management systems
Architect/designer	Earlier visualisations of the design Assists collaboration with other design disciplines Generation of accurate and consistent 2D drawings at any stage of the design Facilitates easier cost estimates at the design stage Improved modelling of energy requirements
Contractor and subcontractor	Design model for the fabrication of components Identification and resolution of design errors and omissions before the construction process begins Improved safety procedures Construction procurement processes

Section III

for their projects. The need for a more sustainable built environment has also placed a renewed focus on the energy use within buildings, a reduction in the physical waste within the construction process and the operation and maintenance of buildings over their lifetime. BIM has the ability and potential to assist in all these key aspects of sustainability.

Government has realised the potential of BIM. The UK government's Autumn Statement (HM Treasury, 2011) reaffirmed the Cabinet Office's commitment to the use of BIM. The purpose of BIM is to provide a common, co-ordinated source of structured information. This initiative follows on from recent reports that have emphasised the benefits of BIM, for example, the report to the Construction Clients Group by the BIM Industry Working Group entitled 'BIM management for value, cost and carbon improvement' which was published in 2011. This report was the result of an invitation by BIS and the Efficiency Reform Group of the Cabinet Office to look at construction and post-occupancy benefits of BIM for use in the UK building and infrastructure markets. This impetus is not limited to the United Kingdom. Such initiatives may be found in many countries. In Australia, a report to government by buildingSMART Australia has recommend a full 3D collaborative BIM based on open standards for information exchange for all Australian government building procurements by 1 July 2016.

BIM maturity levels

For organisations working in a BIM environment, there are different levels of information management, data sharing and collaborative working. These have been termed BIM maturity levels. They show the increased interoperability from 2D CAD to full 3D modelling with internationally recognised standards of data interchange.

The BIM Industry Working Group identified four different 'maturity levels' ranging from Level 0 BIM, the traditional CAD approach, to Level 3 BIM.

These levels are shown in Figure 9.1.

They define these levels as:

Level 0

Unmanaged CAD, information is probably disseminated in 2D drawings, with paper (or electronic paper) as the most likely data exchange mechanism.

Level 1

Managed CAD in 2D or 3D format using BS1192: 2007 with a collaboration tool providing a common data environment, possibly some standard data structures and formats. Commercial data managed by standalone finance and cost management packages with no integration.

	Level	Platform	Standards	Notes
↑	3	3D Models BIM Tools	BS 8541–1:2012 BS 8541–3:2012 BS 8541–4:2012 IFC:BS ISO 16739 IDM:BS ISO 29481:2010 IFD:BS ISO 12006–3:2007	Models, objects, collaboration, integrated interoperable data Life cycle asset management. Full inter-operability achieved
↑	2	3D Models	BS 7000–4 BS 1192:2007 BS 1192–4 BS 8541–1:2012 BS 8541–3:2012 BS 8541–4:2012 PAS 1192–2 CAPEX PAS 1192–3 OPEX	Models, objects, collaboration, integrated interoperable data but not a 'single' model. The methodology of BIM established.
↑	1	3D CAD 2D CAD	BS 1192:2007 BS 7000–4 BS 8541–2:2011	Models, objects, collaboration, integrated interoperable data, 'Lonely BIM' 3D CAD for large infrastructure projects, for the design of M&E etc.
↑	0	2D CAD	BS 1192:2007 BS 7000–4	CAD drawings, lines arcs, text etc. Data exchange by paper and or electronic documents. Unmanaged CAD data.

Figure 9.1   BIM maturity levels.

Level 2

Managed 3D environment with different BIM tools holding the data from different disciplines. An enterprise resource programme manages commercial data. Integration is achieved by proprietary interface or bespoke middleware. The approach may utilise four-dimensional (4D) schedule data and 5D cost elements.

Level 3

Level 3 BIM is a fully open process. Data integration is enabled by IFC/IFD and managed by a collaborative model server. This could be regarded as integrated BIM (iBIM) or BIM (M), building information modelling and management employing collaborative engineering processes. Within such an environment, each contributor to the design process adds his or her own dis-cipline-specific information to the model on an ongoing basis. This includes the client who may request changes in the requirements of the building. The latest data is available to all parties. Data within the BIM model are freely available for export to third-party simulation software. Similarly, data are available for materials procurement, planning and scheduling, costing purposes, etc.

The UK government's aim is that, by 2016, virtually every UK government project above a certain size should be carried out using BIM Level 2 and it is intended that the industry will move to fully integrated BIM (BIM Level 3) thereafter (NCE, 160812. 2012).

The effort required moving from Level 0 BIM to Level 3 BIM should not be underestimated. It will require the development of collaborative working methods with closer collaboration of all the design team. Input will be required from construction experts and specialist suppliers. Systems that are developed will require full interoperability of software and integrated database systems. It will need new procurement routes and forms of contract aligned to new working methods.

The adoption of BIM clearly has advantages for all organisations in the construction industry. Within the context of this book, we shall focus on BIM and its link to planning and scheduling. This is commonly known as 4D CAD.

The development of 4D CAD

Four-dimensional CAD is the term used to cover the use of computer-based tools to visualise the construction plan in a 4D (3D computer model + time) environment (Heesom and Mahdjoubi, 2004).

Four-dimensional CAD adds time to 3D CAD models and provides the planner with the potential to represent the construction phase of the project visually as opposed to in bar chart form. Again, 4D CAD in construction is not new and has been developing since the 1980s when construction organisations engaged in large complex projects began to use 3D modelling to build manual 4D models (Fischer and Kam, 2001). In these models, schedule information was used with 3D CAD models to provide 'snapshots' of each phase of the project over periods of time. These 'snapshots' were built into models to visualise the construction process. Little attention was given to their use for analytical purposes; the output in the form of a visual presentation was primarily for marketing and promotional purposes. If any aspect of the construction process or construction schedule changed, the model had to be rebuilt manually, a costly and time-consuming process. Such early methods of visualisation were enhanced when customised software and commercially available tools evolved in the mid- to late 1990s, facilitating the process by manually creating 4D models with automatic links to 3D geometry, entities and groups of entities for construction activities (Eastman et al., 2011).

These advances have been incremental. In 2004, Heesom and Mahdjoubi reviewed available software products and commented that

"Very few of the packages offer the ability to carry out analytical tasks on the developed simulation and this was often left to the interpretation of the user." (p. 173)

They went on to note:

"The amount of detail in a 4D simulation is still ambiguous. There is a need for improved use of data exchange standards and the automation of linking construction tasks to the 3D model." (p. 174)

Section III

The last decade has however led to a new generation of 4D planning products and the growth of interest in BIM. (Interestingly, Heesom and Mahdjoubi do not use the word BIM in their paper.) Over the last decade, the software companies appear to have targeted construction as a key market, and the term BIM and the claims attributed to its use have grown to the point that it is now considered by many to be oversold as *the* solution to construction productivity.

Currently, there are three main approaches to 4D modelling:

- A manual method using 3D software tools
- Using built-in 4D features in a 3D or BIM modelling tool
- Exporting 3D/BIM data to a 4D software tool and importing a schedule from the project planning software (Eastman et al., 2011)

These different approaches are shown diagrammatically in Figure 9.2 and Figure 9.3.

In addition to the use of 3D images linked to a time schedule, some commercial 3D modelling tools and commercial BIM software products incorporate the facility to exchange data with project management software such as MS Project or Primavera to generate 4D simulations. Alternatively, a specific 4D software tool may be selected for the simulation. Using such a product requires the input of design data from the 3D/BIM model and input of schedule data from project management software. These data are combined to produce the 4D model of the construction.

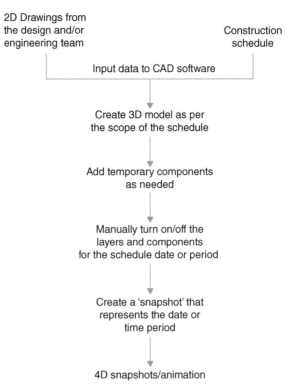

Figure 9.2   Manual/CAD-based 4D modelling process.

Section III

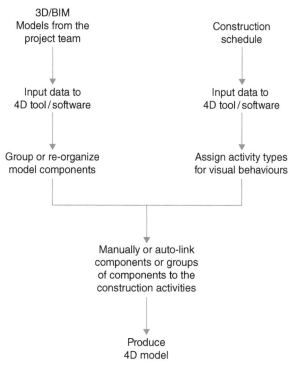

Figure 9.3   Four-dimensional tool/BIM-based 4D modelling process.

Since then software has improved considerably. Current 4D simulation software provides full visualisation of both the construction of the building and the construction process. Software use and the detail included however remains at the interpretation of the user. The example described in Appendix 1 shows clearly the advantages gained by the contractor: detailed building information showing the components of the building and their dimensions, specification details of each building component, data relating to the building performance (e.g. structural/ loading data), and design and procurement data. To this information, the contractor may add details of all the temporary components, scaffolding, formwork, etc., related to the construction. The current interest in BIM has led to the development of a number of new software tools to meet these requirements, and new products are continually emerging to assist the construction planner.

The BIM Handbook (Eastman et al., 2011) lists the following things to consider when evaluating specialised 4D tools:

- BIM import capabilities
- Schedule import capabilities
- The ability to merge/update multiple files into the 3D/BIM model
- The ability to reorganise data after it has been imported
- The ability to add temporary components required for the construction stage
- Animation
- Output facilities
- Automatic linking facilities

Section III

Eastman states that 'The degree to which these features are available within the software product will determine the ease of use and the benefits that may be obtained from the models produced'.

Key to the success of the BIM modelling remains 'interoperability', the ability to share information between different models. Interoperability allows the reuse of project data that has already been created and ensures consistency between different models. With interoperability, there is no need to re-enter data, a costly task with inherent dangers of creating errors.

The opportunities for increased productivity are considerable. Extensive use of 4D modelling presents a new way of planning and scheduling: the use of digital product models, and realistic graphical simulations that address the broader issues of physical and operational planning. This may be termed virtual prototyping or virtual construction.

Virtual construction

Using virtual construction, the planner can examine the construction process digitally by combining building components and checking potential problems with the construction process. This is particularly important/useful where the building construction incorporates customised design and prefabricated components such as structural steel, precast concrete items, mechanical and electrical systems and/ or complex temporary works.

Where the construction team wish to adopt a virtual construction approach, there are typically three phases of virtual construction implementation: the project requirement collection phase, the model building stage and the process simulation phase. Table 9.2 shows the three phases: the tasks; the information involved; and the people who will be required to provide this information.

The requirement collection phase

In the requirement collection phase, it is necessary to identify the major challenges of the construction process that the planner seeks to examine. These challenges may relate to the design of the building, the building components, the design of the temporary works, the construction process, the resources available and the time required for completion. To collect this information, the planner must liaise closely with the architect, the contractor's construction team and specialist subcontractors. Typical requirements for a high-rise residential building constructed from precast components are listed in Table 9.3.

The model building phase

In the model building phase, the planner works with the designers, subcontractors and model building experts to produce a model with the components necessary to simulate the construction process. These components will include not only the

Table 9.2   The three phases of virtual prototyping implementation.

	Project requirement collection phase	3D models' building phase	Process simulation phase
Tasks	Identify the key challenges in the design and construction of the building Define the scope of the work to be covered in the simulation	Create 3D models of individual building components and groups of components for temporary work facilities and plant (e.g. components of tower crane facility)	Identify the different construction processes Simulate each process Validate the construction sequences Identify time–space conflicts Check and optimise resource utilisation
Information required	Architect's drawings of key building components Construction method statement(s) Construction master programme The anticipated cycle for the production of each floor of the building	Detailed floor cycle programme with working bays/areas identified Layouts of building components Workshop drawings Details of temporary work activities, facilities and plant	Safety plan for the construction of each building element Detailed programme for construction Productivity/output rates of different trades in each of main activities
People involved	Architect, engineer, project manager Main contractor Representatives from major subcontractors and major suppliers	Project planning team Suppliers of building components Temporary works designers	Project planning team Representatives from major suppliers and subcontractors

Based on data from Baldwin et al. in Brandon and Kocaturk (2008).

Table 9.3   Requirements for the modelling of a precast high-rise residential building.

	Information required	Source of information
Design information	Co-ordination of the welding joints and reinforcement layout of the precast components	Architect Design team
Temporary works	Co-ordination of the working platform design	Main contractor Specialist subcontractors Supplier/fabricator
Construction process	Prepare the best sequence for installing precast components Scheduling of the tower crane: lifting cycles for precast components, materials and temporary works items Ensure that the tower crane is not overloaded	Main contractor's planning team Specialist subcontractors Tower crane supplier

Based on data from Baldwin et al. in Brandon and Kocaturk (2008).

Section III

components of the building but also temporary works and construction plant items, for example, a tower crane.

Here, it may be necessary to change the model of the building product to reflect the construction process. For example, the designer may have developed the model of the building with only one element to represent the ground-floor slab construction. The planner may decide that this slab needs to be constructed in three separate concrete pours, thus necessitating the need to have the building model amended. A key question here is 'granularity', how much detail should you add to the model. To ensure a safe, viable construction process, the models need to be built to a level of detail that reflects both dimensional and space conflicts and the process to be undertaken. These factors combined with the complexity of the temporary works and safety requirements will determine the overall detail required.

The process simulation phase

The process simulation phase demonstrates visually the planned construction process. Here, the planner explores alternative construction plans, examines the construction sequence, resolves any construction conflicts (including time and space conflicts) and validates the overall construction sequence. This is an iterative process. The planning team and the modelling team must work closely together to develop the necessary models to optimise the construction process.

How will BIM change construction planning and scheduling?

Planners and schedulers are already using 4D CAD and virtual construction. The adoption of these technologies is changing the way that planners work, who they work with and when.

Current use of 4D CAD is dependent on planners working with modelling experts to produce and amend the simulation models. This requires models to be produced on an iterative basis, amending and exploring in turn different aspects of the construction process until the optimum construction sequence is reached.

Over time, even though the adoption of BIM is likely to radically change the way that buildings are designed and constructed, the basic responsibilities of the planner are unlikely to change.

For projects secured under a traditional form of contract, 4D CAD will be used at the tender stage and at the precontract stage. At the pre-tender stage, the use of 4D modelling will be in the production of models linked to a time schedule to demonstrate to the client how and when the facility will be delivered. The main use of virtual construction will be at the precontract stage. Here, the contractor will utilise 4D CAD and virtual construction for a range of tasks depending on the type and complexity of the construction work.

For the clients of construction, other forms of procurement such as Design and Build and Guaranteed Maximum Price enable the use of BIM to bring a wider range of benefits and new responsibilities for the participants. The roles of the professions within the construction project have been established over a long period of time. With the increasing use of BIM, these roles may be expected to change, and new business models will emerge. Where clients fully embrace the use of BIM and seek to utilise its potential throughout the operation of the facility, there will be a greater need for the input of component detail (from specialist suppliers) and the proposed construction process (from the contractor) if design, construction and operation are to be optimised. Therefore, the construction planner and the construction organisation will need to be involved earlier in the project cycle (see Baldwin et al. in Brandon and Kocaturk, 2008).

BIM and the law

The full legal implications of BIM are still being explored and will not become fully evident until such time as the use of Level 3 BIM becomes widespread. It is however unlikely that BIM will radically shift legal risks and responsibilities: indeed, the overall risk to the various parties should in fact decrease if the key issues are identified at an early stage and the parties adopt a collaborative approach. Cleary, contracts must recognise BIM, and it is crucial that the parties define the full extent to which BIM will be used on their project. Currently, the only existing contracts that could be used in their current form on a BIM Level 3 project are JCT Constructing Excellence, NEC3 and PPC2000, all of which employ a collaborative approach and share risk. The use of other standard forms would require significant amendment or 'Z' clauses (Fenwick-Elliott, 2012).

Appendix 3 provides a current overview of BIM and the law.

Key points

- The use of BIM has increased greatly over the last decade and may be expected to continue.
- Four-dimensional CAD or 4D planning as it is also known is already well established.
- The ability to link the building product model to the construction process now extends virtual prototyping whereby not only the time taken to complete construction tasks but full construction process may be examined.
- Moving from computer-aided drafting to BIM involves a paradigm shift from drawing to modelling.
- The role of the planner does not change – only the people with whom the planner must interact with and when this interaction will take place.
- Whilst legal issues do arise from the use of BIM Level 3 (and also BIM Level 2), they are by no means insurmountable and should not hinder the future use of BIM.
- Provided BIM is used properly and accounted for in the contract documents, it should minimise design conflicts at the design phase prior to the build commencing.
- When design defects are discovered on-site, delay can quickly become an issue. The use of BIM may reduce the incidence of professional negligence claims and delay and disruption claims should in turn also decrease.

Section III

Chapter 10
Planning for Sustainability with BREEAM

Samuel Ewuosho

Graduate, Construction Engineering Management Programme, Loughborough University, Loughborough, UK

Background

The need to develop with future generations in mind has been made increasingly apparent since the Kyoto Protocol; governments around the world have made commitments to uphold sustainable development within their shores, and the United Kingdom is no exception. Sustainability is now a key concept in development thinking, and the construction industry is at the forefront of the fight to deliver it. By use of legal and voluntary instruments, consecutive governments have sought to monitor and reduce the waste in the industry. However, it is generally considered that the voluntary measures give the best results. 'Green buildings have attracted international attention as a means of sustainable development within the current energy crisis and the deterioration of the world's natural environment' (Liu et al., 2010). In the measurement of sustainability, the 'greenness' of a building is determined by the context of the performance benchmark set (Burnett, 2007). Arguably the foremost of such measures in the UK is the British Research Establishment's Environmental Assessment Method (BREEAM).

The need for sustainable construction

The construction industry has come to be synonymous with economic growth contributing to almost 10% of GDP but fraught with inefficiencies, the amounts and proportions of waste within the industry having been well documented over the decades. In Rethinking Construction, Sir John Egan (1998) cited 'that up to 30% of construction is rework, labour is used at only 40–60% of potential efficiency and at least 10% of materials are wasted'. He also suggested that these were 'conservative estimates' and concluded that there is 'plenty of scope for improving efficiency and quality, simply by taking waste out of construction'.

Several reports (of which Latham and Egan are most important) have warned of the impacts of the industry on our future generations, prompting subsequent governments to procure a mixture of legal and voluntary means to force the industry to pay attention because it currently contributes to about 40% of the total energy consumption around the world (Wall, 2006).

In outlining a strategy for sustainable construction, the UK government have identified the following themes of action:

- Design for minimum waste
- Lean construction and minimise waste
- Minimise energy in construction and use
- Do not pollute
- Preserve and enhance biodiversity
- Conserve water resources
- Respect people and local government
- Monitor and report (i.e. use benchmarks) (Department for Trade and Industry, 2004)

This brief by the government department suggested that 'these points simply make good business sense'. The argument in the Latham and Egan reports was that the efficient use of resources is key to British competitiveness within the global construction market: effective utilisation of the supply chain, as well as an understanding of the drivers of the sustainability agenda (Edwards, 1998).

Sustainability had emerged as an important issue.

The benefits of green (or sustainable) buildings may be categorised into direct, indirect and global benefits. The direct benefits are lower energy usage and operational costs, market advantages for the building developer and higher indoor environmental quality (therefore higher productivity of inhabitants). The indirect benefits include 'a psychologically and mentally more pleasant indoor environment, due to the utilization of natural lighting and ventilation and they improve the image of the company'. The global benefit is the contribution to the overall effort to tackle environmental problems on the large scale such as global warning, loss of biodiversity, ozone layer depletion and the increasing consumption of scarce resources (Prior, 2007).

Drivers of sustainable construction

Legislative drivers

The Kyoto Protocol commits industrialised countries and the European Community to the reduction of greenhouse gas emissions. 'Beyond the EU, the most important international influence on the UK built environment is undoubtedly the UN Framework Convention on Climate Change (UNFCCC)' (Ekins, 2008). That said, some researchers have argued that legislation can only go so far in enhancing sustainability and that voluntary measures must be brought to bear in order to change the attitudes in industry (Lee and Yik, 2004). Lee goes on to argue that 'due to the high cost implications to building owners for compliance, opposition to enforcement of regulatory control is likely to occur and if enforced, could incur high costs for policing and prosecuting violations, rendering control unenforceable. This explains why only moderate targets can be set and only moderate results achieved through this type of instrument (regulation)'. Notwithstanding, sustainability is now a legal and planning requirement.

Client (market-led) requirements

In addition to legislation, the government estate and government-related organisations aim to set the standards as lead clients in sustainability through more voluntary measures. For example, in the United Kingdom, the Housing Corporation must achieve a Level 3 rating under the Code for Sustainable Homes, 'The Ministry of Defence (MoD), the Prison Service, the Department for Education and Skills (DfES) and the National Health Service (NHS) all have their environmental assessment methods (based on BREEAM) that must be undertaken for all projects' (See Introduction to Sustainability, Cheshire, 2007). In following the lead from these government bodies, many private-sector organisations search for 'green labels' as part of their building procurement strategies. This will either have the benefit of increasing the value of their investment or reducing the running costs of the building: 'If the eco-label can truly reflect enhanced environmental performance of buildings that would be treasured by prospective buyers or tenants of properties, such schemes would encourage building developers to target buildings with higher environmental performance' (Lee, 2004). Certification and labelling codes may now be seen as an essential part of the mechanism to increase the supply of environmental public goods.

Professional responsibility

Aside from pressures of legislation and client expectation, more construction organisations are becoming conscious of a professional responsibility (CSR), as ambassadors for the industry, to show a 'duty of care' not just to their clients and shareholders but also to the community in which they are working (in terms of noise and air pollution) and the wider society (the environmental and climate change). For example, 'The Considerate Constructors Scheme provides a creditable scheme and guidance on community affairs during construction' (Brownhill and Rao, 2002). It is through such voluntary schemes and various other CSR initiatives that companies are shown to carry out their professional responsibility to 'the triple bottom line'. Building sustainably may increase the corporate ethical credentials of a company, and many organisations believe that commitment to sustainability will bring corporate benefits.

Competitors

Finally, given the increasing legislation, intensifying targets and higher demands from clients, construction companies that are seen to grasp and engage with the sustainability agenda in the most positive and innovative ways will get ahead of their contemporaries. CIBSE London (Cheshire et al., 2007) put it this way, 'The drivers for the sustainable development are likely to become stronger and those who embrace and understand this agenda are likely to see an increased demand for their skills'. This in itself is a driver because market forces mean construction companies increasingly need to prove their competencies and commitments with respect to sustainability in order to remain competitive. If 'competitiveness is the ability to provide products and services as or more effectively and efficiently than relevant competitors' (Blunck, 2006), then 'the definition of competitiveness as well as the definition of development sustainability requires adequate interpretation and quantitative assessment' (Rutkauskas, 2008). In setting the pace for sustainable development, the British government has also given an interpretation and a quantitative assessment method of sustainable construction, thereby giving a measure of competitiveness in the same arena (as required by Rutkauskas): the BREEAM. To reduce environmental impact, it is necessary to measure environmental performance.

The notions of sustainability – including the themes of action set out by the government – include plans to 'minimise', 'preserve' and 'conserve'. Without clear methods of measuring, reporting and benchmarking performance, these notions cannot progress. Benchmarks help enforce a reduction in the environmental impact of construction activities by providing the necessary yardstick for project-by-project comparisons (Crawley and Aho, 1999). In the United Kingdom, BREEAM is the main measurement and benchmarking tool to measure sustainability.

BREEAM

> "BREEAM (Building Research Establishment's Environmental Assessment Method) is the world's leading and most widely used environmental assessment method for buildings, with over 115,000 buildings certified and nearly 700,000 registered. It sets the standard for best practice in sustainable design and has become the de facto measure used to describe a building's environmental performance. Credits are awarded in nine categories according to performance. These credits are then added together to produce a single overall score on a scale of Pass, Good, Very Good, Excellent and Outstanding." (BRE, 2008, p. 5)

Introduced in 1990, the BREEAM was the first assessment tool for buildings and has evolved over the years to cover different kinds of buildings (schemes). Its operation is overseen by an independent Sustainability Board an independent body of industry experts representing a wide cross-section of construction industry stakeholders. BREEAM provides a credible environmental label for buildings and enables buildings to be recognised according to their environmental benefits. It is hoped that the use of BREEAM stimulates the demand for sustainable buildings and mitigates the impact of buildings on the environment.

In carrying out their aims, the establishment scores each registered building according to a set of categories (subthemes): management, health and wellbeing, energy, transport, water, materials, waste, land use and ecology, pollution and

Section III

Table 10.1 BREEAM section and weightings.

BREEAM section	Weighting %
Management	12
Health and wellbeing	15
Energy	19
Transport	8
Water	6
Materials	12.5
Waste	7.5
Land use and ecology	10
Pollution	10
Innovation	10
Total	110

Based on data from BRE, (2008).

innovation. Table 10.1 shows these categories together with the main items and the category weightings for new build, extensions and major refurbishments.

These weightings give the relative importance of each category based on the research by the BRE.

It is important to note that the summation of the weightings equals 110% – unless the innovation weights are subtracted. 'Innovation credits provide additional recognition for a building that innovated in the field of sustainable performance, above and beyond the level currently being recognised and rewarded within the standard BREEAM…they therefore enable clients and design teams to boost their BREEAM performance, help support the market for new innovative technologies and practices' (BRE, 2008, p. 6).

The credits achieved in the categories are multiplied by the weights and added together to give a percentage score and therefore a BREEAM rating:

- Unclassified   Less than or equal to 30%
- Pass   Greater than or equal to 30%
- Good   Greater than or equal to 45%
- Very Good   Greater than or equal to 55%
- Excellent   Greater than or equal to 70%
- Outstanding   Greater than or equal to 85%

Although some credits are tradable, gaining certain BREEAM ratings requires minimum credits to be achieved in some of the subcategories. For instance, in order to be considered 'outstanding', the project must gain both credits under the Considerate Constructors Scheme. Apart from scoring ≥85% on the BREEAM, an 'outstanding' building requires that 'a good quality case study is produced that design teams can refer to' (BREEAM, 2008).

BREEAM has two stages of assessment. In the design stage (DS), an interim assessment is made on the basis of the design and the involvement of a BREEAM assessor. At the post-construction stage (PCS), the assessor certifies the 'as built' performance and the BREEAM rating. This 'serves to confirm the interim

BREEAM rating achieved at the DS in accordance with the reporting and evidential criteria of the technical guidance' (BRE, 2008).

The items within BREEAM categories are scored and totalled for each section and then the weightings included to produce a single score from which a rating is awarded.

In 2008, the BREEAM scheme underwent revisions, and republished. It is this version that formed the basis of the research as most of the projects under review and most industry experience related to this version of BREEAM. BREEAM 2011 was introduced in July 2011, and the full ramifications of this new version are as yet unknown to the industry. Fundamentally, there is no real change to the ethos of the programme in 2011, except that where there were different schemes (BREEAM courts, education, industrial, etc.) in previous editions, BREEAM 2011 brings them back under the one umbrella, whilst making it more challenging to achieve certain ratings.

BREEAM sections

The BREEAM sections include management, health and wellbeing, energy, transport, water, materials, waste, land use and ecology, pollution and innovation. It is important to note that the categories are weighted differently – that is, the water section is about 6% of the entire weighting as compared with the energy section, which is nearly 20% of the rating.

Management

This category assesses the management of the building works itself, as well as the quality of community interface, the use of external consultation and the maintenance of the site. It assesses whether the builder is a considerate constructor or a good corporate citizen and the extent to which they have the considered the life cycle cost of the project – including the end-user building management.

Health and wellbeing

This category is concerned with the comfort of the anticipated users within the building; the thermal comfort, acoustic performance, the air quality and day lighting factors are all addressed in this category, including the view to the external landscape. '…Existing literature contains moderate to strong evidence that characteristics of buildings and indoor environments significantly influence rates of communicable respiratory illness, allergy and asthma symptoms sick building symptoms and worker performance' (Fisk, 2000).

Energy

This category deals with the design for control and efficiency of energy usage and CO_2 emissions within the building. The most strongly weighted category directly addresses the problem of climate change within the construction industry, with

Table 10.2 BREEAM credits and CO_2 index.

BREEAM credits	CO_2 index (EPC) rating	CO_2 index (EPC) rating
	New build	**Refurbishment**
1	63	100
2	53	87
3	47	74
4	45	61
5	43	50
6	40	47
7	37	44
8	31	41
9	28	36
10	25	31
11	23	28
12	20	25
13	18	22
14	10	18
15	0	15
Exemplar credit 1	<0	<0
Exemplar credit 2	True zero carbon building	True zero carbon building

Based on data from BRE, (2008).

specific respect to gas emissions. Credit is given for measures such as 'low or zero carbon' technologies, the use of energy efficient equipment within the building, based on the Energy Performance Certificates (EPC). More specifically, there are 15 credits associated with the reduction CO_2 emissions; Table 10.2 shows the credits given for different EPC ratings for a new or refurbished healthcare unit.

Transport

Mainly considered in the DS of the project, this category aims to recognise and encourage proximity to pedestrian, cycle and local transport facilities. By so doing, the development can help to reduce transport-related emissions and traffic congestion, reducing the need for extended travel or multiple trips, 'encouraging the use of alternative means of transport other than the private car' (BRE , 2008). One of the credits is available for considering the safety and the access to and from the site. This is specifically crucial in a hospital; vehicles bring emergency cases into the hospital, whilst other vehicles may be on standby for fast response to call-outs.

Water section

This section ensures thought has gone into an ecological design for water usage and handling of foul/storm water. As mentioned, this is the least weighted section in the scheme, but it forces the design team to address measures to control and assess methods to minimise wastage.

Materials

This section is to encourage the use of environmentally friendly, sustainable or recycled materials in the design of the project. It is concerned with the materials specified by the client/architect and its environmental impact over the life cycle of the building; 'The Green Guide to Specification' is the mechanism BREEAM uses to measure the environmental impact of specifications for key building elements.

Waste

This category seeks to promote the responsible use of resources and the appropriate management of construction site waste. In conjunction with the 'Half Waste to Landfill' scheme of the government, this gives added incentive for the design team to think of ways to incorporate sustainable, recycled and reusable materials and processes, in order to divert the waste.

Land use and ecology

This is mainly concerned with the improvement of ecological value of the land, vis-à-vis consultation with ecological, wildlife and local community representatives. It focuses on the most effective use of the land available, whilst mitigating the ecological impact by protecting the ecological species present. To achieve these credits, the project team must show an appreciation for the long-term impact of the project on the biodiversity of the area.

Pollution section

This section caters for the design consideration of water light pollution, NOx emissions, refrigerants and insulation materials. Any pollution – noise, light, water, heat and leakages – are all addressed in this category to ensure that appropriate care is taken not to cause nuisance and/or destruction to the locality.

Innovation

In addition to the initial ten categories of sustainability, BREEAM awards up to 10 extra points to projects that innovate to meet exemplary performance. 'Exemplary performance' is demonstrated by meeting specific criteria for which key credits are awarded. BREEAM issues which award key credits include:

- Man 2 – Considerate contractors
- Hea 1 – Daylighting
- Hea 14 – Office space (BREEAM retail and industrial schemes only)
- Ene 1 – Reduction of CO_2 emissions
- Ene 5 – Low or zero carbon technologies
- Wat – Water metre
- Mat 1 – Materials specification
- Mat 5 – Responsible sourcing of materials
- Wst 1 – Construction site waste management

Section III

Industry response to BREEAM

All business organisations recognise the need to ensure that they have the capabilities to remain competitive in this nature and advance with the times. All construction companies are aware of the industry's image and the failure to meet sustainability targets. Published in May 2011, the Sustainable Times claimed that 'our buildings are using 30% more energy than they did 30 years ago, the impact on the natural environment is getting worse and ultimately, we are failing to prepare for the huge and imminent consequences of climate and resource depletion. It is time for a rethink of what we are doing, where we are going and how we should move forward'. Why have efforts to make the industry more sustainable not resulted in substantial improvement in real terms?

The root of most problems in construction lies in the fragmentation of the industry – the implementation of BREEAM is no different.

Different industry parties and stakeholders believe that certain sustainability responsibilities lie elsewhere and naturally; these stakeholders believe that certain aspects of sustainability are more important (therefore deserving of more attention) than others. Stakeholders in construction rank each aspect of sustainability differently: the developers/investors are most interested in their economic return; the local authorities (those who grant planning permission) are careful to foster socially sustainable societies; and the construction professionals are most worried about their effect on the environment. Differences in interests and requirements of the stakeholders (Cole, 1998) often characterise the sector and when left to market forces either lead to compromise (where the environment suffers for economic gain) or can lead to the boom and bust problems as economic performance interests investors, whilst health and comfort are key for tenants – for example. Government intervention not only as a legislature but more importantly as (the client) an active player in the market is a principal prerequisite not only in the sustainability agenda as a whole but also in the success of BREEAM: '…the client, design team, contractor…have an important role to play if the desired performance level is to be achieved and reflected through the certified BREEAM rating. However, the onus of orientating the brief towards sustainability needs to come first and foremost from the client' (BREEAM Technical Manual, 2011).

Because of the ongoing shift in the way that construction is procured in this country, more projects are on a 'design and build' (D&B) basis, as opposed to the 'traditional' route, and as such, the design risk is also with the main contractor. In traditional procurement, the client would present a completed design to the contractor. Here, the design risks lie with the client. Along with the increased risk from D&B comes the responsibility for sustainable design and construction, which more often than not equates to contractors being set a BREEAM target to achieve – and this is certain to rise.

Of course, taking waste out of construction is hardly a 'simple' affair, and as far as many in the industry are concerned, waste is simply a by-product of the process. Although this is a less publicised point of view, informal interaction with some construction site staffs indicates a perception that waste management had become a burden, a bureaucratic process that sought to jump hoops rather than to actually reduce waste.

It has since become accepted in industry that just like we can 'design out' health and safety risks to minimise the implications on-site, we can also 'design in' sustainable methods of construction – such as specifying precast members of concrete or steel, thereby cutting down on the waste of unwanted material. Attention has turned beyond the *sustainable qualities* of the building and the business model to the whole life cycle of the project. In fact, some contractors are using BREEAM IN-USE to 'address the challenges posed by existing building stock…allowing the facilities managers and occupiers of a building to obtain an indication of its sustainability performance…' (Hughes, 2011). Through the transparency of the credit system, BREEAM is helping the industry to see that sustainability is everyone's responsibility.

Case study analysis

Research at Loughborough University in the United Kingdom examined three case studies to review how BREEAM was implemented on similar projects. All three of the projects were for the provision of hospital facilities. They were procured under different methods of procurement.

In the three case studies, each of the BREEAM sections (management, health and wellbeing, etc.) was broken down to decide for each project: how many of the possible points could be achieved, how easily (resource intensive) they could be achieved and which stakeholder was responsible for achieving it. The codes A to D show the degree of difficulty associated with each credit, with (A) being easy and/or routine, (B) meaning that cost allowances must be set aside to achieve these credits, (C) meaning that the costs of these credits and efforts required are relatively high and (D) meaning that the credit is uncertain or almost impossible without specific attention from the client.

The BREEAM targets and performance were established and monitored, and the overall process was managed using extended spreadsheets. Other columns of the spreadsheets show the organisation responsible for the work to secure the credits (client, principal contractor, work package holder, etc.) and the individual responsible for ensuring that the task is completed. Summary spreadsheets were then produced to total the number of credits likely to be achieved and examine different scenarios, thereby ensuring that the desired rating is achieved. Tables such as these should show clearly the original analysis of the potential credits in the management section of the analysis together with those potential credits where there is the likelihood of a high cost to achieve and also indicated who was responsible for ensuring the credits are secured. It is also helpful to present a final summation table and analysis for three scenarios. The first table should show the potential, or 'likely', credits. The second table should show the credits that can be achieved with minimum additional effort or cost. The final table should illustrate the scenario where accepting all the potential credits together with those that are uncertain to secure then that an overall rating of 'excellent' could be achieved.

Different projects produce different management situations

First, in Hospital A, the design team was responsible for 51 credits – over 40% of the total credits available on that job – which also equated to 40% of the total weightings. However, almost half of their allocated credits appear in the 'C&D'

zone, which meant they required a lot of effort to achieve. In contrast, although the architect credits were far less (in number and weighting), a larger percentage of them fell in the easy (A&B) category – this indicates that on this job, the design concerns were crucial and required careful planning and execution.

A similar situation occurred in the case of Hospital B; the design team had considerable responsibility, but again, over 40% of their credits lay in the 'C&D' zone, a situation which could jeopardise the rating especially when one considers that a few of their credits are mandatory.

From observation, a frustration on the part of the main contractors was that in many ways, 'the buck stopped with them' because it was their responsibility to translate the efforts of the project team into the building the client desired within the allocated time. Throughout the project life cycle, the responsibility (risk) for sustainable construction is seen to come down from the government through the client, through their consultants and through the design team and falls squarely on the lap of the main contractor, having been multiplied several times through the chain.

The contractor is often waiting on a design or an amendment before being able to build (which can slow the progress of the project) and then waiting for the correct documentation post-construction to prove the building credentials to the BRE assessor.

This frustration was evident in the cases of Hospital A and Hospital B, but the difference with the latter case was that because 100% (17/17) of the client's credits were in the 'A&B' section, the rest of the project team were able to produce a higher BREEAM rating. This echoes with interviewee claims that when the client takes charge from the outset of the project, achieving a BREEAM rating is much easier.

Finally, in Hospital C, an example of a D&B contract placed a big responsibility on the contractor to deliver 50% of the BREEAM credits. The responsibilities of the architect diminished and the design team subcontracted. Although this partly solved the problem of having a long chain of responsibility as the previous two case studies, it tended to increase the number (and weight) of contractor credits in the C&D category, as the contractor had less mechanical and electrical expertise than the design team.

For this reason, the D&B form of contract is mainly used for less technically demanding BREEAM projects in healthcare, whilst the more technically challenging projects are executed in the traditional way. Again, the client's credits weren't considered straightforward, and this had an adverse effect on the total percentage of BREEAM credits achieved.

Individual perceptions of sustainability and BREEAM

In interviews with industry personnel, it was agreed that BREEAM was the most widely known and used method of measurement in the United Kingdom. However, a problem of measuring against set criteria is that the benchmarks will prioritise one set of metrics over another; parties may chase the favoured metrics at the expense of the less easily categorised. The interviewees noted that the consequences of using BREEAM were not always positive.

One interviewee offered that '…I think they the system looks at the small details and not the bigger picture', showing one of the principle problems with BREEAM: that many feel it is a tick box, a simplistic approach to sustainability. Another agreed to an extent, saying that even if the building meets a sustainable standard, there is a substantial challenge in marshalling the paperwork and evidence required in such a fashion suitable for the BREEAM assessors. There is a feeling that the bureaucracy in itself becomes an issue, sometimes overshadowing the point of the exercise.

Describing it as 'onerous', another interviewee said that the burden of proof is too heavy; the wording of the guidance is open to several interpretations and the process too arduous. At worst, the net effect of this is often a finger-pointing and blaming culture within project teams as deadlines close in and the clients mount the pressure. In difficult economic times it is the belief of some that sustainability will still be present in terms of paperwork and public perception, but the actual delivery of sustainability will suffer. Others thought that if the building was designed properly in the first place with BREEAM in mind from the beginning, then the extra cost implications might be minimised. From another perspective, achieving BREEAM increasingly means derogating against operational necessities. In many ways, the ultimate sacrifice has been involved in hospital projects whereby to get the funding required for a building project, BREEAM thresholds must be met, and in a 'Catch-22' scenario, these thresholds require funding. It may be argued that the overall effect is that the hospital is forced to operate at a lower capacity in terms of the tasks the hospital was meant to achieve in the first place, in the name of BREEAM.

On the whole, the research revealed that BREEAM is not perceived to add to the economic value of a development. Although there is tacit agreement that it is necessary to monitor carbon emissions to slow climate change, BREEAM is seen to focus too rigidly on the construction process, without the due attention to the life cycle of the building; client education as to effective operation and maintenance of the facility is seen to be neglected, with BREEAM as a badge of environmental sustainability. The main contractor is expected to balance the expectations of all stakeholders to deliver a BREEAM project, but the bureaucratic nature of the process has been observed to have an adverse impact on the dynamic of the project teams, with the client rarely satisfied with the capabilities of the facility, having had to trade-off against functional utilities. An engaged client should drive the BREEAM agenda from the very start of the project even though sustainability is the responsibility of every stakeholder.

Section III

Key points

- The BREEAM is a voluntary measurement rating for green buildings that was established in the United Kingdom by the BRE.
- The use of BREEAM is now widespread throughout the world.
- The BREEAM uses a range of weighted criteria to assess the impact of a construction project on the environment throughout its life cycle. These weightings enable an overall rating for the building to be produced. This rating is based on six levels, from 'unclassified' to 'excellent'.
- The BREEAM is now an accepted part of the drive for sustainability.
- Some practitioners however query its effectiveness in producing sustainable buildings and argue that systems such as BREEAM are divisive and place too much of a burden on the contractor and that truly sustainable buildings will only be produced if the client is fully committed to sustainability and the design team respond accordingly.

Chapter 11
Planning for Waste Management

Sarah-Jane Holmes[1] and Mohamed Osmani[2]

[1] Graduate, Architectural Engineering and Design Management Programme, Loughborough University, Loughborough, UK
[2] Senior Lecturer, Architecture and Sustainable Construction, School of Civil and Building Engineering, Loughborough University, Loughborough, UK

Background

In 2004, the UK government published requirements for Site Waste Management Plans (SWMPs). These were as a voluntary code of practice. In 2008, they became mandatory regulations for construction projects costing more than £300 000. The aim of SWMPs is to divert waste from landfill by increasing on-site reuse and recycling rates. The onus is on the client or principal contractor to make sure the project meets SWMP regulatory requirements through effective implementation and monitoring.

Due to the statutory requirement for SWMPs, UK construction contractors have fully embraced an important vehicle for reducing construction waste. Little information had been published on contractors' current SWMP implementation practice and associated challenges. This chapter reports on research that set out to carry out a gap analysis on the limitations and missed opportunities of the implementation of SWMPs. The research found that the main implementation difficulty for those producing SWMPs is the estimation of waste quantities. To enable systems to be implemented effectively, space for on-site skips should be designed into the project for waste segregation. Depending on the type of project and space available, there may be the need for the increased use of external waste contractors. Subcontractor involvement with SWMPs may be limited. As key participants in the construction process, they should be given part ownership in the management of the system and feedback provided to them. Unless there is the involvement of the design team in the implementation of SWMPs, there is unlikely to be significant reductions in the levels of construction waste.

Construction waste is a material that arises due to damage, due to non-use or as a by-product of the construction process that needs to be either discarded off-site or recycled and reused on-site (Ekanayake and Ofori, 2000). With the increasing

Section III

focus on environmental problems, the UK government has become increasingly aware of the substantial waste dilemma, particularly in concern with the Construction and Demolition (C&D) industry. Each year, the UK C&D industry generates around 120 million tonnes of waste in utilising over 400 million tonnes of materials (Jones, 2008a). Of this total, 13 million tonnes comprises materials classified as damaged or defective resulting from poor materials management. This equates to a third of all waste in the United Kingdom. Clearly, there is a significant potential for minimising construction waste. The construction industry is not as advanced as other industries in reducing the significant quantities of waste produced. Much of the waste produced by the construction industry is both useable and avoidable. This view is fully supported by the UK government who have initiated a combination of voluntary policies and mandatory legislation to help reduce the quantities of construction waste, including the target to halve construction waste to landfill from the 2008 figures by 2012 (WRAP, 2009c). To achieve this target, SWMPs came into force in April 2008.

Construction waste causes and origins

Construction waste arrives throughout all stages of a construction project: design, materials procurement and site operations. Table 11.1 summarises some of the main ways that construction waste arises.

Materials procurement

Waste through materials procurement waste arises in two ways: firstly, through the contractual obligations set out by the client during the principal stages of

Table 11.1   Construction waste causes and origins.

Project phase waste origin	Causes of waste
Materials procurement	• Ordering errors (not ordering materials in compliance with specification) • Over-ordering (poor design information causing assumption of quantities) • Supplier errors
Design	• Design changes • Detailing errors in design • Poor specifications • Lack of design for standard-sized materials • Lack of coordination and communication (slow response from architects and clients)
Site operations	• Unskilled or poorly trained labour • Inappropriate equipment available • Inappropriate storage areas • Poor material handling • Unused materials

Section III

a construction project and, secondly, through the ordering and supply of materials on-site. The design and works procurement stage establishes a great opportunity to design and implement waste minimisation techniques (WRAP, 2007a). Waste minimisation should be a consideration from the initial concept of the project. Ideally, requirements for materials' waste management should be included within the tender documentation. The ordering and supply of materials is a critical part of the construction process. Osmani et al. (2008) argue that the over-ordering of materials is a high contributor to site waste. This is often due to poor design information, causing contractors to assume the quantities of materials required. Coventry et al. (2001) found that 10% of materials on-site were wasted through not being used in the construction works. Materials procured and unused are then wasted if they cannot be returned or sold on.

Design

There is a general consensus amongst construction contractors that the main cause of construction waste is due to design problems or the lack of waste mini- misation strategies during the design. Osmani et al. (2008) argued that waste is generated on construction projects throughout the entire process from inception to completion, with the phase prior to construction contributing a vast propor- tion of waste. Jones (2008a) illustrated the significance of employing waste man- agement practices at the initiation of a construction project by arguing how vital the design stage is in reducing the quantities of waste produced on-site; an exam- ple of this is the provision of key performance indicators in assessing design opportunities. Other investigations have produced results where significant causes of waste have been attributed to the design process (see and Faniran and Caban, 1998; Ekanayake and Ofori, 2000).

Furthermore, it has been argued that the designer has the best possibility to increase recycling of construction waste as they specify the materials used within the construction (Price et al., 2009). This has been reinforced by Jones (2008a) who confirmed that the specifications produced during the design stage may have significant impacts on waste reduction, such as the specification of products with recycled content. Interestingly, contradicting research claims that architects them- selves believe that the majority of construction waste develops in the course of site operations and is seldom produced during the design stages (Osmani et al., 2006). However, some designers have helped to significantly reduce waste produced dur- ing construction projects by implementing measures to design out waste during the design stages (Langdon, 2009).

Site operations

Contractors acknowledge that on-site construction waste could be reduced through better selection of qualified site personnel as well as the correct equip- ment to reduce damage caused through material handling (Ekanayake and Ofori, 2000). Other researchers have found that untrained and unskilled labour were a major cause of waste on-site (Osmani et al. 2006). Waste may be reduced by better storage facilities for materials. Poor weather can cause damage to materials and therefore waste if the appropriate protection of materials is not supplied. Other external factors such as theft and vandalism can contribute to the waste on a construction project. Measures to prevent this type of waste will be dependent on the type of project and its location.

On-site waste management practices

Waste management is an effective methodology in reducing the amount of construction waste going to landfill. It is now considered as a strategy of resource management throughout the course of construction, a key part of construction project management. It should be employed from the onset of the project to achieve the greatest benefits (Coventry et al., 2001; Tam, 2007). Waste management should focus on reducing the quantity of waste produced prior to reuse, recovery and disposal.

WRAP (2007a) and Keys et al. (2000) argue that waste management involves the identification of waste streams and setting targets to achieve waste minimisation goals. The monitoring of waste minimisation is vital to understand the benefits and successes of the waste management practice (Griffiths, 2004). Once waste arisings have been measured, they can then be used as a benchmark for further progress (McGrath, 2001). Lingard et al. (2001) observed that goal setting and feedback systems help to both reduce waste at source and increase material reuse rates but had little impact on the level of recycling achieved.

Poon et al. (2004) state that on-site waste segregation should be an integral part of an SWMP as it helps to open up business opportunities for recycling organisations. Shen et al. (2004) argue that waste, which is either recyclable or reusable, should be classified and separated as soon as the waste is created. Segregating waste on-site with the clear labelling of skips has been proven to be extremely beneficial (WRAP, 2008). By segregating inert waste from the rest of the waste streams, this itself can be disposed of in a lower tax band and thus present cost savings (Coventry et al. 2001).

When all staffs on-site are given an opportunity to contribute to the waste minimisation process, there is a positive impact on waste minimisation (Griffiths, 2004). Jones (2009) illustrates how WRAP has been working with contractors to initiate subcontractors in identifying the waste produced by their work and how they can contribute to waste reduction practices. As well as this, Coventry et al. (1999) explain how, in order to maximise the implementation of waste minimisation techniques, commitment needs to be demonstrated at the highest corporate level in order to illustrate the importance to all personnel. This view is supported by Tam (2007), who noted that the environmental awareness of a project can be enhanced with the support of senior management. Such support is not always forthcoming. Teo and Loosemore (2001) found that waste management was considered as a low priority by many managerial staff on construction projects, and thus, support within the construction team was rarely established.

On-site waste management techniques

Different waste management techniques may be adopted dependent on a company's philosophy, as well as the financial and time input available. Different companies approach on-site waste management differently. Some of the key techniques used are explained in the following.

Mcdonald and Smithers (1998) studied different methodologies used for waste management plans on-site and found that with the development of a waste collection system on-site, it provided the subcontractors with a convenient and functional method to deposit waste. For good practice, the waste storage facilities on-site should be centralised and well managed in order to gain maximum benefits (WRAP, 2008). The centralisation of the waste facilities is also supported by Shen et al. (2004) as it will help to reduce the supervisory effort required for waste collection.

A trade-based waste collection system helps create the philosophy that the different trades are responsible for the collection and deposition of their own waste (Mcdonald and Smithers, 1998). Nevertheless, a single person needs to be identified and given overall responsibility for the management of waste on-site. This has been found to give immediate improvements for waste segregation and record keeping on-site (WRAP, 2007a). A simplified skip colour-coding system has minimised errors with the separation of waste at source (Jones, 2008b). By separating waste materials at source, a reduction in the current tax duty of up to 90% can be achieved (Jones, 2008b). However, research by Trufil-Fulcher et al. (2009) has shown no correlation between employing a 'waste champion' and achieving any cost savings.

Dainty and Brooke (2004) identified that toolbox talks were a strategy highly used on construction sites to increase the knowledge of the benefits of waste minimisation on-site. In addition, WRAP (2007a) suggested training for personnel at each level of the waste supply chain to improve awareness of the waste minimisation methods. Motivation for improved behaviour in terms of waste management with subcontractors can involve incentives for segregation and penalties for waste contamination (WRAP, 2007a). This has been supported by Lingard et al. (2001) who found increased recycling efficiency on projects which had in place an incentive scheme; however, in addition to this, the research showed that site managers were apprehensive of this strategy as some found increased quantities of useable materials in the recycling bins.

NetRegs (2007) explained how ordering materials to specification will help to reduce offcuts of materials and save workman's time. Take-back schemes by manufacturers can also help to reduce the amount of waste produced on-site (WRAP, 2008). Also, 'just-in-time' delivery is a key methodology used for improved waste minimisation (WRAP, 2007a). Failing this, the physical size of waste on-site can be reduced by the use of compaction systems such as balers (WRAP, 2008). This is beneficial as void space in skips can be up to 40% (WRAP, 2008).

The SMARTWaste auditing tool by BRE is used as a means of recording and generating data on the quantities and types of waste produced (WRAP, 2007a). By auditing generated wastes, these can be used as a benchmark for waste control on future sites (Shen et al., 2004). The SMARTWaste tool can be used by a company to create separate waste plans for each individual site and make all the information on the project available company-wide for benchmarking benchmark waste control on future sites (BRE, 2009).

Jones (2008b) explained how SABRE, a software tool designed by BRE, can be portably used to convert the quantity of waste arising surveyed into a tonnage mass. The SMARTWaste tool has been used as a means of connecting the construction process with the waste hierarchy and so increasing material reuse and recycling (McGrath, 2001).

WRAP has developed the recycled content toolkit which helps project teams calculate the amount of reclaimed or recycled materials which are used in new construction projects by value (Addis and Jenkins, 2008). As well as this, Langdon (2009) described the WRAP Net Waste Tool which helps to identify where waste will occur on-site, thus ascertaining the prospects for reducing waste disposal and material costs (Moon, 2008). In addition to this, WRAP has produced a Reporting Tool which helps to monitor the progress being achieved by personnel within the industry against the government target to halve waste to landfill by 2012 (Moon, 2008).

Finally, SWMPs may be produced in-house using proprietary software products or by the use of third-party auditing tools. Such auditing tools are frequently made available within a suite of software products designed to assist the project team to monitor the environmental performance on the project.

Site Waste Management Plan (SWMP) requirements

SWMPs require the client, the client's representative or the principal contractor to take overall responsibility for the preparation and implementation of the plan. They necessitate the production of an SWMP that describes each waste type expected to be produced in the course of the project; estimation of the quantity of each different waste type expected to be produced; and identification of waste management action proposed for each different waste type, including reuse, recycling, recovery and disposal. The SWMP needs to identify the subcontractors producing the significant waste streams and set realistic targets on the amounts of waste produced which could be recycled, reused or safely disposed (NetRegs, 2007). The principal contractor and client should declare that all waste will be processed with the Environmental Protection Act 1990 and the Environmental Protection Regulations 1991 in accordance with the waste duty of care. In addition, special arrangements need to be provided for the disposal of hazardous waste (NetRegs, 2007).

The main actions involved in SWMPs are waste segregation, recycling and reuse of the waste produced, plus the segregation of hazardous waste from other waste streams on-site to avoid contamination (WRAP, 2008). When waste is removed from the site, the types of waste being removed, the identity of the waste remover and the location to which it's being taken need to be recorded. The registration number of the waste remover needs to be recorded. For all but the smallest construction projects, there will be a need to update the SWMPs at regular (say, 6 monthly) intervals to reflect the progress of the construction work and the types of waste being produced. The principal contractor must supply information about SWMPs to site personnel through methods such as project induction or toolbox talks, and at the end of the project, the SWMP must be reviewed and any deviations from the plan must be explained. The

Section III

principal contractor should keep the SWMPs for 2 years after the completion of the project. Within 3 months after project completion, a comparison between the estimated and actual quantities of each waste type should be produced together with details of the cost savings. This will provide useful information for the management of waste on future projects.

A study carried out by WRAP (2009d) indicated that there is confusion on who is responsible for implementing SWMPs and at what stage. The same study reported that contractors have been slow to adopt SWMPs and concluded that over 51% were still unaware of their legal obligations. On the other hand, NetRegs (2008) stated that a quarter of organisations that had used SWMPs believed the Regulations had helped them to win new contracts. They suggest 'having environmental credentials could be the route for construction businesses to differentiate themselves from competitors and win new business in a shrinking market'. Osmani (2011) endorses this view. He suggests that clients are starting to promote waste minimisation as they seek higher levels of corporate social responsibility engagement.

How the research was undertaken

A 'triangulated' method of data collection was employed to capture contractors' views on the limitations and missed of SWMP Regulations 2008. A comprehensive literature review was undertaken with a view to gather reliable, insightful information to bring clarity to the waste predicament in the construction industry, whilst also examining current site waste management practices as well as exploring the legislative requirements of SWMPs. The literature review findings informed the design of an in-depth questionnaire, which was distributed to the top 100 contractors rated by turnover and profit in the United Kingdom. This was followed by semi-structured interviews with ten contractors. The data collected by each method were then analysed, and individual findings confirmed by cross-checking against data from other sources.

Research results

Both the questionnaire survey and follow-up generated valuable insights into contractors' SWMP implementation practices and challenges and captured their views on possible amendments to the Regulations. Of the 100 contractor organisations invited to take part in the survey and sent survey documents, 53 completed questionnaires were received, representing a 53% response rate. The participating contractors' quantitative and qualitative responses emanating from the questionnaire survey and the ten follow-up interviews are discussed in the following.

Construction waste origins

Respondents were asked to rate on a scale of 1 to 5 the key on-site construction origins (1 = not a waste origin, 5 = major waste origin); their responses are summarised in Figure 11.1 Respondents rated 'design changes during construction

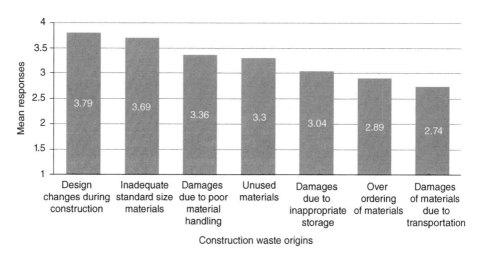

Figure 11.1   Construction waste origins.

stage' and 'inadequate standard size material' as the most significant on-site waste origins with mean value of 3.79 and 3.69, respectively. This would suggest that design and specification of materials indirectly impact on on-site waste generation. Similarly, damaged materials and materials delivered to the site but not used have been rated as significant on-site waste generated with mean value ranging from 3.36 to 2.74.

Waste production and potential waste minimisation across projects' life cycle

Respondents were asked to rate on a scale of 1 to 5 which phases in a construction project tend to produce most waste (1 = no waste generation, 5 = major waste generation). They were also asked to rate which project stages have the highest potential for waste minimisation (1 = no potential, 5 = major potential). The responses can be seen in Figure 11.2, which indicate that the design phase has the greatest potential for waste minimisation (mean value of 4.49); however, the construction stage was opined by respondents (mean value of 4.04) to be the phase that produces the largest amount of waste.

Respondents were asked to rate on a scale of 1 to 5 the challenges that are currently affecting the implementation of SWMPs (1 = not a challenge, 5 = major challenge). Their responses are summarised in Figure 11.3. The majority of respondents saw 'lack of designers' involvement' (mean value of 4.21) as a major impediment to in the planning and implementation of SWMPs. This was equally followed by 'no client's responsibility allocation'; 'difficulty estimating expected waste production'; and 'limited on-site space for waste segregation'.

Respondents were also asked to rate on a scale of 1 to 5 amendments that could potentially improve the implementation of SWMPs (1 = no potential, 5 = major potential). Their responses are summarised in Figure 11.4 which show a strong call from participating contractors to make SWMPs as a legislative requirement during

Section III

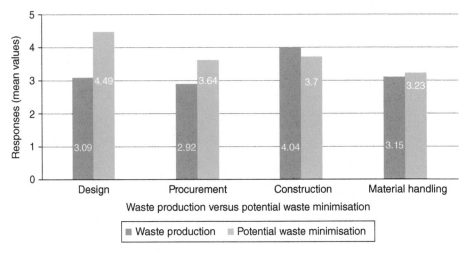

Figure 11.2   Waste production and potential waste minimisation across project stages.

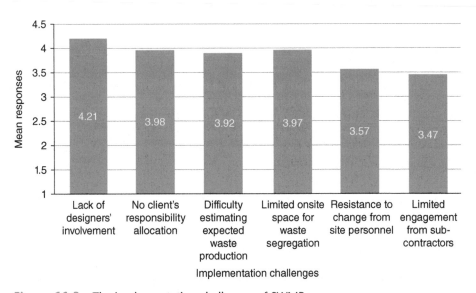

Figure 11.3   The implementation challenges of SWMPs.

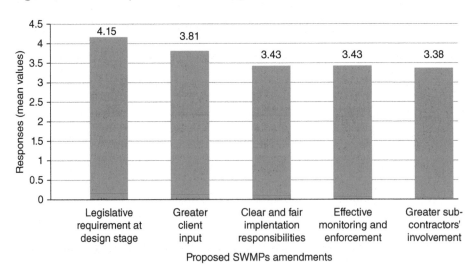

Figure 11.4   Potential amendments to SWMPs.

the design stage (4.15) and asked clients to be fully engaged in the implementation of SWMPs (3.81). They also opined that clearer responsibilities; effective enforcement and enforcement processes; and greater subcontractors' input should equally be included in the future SWMP amendments.

The interviewees who argued that designers have a pivotal role to play in curbing waste generation reiterated this view pointing out that designers are not legislatively required to do so. They went further to explain 'when producing SWMPs before construction starts, it is too late to implement some measures as these were missed during the design stage'. When probed for an explanation of the designers' lack of involvement, there were a number of reasons expressed. Typical comments were as follows: there are no client's requirements for waste reduction during design, designers do not see the cost element of wastage, and the design team are not obligated to comply with SWMPs. As such, there was a consensus amongst the interviewees to include the design team in future amendments of the SWMP Regulations. Some interviewees opined that this could be incorporated into the CDM Regulations since designers are already responsible for the health, safety and environment of their design and could practically be revised to include waste reduction. Furthermore, increased training could help in developing the design teams' awareness of SWMPs and improving communication between all parties in relation to designing out waste issues.

The interviewees reported that by and large clients are not aware of their responsibilities and have no clear understanding of what was required of them. As such, they usually assume that SWMP implementation is the sole responsibility of the contractor. That said, they acknowledged that client's involvement, particularly public bodies, has increased in their recent projects. In terms of potential amendments of SWMPs, it was emphasised by all interviewees that the client should be more involved in the design stage and a meeting should be incorporated early on with all the major stakeholders with SWMPs on the agenda, therefore increasing the clients' awareness. They went on to argue that major flaw with the current legislation was the limited guidance and training, which were not properly implemented before the legislation enforcement, therefore causing a lack of awareness amongst certain stakeholders.

The interviewees recognised that skip space layout should be designed into the project for waste segregation; however, they argued that the ease in which this could be done was dependent on the geographical location and the size of the project. When probed about possible amendments to SWMPs to improve on-site waste segregation, all interviewees suggested the employment of external waste contractors for the segregation of waste, since they are far more stringent with their segregation process, thus increasing efficiency of the SWMPs. However, one of the respondents felt that the use of external waste contractors reduced the main contractors' control in terms of assurance that all the waste was actually getting streamed.

Discussion

The research found that there was a lack of awareness throughout the industry concerning the workings and features within proprietary software tools such as SMARTWaste and the WRAP Net Waste Tools. Most construction contractors preferred to develop their own software tools for waste management based on commonly available applications software such as Microsoft Excel. There were few technical challenges in the production and monitoring of SWMPs. The main difficulties centred on the challenges of setting waste targets prior to the commencement on-site.

Sometimes, a major problem was a lack of space on-site for waste segregation and an increased amount of time and cost involved in the implementation of SWMPs. Subcontractors are limited to their involvement with SWMPs; however, they should be given part ownership in on-site meetings where feedback is given to them.

Many consider that the greatest impacts on waste production are design changes during the construction process and inadequate standard-sized materials. The design phase has the greatest potential for waste minimisation:

- Allegedly architects at present have low to no responsibility for waste management but should potentially have a significant to high responsibility.
- Neglected design team involvement is seen as the predominant challenge for SWMP implementation.
- It appears that the client delegates responsibility for SWMP implementation.
- There is a difficulty in estimating the expected quantities of waste for the requirements of SWMPs.

Key challenges associated with implementing SWMPs

Our research identified several recurring themes with respect to improving waste reduction through SWMPs:

- Non-involvement of the design team in SWMP implementation is having significant impacts on the amount of waste contractors can reduce.
- Clients are not always fully engaged in SWMP implementation process.
- The need to formulate SWMPs in the planning stage of a project. This ensures the engagement of the client in the process, as well as increasing the possibilities of waste minimisation through design.
- The initial estimation of waste quantities in the formation of the plan was also seen as a major challenge due to a lack of best practice dissemination and of benchmarks throughout the industry.

Waste is not only created through the construction phase of a project but can be expelled throughout all phases and therefore involves many stakeholder responsibilities. SWMPs should encompass all of the stakeholders who contribute to the production of construction waste and the effectiveness of the plan. Involvement of the design team is critical and has a significant impact on the levels of waste produced on-site. Osmani et al. (2008) consider that architects have a primary role in the reduction of construction waste. Similarly, clients are frequently not directly engaged with the implementation of SWMPs. Client commitment is essential if waste is to be minimised.

The research supported the views of other researchers: the design phase has a significant impact on the waste produced in a construction project, and the current UK waste legislation neglects to enforce designers to comply within the regulations.

Throughout literature, there has been a general consensus that there needs to be a focus on waste reduction in the design phase in a construction project, as this is a major origin and cause of construction waste. Legislation is lacking to impose responsibilities on architects to minimise waste, and as established throughout the literature, it is by far easier to reduce waste at the onset of a project through design, rather than implementing waste minimisation measures later on during construction.

In December 2013 after consultation with construction contractors, private business, and Health and Safety Officers, Local Authority representatives, and construction clients the UK government repealed the Site Waste Management Plan Regulations (2008). They concluded that the consultation had shown that, 'opinions on the proposal to repeal were quite mixed with an even split between those in favour and those against. A slight majority indicated that they were in agreement with the impacts identified in the impact assessment that had been produced, and significantly, respondents were unable to provide sufficient evidence that there would any additional impacts and no evidence that impacts would be very different to those expected by Government. The majority of respondents also indicated that they would continue to use SWMPs or a similar process if the Regulations were repealed, which gives more substance to that expectation'. (DEFRA, 2013)

Key points

- Current waste management practices have had a substantial effect on minimising waste sent to landfill by incorporating the philosophy of the waste hierarchy into company policy and thus increasing the levels of recycling and reuse.
- Architects and designers need to be more involved at the start of a construction project to reduce the waste, which is the main goal of the waste hierarchy.
- SWMPs are only the start for reducing waste in construction; in order to achieve the maximum waste minimisation for a construction project, the materials resource efficiency needs to be focused upon at the onset of the design process.
- In order to maximise the benefits of SWMPs, more attention needs to be given to incorporating waste reduction in the preconstruction programme and contractors required to include details of their strategies for waste minimisation and confirmation of their SWMPs in the tender documentation.
- Although the UK government has now repealed the Site Waste Management Plan regulations (2008) the industry representatives and clients believe that, irrespective of legislation, waste management plans have an important role in reducing construction waste and they would ocntinue to use them.

Section III

Chapter 12
Planning for Safety, Health and Environment

Alastair Gibb

ECI Royal Academy of Engineering Professor of Complex Project Management, School of Civil and Building Engineering, Loughborough University, Loughborough, UK

Background

The safety, health and environment (SHE) management model provides a framework for setting policies, standards and procedures. It considers the status of the project, develops goals and provides a basis for planning which takes regard of the risks inherent in the proposed construction methods. It extends beyond the assessment of risk into the development of a safe system of working. Having established the necessary controls, it develops procedures for organising, implementing and measuring health and safety performance. The occupational SHE model has been produced by the SHE task force of the European Construction Institute (ECI). It follows international standards such as EN ISO 14001:2004 and EN ISO 9001:2000 and relevant national guidance. In this chapter, we focus on the planning aspect of the model. Full details of the framework are provided in the ECI guidebook (ECI, 2013).

SHE management model: An overview

The SHE management model (SHE-MM) provides a goal-setting framework, within which the project management organisation (PMO) can develop detailed policies, standards and procedures and incorporate them within their system. This can be applied from the start to the finish of the project. In addition to a framework, this guide provides exemplars of key procedures and documentation. The main sections of the model are:

Status review: Establishing the current status of the project and its management systems
Developing: Developing the policies, plans and procedures to achieve the desired objectives

---

A Handbook for Construction Planning and Scheduling, First Edition. Andrew Baldwin and David Bordoli.
© 2014 John Wiley & Sons, Ltd. Published 2014 by John Wiley & Sons, Ltd.

Section III

Goal setting: Setting objectives and measurable performance indicators
Planning: Planning for SHE management
Organising: Organising people and resources to accomplish the objectives
Implementing: Implementing policies, plans and procedures
Measuring: Measuring, auditing and reviewing of performance and outcomes

The *status review* section establishes the current status of the project and its management systems. Each project stakeholder will have their own management systems; however, it is essential that a project-specific management system is agreed by all parties to ensure effectiveness and consistency. Both the initial status review and periodic status reviews (PSR) are required to develop and maintain management systems that are suitable for the project.

The *developing* section covers the policies, plans and procedures to achieve the desired objectives. A project policy, endorsed by the project stakeholders, must be developed from the initial status review and the corporate policies of the stakeholders. This policy describes the overall aims that have been identified and the commitment to a continual improvement SHE performance approach.

The *goal setting* section establishes objectives and measurable performance indicators. Some of these objectives and indicators will be leading or proactive, and others will be lagging or reactive.

Planning for SHE management builds upon the initial status review and continues to reflect the changing hazard and risk profile of all activities as updated through the PSR. It is a live and fully updated reference and outline plan for the management of all activities appropriate to planning and management. The section covers hazard/ risk assessment and control, the SHE plan, assessing and managing constriction risk, constructability reviews, method statements, job safety analyses, environment and emergency preparedness.

Organising is a key activity building on the initial status review, particularly the resource implications and the identified required arrangements to accomplish the project objectives. The policy and performance indicators are assigned organisa- tionally. This organisation activity also facilitates the application of the SHE plan principle of right information, right people and right time.

Implementing policies, plans and procedures requires the realisation of the objectives and plan within the project organisation, with appropriate and adequate control. The PMO should execute the SHE plan and related contingency plans: arrange audits and PSR; motivate all employees; assess, select and control contractors and subcont- ractors; provide competent and effective management and supervision; prepare and maintain sufficient documentation; and communicate, coordinate and consult.

The *measuring* section covers the measurement, audit and review of performance and outcomes against the previously agreed performance indicators. Proactive meas- urement and audit is recommended in addition to a formal investigation procedure to respond to significant events during the project.

The SHE-MM provides a framework, within which the PMO can develop detailed policies, standards and procedures and incorporate them within their system. In order to achieve this, this chapter states objectives that should be achieved, rather than pre- scribing detailed measures to be adopted. In this way, it provides a goal-setting format that can be applied right at the start and throughout the life cycle of the project.

The main sections of the ECI's SHE-MM are illustrated in Figure 12.1.

Section III

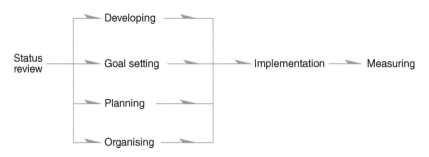

Figure 12.1   The SHE-MM.

Phase	Activity	SHE process
Definition phase	Comply with SHE legislation standards / specs & codes	SHE policy
Engineering plan	Determine risks / hazards constructability reviews	Hazard and risk register
Project team activity	Establish risk / hazard mitigation measurements	
Project approvals	Detailed engneering review / design solutions	HAZOP / HAZCON
		SHE accountabilities and responsibilities
Implementation phase	Management / client review	SHE tasks and roles
Field activity plan	Establish objectives and resources	Risk awareness programme
Build SHE culture	SHE training and education	Field specific she plans
Site mobilisation	SHE programme	
Field activity phase	Monitor and control	SHE programme

Figure 12.2   An example of a PMO SHE management process showing activity phases.

Figure 12.2 provides an example of a PMO SHE management process showing activity phases.

Planning

Planning is an integral part of all business activities and objectives defining the full scope of work, from project conception through detailed design, scope definition, planning, construction and execution to final commissioning and handover to the eventual client or end user.

Planning builds upon the initial status review of safety health and environment and continues to reflect the changing hazard and risk profile of all activities as updated through the PSR. It is therefore a live and fully updated reference and outline plan for the management of all activities appropriate to planning and management.

We now consider hazard/risk identification and control, developing the SHE plan, coordination and consultation.

Hazard/risk identification and control

Hazard identification, risk assessment and control are based upon the principle that all hazards must be understood and suitably controlled.

The following essential steps must be carried out to successfully manage a hazard:

- Hazard identification
- Risk assessment
- Control measures

The main principles embodied in this process should have been applied during the design phase for the identification and elimination of hazards during the period of construction, commissioning, use and maintenance of the project.

The identification of risks in terms of hazards and potential effects and the subsequent risk control is a line management responsibility and key element in the SHE-MS and concerns all positions, functions and levels within the entire PMO organisation.

Risk control measures

Establishment of the level of risk requires that the identified hazards shall be:

- Assessed and evaluated in terms of potential effects and probability (qualitatively or quantitatively)
- Evaluated for existing control measures to their efficiency
- Related to the appropriate acceptance criteria
- Demonstrated that the risk is within the performance standards and acceptance criteria

If the assessment identifies that the risk does not meet the relevant acceptance criteria then, additional control measures are required to be taken. The priority for the implementation of risk control measures is as follows[1]:

- Avoid, eliminate and design out the hazard at source
- Reduce severity and/or probability of potential effects
- Adapt activities and design to the individual and environment
- Allow for current technology
- Give collective measures priority over individual measures

Risk control measures are to be integrated including technical, organisational, human and external aspects.

It is essential that the process is taken beyond the assessment of risk and developed into a safe system of work. A safe system of work is a formal process involving a systematic examination of a task to identify all the hazards. It goes on to define safe methods of working so that hazards are eliminated or risks minimised.

Various techniques for the systematic identification of hazard, risk assessment and control are available for use within the PMO organisation. A typical approach (endorsed by the IChemE[2] and CIA[3]) is detailed in Table 12.1. This is a six-stage

[1] Based on the principals of prevention from the EC Framework Directive – 89/391/EEC.
[2] IChemE – UK Institution of Chemical Engineers.
[3] CIA – UK Chemical Industries Association.

Section III

Section III

Table 12.1 SHE six-stage approach.

Stage	Description	PMO activities			Responsibility	
		Design reviews	Safety documents	Safety calculations	PMO	Client
One	Conceptual	*Initial review*	Safety scope of work (hazard identification)		Primary	Attend reviews Review documents
Two	Basic engineering	Preliminary hazard analysis Plot plan Review *Constructability review*	Project quality plan Environmental management plan Safety plan (design) Process basis of design Basis of systems Design process safety (overpressure protection) specification	Relief and blowdown calculations		
Three	Detailed design	SIL analysis HAZOP study Design Model Reviews Process Hazard Analysis	Noise control specification Fire protection specification Fire and gas detection specification Hazardous area classification Health and safety plan (construction)	Radiation, dispersion and noise modelling		
Four	Construction/ Design Verification	Pre-Startup Safety Review	Operating procedures Construction safety audit Construction quality programme forms STAR construction project turnover package			
Five	Pre-commissioning Safety Review	As per client requirements			Attend reviews	Primary
Six	Project closeout/ Post-Start-Up Review	As per client requirements				

Items in italics are not specifically related to SHE and address a wider range of design issues.
SIL, Safety Integrity Level; STAR, Stop–Think–Act–Review.

Table 12.2 Design hazard identification and risk assessment.

	Hazard	Aspects of hazard	Mitigation/actions	Controls
1	Access/egress	To specific work areas onto site	Provide safe means of access to all work areas at all times Designate routes and methods for access, traffic management plan Identify key routes and methods in preconstruction SHE information, SHE file/manual (including any planning restrictions)	
		Interfaces with plant/traffic	Specify arrangements in tender documentation for warning staff at locations where they leave one site area and enter another site area with differing operational rules and/or constraints Specify design arrangements in initial proposals/plans for enabling plant/traffic access for maintenance activities	
2	Asbestos	Contaminated land Demolition Refurbishment of structures	Where asbestos is likely, carry out survey to determine nature and location of asbestos Inform statutory authorities Avoid or minimise disturbance to asbestos Provide information on asbestos in preconstruction SHE information Ensure disposal to a registered landfill site using licensed waste carrier with sealed containers Provide information on asbestos and residual risks for maintenance and demolition in SHE file/manual	
3	Biological/ chemical health hazards	Anthrax Asbestos Leptospirosis Isocyanates Solvents, etc.	Liaise with appropriate authorities Provide information on uncommon risks in preconstruction SHE information Provide information on uncommon residual risks in SHE file/manual Where possible, avoid disturbance Otherwise, minimise time and personnel spent working in the environment Ensure appropriate disposal	
4	Cleaning >2 m	Access/egress Safe system of work Working at height	Design structures to enable safe access and egress to areas requiring cleaning Design structures so that there will be a safe system of work for cleaning operations Where possible, design structures so that permanent edge protection exists for cleaning purposes Provide information on access/egress, safe system of work and residual risks for cleaning in SHE file/manual	
5	Confined spaces working	Asphyxiation, explosion, flooding, heat, humidity in: Existing confined Spaces Confined spaces to be constructed/ maintained	Wherever possible, eliminate confined spaces from design Reduce the need for entry into confined spaces during construction and maintenance Provide information on residual risks in preconstruction SHE information, SHE file/manual Consider control measures, method statement, risk assessment, permit to work, tag system	

Section III

(Continued)

Table 12.2 (Continued)

	Hazard	Aspects of hazard	Mitigation/actions	Controls
6	Contaminated land	Contaminated ground	Examine ground investigation reports to determine nature and location of contamination	
		Ground gas	Check local authority and client records to determine nature and location of contamination	
			Avoid or minimise disturbance to contaminated ground	
			Provide information on contamination in preconstruction SHE information, SHE file/manual	
			Ensure disposal to an appropriate landfill site, using appropriate licensed waste carrier with sealed containers	
7	Demolition and site clearance	Existing structures	Structural survey and study of existing construction details, including fragile roofing	
			Consider stability of partially demolished structures, for example, arches when planning work sequences	
			Plan and phase work to minimise effect on public (spread of dust, traffic routes)	
			Provide information on structure in preconstruction SHE information	
			Specify site fencing and security around demolition area in preconstruction SHE information	
			Ensure fencing does not block traffic routes, for example, maintenance traffic access, third party access to their land	
		New structures	Design structures to facilitate a practical sequence for demolition works	
			Provide information for subsequent demolition of structures in design statements, SHE file	
			Ensure demolition does not interfere with third parties, for example, owner, general public	
8	Earthworks	Ground movements	Examine ground investigation reports for design	
			Minimise extent of earthworks	
			Consider effect of earthworks on stability of existing structures and foundations – review by applicable design discipline where appropriate	
			Provide information in documentation to allow contractor to design all necessary temporary works, monitoring systems, etc.	
		Contaminated ground, ground gas	Refer to contaminated land	
9	Excavation	Collapse	Examine ground investigation reports for design	
		Falls into	Minimise extent of deep excavation	
		Ground movements	Consider effect of excavation on stability of existing structures and foundations – review by applicable design discipline where appropriate	
			Provide information in documentation to allow contractor to design all necessary support works, monitoring systems, etc.	
			Consider early installation of site roads/paving/ground conditions	
		Areas prone to flooding	Liaise with authorities to predict possibilities of flooding	
			Provide information in documentation to allow contractor consider safety in case of flooding	
			Provide information on residual risk in preconstruction SHE information and tender documents	
		Contaminated ground, ground gas	Refer to contaminated land	

10	Fire/explosion	General construction Confined spaces and tunnels	Design structures using materials that are non- or less combustible and do not produce noxious fumes
			Avoid specifying or designing in the need for site welding, cutting or hot-working methods where reasonably practicable
			Specify the use of enclosed plant and/or fire suppressant systems where appropriate
			Specify compliance with relevant fire legislation (e.g. in the United Kingdom – the Fire Precautions (Workplace) (Amendment) Regulations 1999)
			Provide information on control of major hazards sites within influencing distance of the construction works in preconstruction SHE information and tender documents
			Refer to demolition and site clearance
11	Fragile surfaces/ glazing	Demolition of existing structures Structures to be constructed	Avoid the use of fragile surfaces
			Design structures to enable safe access/egress and a safe system of work to maintain or replace glazing units at height
			Where possible, design structures so that permanent edge protection exists to allow maintenance or replacement of glazing units at height
			Provide information on access/egress, safe system of work and residual risks for maintaining/ replacing glazing units in SHE file/manual
			Refer to cleaning above-ground level
12	Lead	Demolition of existing structures Grit blasting of lead-based paints	Specify methods to minimise generation and spread of dust
			Provide information on residual risks in preconstruction SHE information
13	Mechanical lifting operations	Impacts	Design works to minimise possible interaction with existing structures, highways, railways and services – review affected interface where appropriate
		Loads	Design works to minimise number of lifting operations
			Design structural elements to include lifting points
			Provide information on casting and handling requirements in documentation

(Continued)

Section III

Table 12.2 (*Continued*)

	Hazard	Aspects of hazard	Mitigation/actions	Controls
14	Maintenance	Access/egress Working environment, for example, working space	Incorporate maintenance requirements into design Consider the use of high-durability, low-maintenance materials where appropriate Consider access and egress for maintenance, safe systems of work, permit to work, isolations Allow sufficient working space for maintenance, including proximity to overhead live equipment Provide information on residual risks for maintenance in SHE file/manual and maintenance manuals	
15	Manual handling	Weight Size Availability of secure handholds Hazardous surfaces (rough/slippery) Unstable Confined spaces Difficult travel routes (e.g. corners, ramps, stairs) Temperature and humidity extremes	Where possible, specify systems which promote mechanical handling and minimise manual handling Where manual handling is necessary, select and/or design materials and components which meet requirements of current legislation Ensure unit weights and sizes of materials (e.g. cement bags, building blocks, kerbs) are reduced to acceptable levels where manual handling is unavoidable. Specify easily achievable tolerances where possible	
16	Noise and vibration	General construction	Where possible, specify construction techniques/equipment that produce reduced levels of noise, for example, silent piling, plant with lower noise emissions, muffling of jackhammers and compactors Specify in preconstruction SHE information limits in accordance with legislative requirements	
17	Pressure testing, hydrostatic/pneumatic/vacuum	Commissioning Possibility of overload Possibility of sudden release	Plan for steam blows and venting to be carried out outside work periods whenever practicable Design foundations/structures to support the combined weight of the equipment and test medium Where possible, specify pressure test systems to maximise separation between active test area and other activities Specify nature and location of security fencing where critical Design structures so that effect of subsequent maintenance work is minimised Where appropriate, assess risks to site personnel and public, evacuation and emergency services access; include in preconstruction SHE information and tender documents	

18	Public safety		Phase construction works to minimise disruption to public (pedestrians and traffic) and to reduce conflict between adjacent construction sites
			Design construction works to maximise separation between active work area and public
			Specify nature and location of security fencing where critical
			Design structures so that the effect of subsequent maintenance work on public is minimised
			Where appropriate, assess risks to public, evacuation and emergency services access; include in preconstruction SHE information
19	Services	Overhead services	Implement colour coding for service/product lines of electricity, water, steam, gas, hydrocarbons, telecom cable, etc.
		Underground services	Agree positions of services with client/utility providers at design stage, arrange disconnection if possible
			Specify the use of cable location device to locate underground services prior to and during construction
			Avoid service diversions and the need to work in the vicinity of services where possible
			Identify requirements and/or constraints of utility provider and incorporate measures for safe construction, maintenance and demolition into design
			Provide information on services and residual construction risk in preconstruction SHE information
			Provide information on services and residual risks for maintenance and demolition in SHE file/manual
			Always treat utility as gas, hydrocarbons, steam or electricity until otherwise proven
			Liaise with client/service provider
			Remove all personnel from the area and secure the area
			Warn occupants of potentially affected buildings
20	Site plant/ traffic	Emergencies, collisions, damage	Secure sufficiently large site to allow for safe movement of plant with segregation of pedestrians
			Phase the construction works to reduce the plant/personnel interface
			Design works to maximise separation between operational plant and personnel
			Where appropriate, review by owner interface
21	Substances hazardous to health	Carcinogenic diseases	Identify substances hazardous to health during the design and control so far as reasonably practicable using the following hierarchy
		Respiratory injuries	Eliminate the hazard, by choosing another method
		Skin diseases	
			Avoid specification of known carcinogenic materials and substances where possible.
			Where no alternative, ensure that adequate information is available and provided in preconstruction SHE information
			Substitute with a less hazardous alternative
			Reduce the exposure to the work group
			Issue personnel protection equipment as a last resort

Section III

(Continued)

Table 12.2 (Continued)

	Hazard	Aspects of hazard	Mitigation/actions	Controls
22	Temporary stability of structures	Existing structures affected by works	Consider effect of construction works on stability of existing structures – review by applicable design discipline where suitable	
			Provide information in documentation to allow contractor to design all necessary support works, monitoring systems, etc.	
		Existing structures to be demolished	Refer to demolition and site clearance	
		New structures to be constructed	Design structures for stability at all stages of construction	
			Design structures for prefabrication of elements	
			Ensure contractor has sufficient information to produce suitable temporary works support system in preconstruction SHE information	
		New structures to be maintained	Ensure owner has sufficient information to design alterations to structures in design statements, SHE file/manual	
		New structures to be demolished	Refer to demolition and site clearance	
23	Unexploded ordinance	Bombs	Obtain and inspect records held by local authorities and other sources	
		Shells	Survey sites to locate any recorded suspected ordinance	
		Cartridges	Prior to main construction works, employ specialist contractors to clear site where needed	
			Provide information and warning in preconstruction SHE information	
24	Working at height	Falls of persons from height	Design works to maximise prefabrication at ground level	
			Design works to minimise time working at height	
		Falls of materials from height	Design works to enable safe access and egress	
			Design structures so that permanent stairs, walkways and edge protection are constructed as early as possible	
25	Working over or adjacent to water	Falls of persons into water	Design works to minimise time spent working over or adjacent to water	
			Design works to enable safe access and egress	
			Design structures so that permanent stairs, walkways and edge protection are constructed as early as possible	
26	Working with or near to electricity	Services	Refer to services	
		Installation of electrical supplies	Design so that live work is never necessary	

approach. Each stage verifies that the actions of the previous stages have been carried out and signed off and that the relevant SHE issues have been identified and are being addressed in a timely manner. The various stages and components can be adopted or modified as required to meet client and PMO requirements and take into account the contractor scope of work. The specific approach for a particular project should be defined in the project SHE plan (design). A typical list of hazards and risks can be found in Table 12.2. This shows hazards and the relevant mitigation actions. Control measures are to be added to suit the specific project.

Developing the SHE plan

Planning is essential to achieve project SHE targets. To be effective, the SHE plan needs line management and supervision commitment from all parties involved, with the advice and support of SHE personnel.

It is essential that designers and the coordinators gather SHE information. This will enable the plan to be drawn up at the early stages of the project prior to construction work commencing.

The plan should set out the specific arrangements, resources, responsibilities and measures to be implemented.

The plan incorporates system of allocation of responsibilities and accountabilities for the carrying out the policy and the arrangements as to how these are to be executed.

An effective plan should set out what is to be achieved and in what timescale and should include but not be limited to:

- Project description and scope
- A list of SHE responsibilities and accountabilities for managing the works by everyone involved in the project
- A register of hazards and risks
- Preventative and precautionary measures
- Arrangements for managing the work
- Procedures for monitoring compliance
- Requirements for emergency preparedness
- Training of staff to carry out their responsibilities for the works so as to achieve the stated targets
- Procedures for monitoring and assessing the implementation of these requirements

A typical SHE plan is developed under the following headings:

- Project management system structure
- Policies and significant authorities and references
- Project scope and outline definition
- Roles and responsibilities
- Key interfaces and coordination with all project stakeholders
- Applicable standards, procedures and legislative references
- Key training and competency requirements
- Support and technical resource
- Hazard inventory and risk profile
- Promotion and communication and meetings
- Reporting and measurement

Table 12.3 Typical contents of a SHE plan.

Item	Description
Plan structure	Critical areas of the plan should include the following key items as a minimum.
Project management system structure	The plan will outline the main structure of the management system; its relationship with other company/organisation arrangements (e.g. the quality plan, project execution plan); its relationship to and with other key stakeholders management arrangements, in particular the system as implemented by the client organisation; and finally its relationship to and with the application of key international management systems and arrangements.
Project scope and outline definition	Clearly specified within this section shall be all aspects of the project, the details of all applicable parties and responsibilities of those companies/organisations contracted to undertake work, critical timelines, geographic/location references, pricing structure, prime function of project, contractual relationships.
Roles and responsibilities	The organisation and roles and responsibilities of all parties and project personnel shall be identified with particular emphasis on the achievement and daily management of the aims and objectives of the project. The plan will specify not only the legal requirements of all parties but also those requirements as specified by the client and how these are to be achieved, the contractual responsibilities of all individuals in the employ of the organisation/project, the expectations to be applied by all contractors and suppliers and the disciplinary arrangements in the event of any failure to comply with the roles and responsibilities defined. Also specified shall be the core requirements of each and every person employed on, for and through the project from conception to final completion and handover.
Key interfaces and coordination with all project stakeholders	Following on from the key roles and responsibilities section shall be the outline details of all organisations connected with the project at all stages of the project and the arrangements by which all interfaces shall be effectively managed. This will include arrangements for risk communication and control, period reporting, activity planning and execution etc. Critical also to this section shall be the arrangements for liaison with all external stakeholders including local communities, private organisations, enforcement authorities, key action groups, local and other governmental authorities and ongoing communication and liaison with the general public.
Applicable standards, procedures and legislative references	A summary and detail as necessary shall be made of all key controls in the execution of all project activities, including critical risk management arrangements through design, construction and commissioning activities. Reference shall be made to the application as necessary of all client minimum expectations, national and local legislative control and authority plus those details of all standards, procedures, instructions, etc., as developed and implemented by the project.

Table 12.3   *(Continued)*

Item	Description
Key training and competency requirements	A full and detailed inventory of all minimum training and competency requirements shall be specified and which shall be standard and applicable across all organisations including the arrangements for project induction, management and supervisory training, technical-based training, specific risk control and mitigation training and the training of all personnel and organisations in the requirements of the overall plan, the policies and minimum expectations and any particular, significant and key standards and other controls as identified for the control of any specified site activity. The behavioural expectations and training plan as appropriate shall also be detailed. The minimum competencies of all disciplines and key personnel shall be specified with a robust and ongoing assessment process defined. All training arrangements in the planning and execution of all emergency arrangements and controls shall be specified.
Support and technical resource	Full details of the organisation in place to provide competent information, support and counsel shall be specified and which shall make reference to the minimum competency expectations of this resource. This will include and make reference to project resource at a local level, the corporate resource available plus the expectations in terms of competent resource within other key stakeholders and project parties, in particular the support available to other contractors and suppliers. How these interfaces are managed shall also be specified. Detailed also shall be the arrangements by which personnel can obtain technical data through electronic or other media applications
Hazard inventory and risk profile	A full and detailed profile shall be available identifying and recording all key and significant hazards and risk mitigation and links to training and key reference and control criteria. Arrangements for updating the inventory shall also be specified and communicated.
Promotion and communication and meetings	The arrangements by which issues are identified, communicated and briefed shall be clearly detailed. This will include the types of media used, the ongoing briefing process, links to key activities and organisations and the role that all individuals and organisations are expected to play and participate in the promotion and communication of issues shall be specified The regular and ongoing meeting and interface management arrangements shall be specified in a matrix. This will include the key attendees and stakeholders in each meeting, the aims and expectations, key agenda issues, the management and action tracking of key meeting outcomes and the cascade process for the escalation of all significant meeting issues and subsequent follow-up and debriefing arrangements.
Reporting and measurement	The regular and minimum reporting requirements shall be specified across all project and activities and organisations including a full suite of minimum criteria including all relevant client, project and other organisation key performance indicators leading and trailing metrics, key scheduling for all reports and the dissemination criteria of all relevant data with reference to the client and project document control arrangements.

Section III

Table 12.3 (*Continued*)

Item	Description
Audit and review	The regular scheduling of reviews and audits shall be specified in a project matrix showing key interface and liaison requirements and interfaces across other stakeholders and organisations The key competency requirements of all personnel involved in the audit and review process shall be detailed. Also included will be the reporting arrangements and tracking process for all audits/reviews undertaken. Critical to the audit and review process shall be its alignment to the risk profile and significant activities of the project
Contractor and supplier management	The minimum expectations selection criteria for subcontractor/supplier assessment both precontract award and post-contract award shall be specified. Details shall be provided of the ongoing monitoring arrangements of the performance of these organisations and the contractual disciplinary arrangements for failure to meet the expectations of the client and the arrangements as specified by the project through the plan.

- Audit and review
- Contractor and supplier management

An example is provided in Table 12.3.

The SHE plan must not be limited to safety issues alone, but should include occupational health and environmental considerations. Further guidance is available from the ECI.

Programme for occupational health

This programme should include regular monitoring; reduction of exposure through control measures; selecting, providing and maintaining suitable PPE and supervising its use; preemployment and periodic medical examinations of exposed workers; and training and education of personnel.

It is not intended that legal requirements and internal company processes and procedures are incorporated into the SHE plan – these should be referenced to specific organisations' documentation.

The scope and detail of the plan depend on the complexity and risks of the project as defined in the initial status review. It may range from a simple statement from the PMO or contractor and their lower tier subcontractors identifying who is responsible for implementing and coordinating SHE matters to more detailed documents covering legislative and contractual responsibilities of the worksite.

The SHE plan is not necessarily contained in a single document. Relevant sections of the plan must be incorporated in all the contractual documents through all tiers of contractors, subcontractors and suppliers and must be available on the construction site before the works begin.

Right info, right people, right time

Each project employer and self-employed person has to comply with the agreements/measures of the plan. All managers, supervisors and workers should be made aware of the plan and the role that they play in its implementation. Careful consideration must be given to ensuring that the right people receive the right information at the right time.

Note: UK Legislative requirements mention just one SHE plan that encloses the entire cycle starting from the design phase through the delivery. However, many countries have two parts of the SHE plan,[4] one for the design phase and one for the construction phase. If the choice is made to develop separate SHE plans (engineering/design and construction), there must be a clear relationship between both plans, to maintain a consistent build of the hazard/risk-reducing measures during the design and also during construction.

As the project progresses, the plan should be reviewed and modified according to experience and any information received through, for example, PSR.

The following techniques[5] can be used to develop the plan such that it is relevant to each particular section of the project and to ensure that it remains current:

- Construction risk assessments
- Constructability reviews
- Method statements
- Job safety analyses
- Environment analyses

Construction risk assessments

Following action by designers to eliminate and reduce risks, the residual construction hazards should be clarified. These could be identified by reviewing information in the project documents, existing drawings, site survey investigations, design stage and constructability reviews or could emerge during the construction or commissioning phases.

Once potential hazards are identified, the risks to SHE must be assessed, and the contractor(s) as part of the working procedures should develop a safe system of work.

Constructability reviews

Constructability reviews consider the 'big picture' and form part of the iterative process to refine the design through construction expertise. Whilst these reviews consider project-wide issues, they form an integral part of developing

[4]   92/57 EEC Minimum regulation concerning health and safety for temporary and mobile construction sites, 24 June 1992.
[5]   This is not an exhaustive list.

the SHE plan and should be considered alongside other issues such as logistics and coordination.

Method statements

Detailed method statements should be prepared by the relevant contractor(s) for all work that has risks to SHE,[6] as determined by the project risk assessment. These method statements should be reviewed by the PMO.

The method statements will describe the work procedures, sequence of operation and the controls necessary to achieve effective management of risk and efficient execution of the works. This requirement applies equally to demolition, refurbishment and maintenance activities as to new build.

A detailed job method statement could include:

- Details of the job to be undertaken.
- The individual activities required to complete the job.
- The sequence of the various activities in the execution of the work.
- Plant, equipment and tools to be used in each activity.
- Substances/chemicals to be used in each activity.
- Hazards from existing buildings and facilities.
- Control measures and procedures to be used for each activity.
- Personal protective equipment required.
- Emergency resources required, for example, fire-watchers, fire extinguishers and fire blankets.
- Name of supervisor for each activity.
- Name of person in overall change of the job.
- Novel procedures may require demonstration or training exercises.

Job safety analysis (JSA)

A job safety analysis (JSA) is effectively dynamic risk assessment. This is a procedure used 'day to day' to review job methods by the supervisors and/or workers at the work face. It aims to uncover hazards that may have been overlooked in the layout or design of the equipment, tools, processes or work areas that may have developed after work started or may have resulted from changes in work procedures or personnel.

One simple method to achieve this is STAR an acronym for STOP, THINK, ACT and REVIEW. STAR is one of several safety techniques that emphasise the need to stop and review safety before commencing any new work and before beginning to carry out an existing procedure. STAR demands action at both an individual level and a team level to ensure that it is safe to proceed. A conservative

---

[6] Detailed job method statements should always be prepared for all SHE critical tasks such as novel techniques, piling, heavy lifts, general rigging and lifting, vessel entries, live line welding (hot taps), excavations, grating and handrail installation/removal, heat treatment/stress relieving and any erection work with specific risks.

approach is necessary, particularly when there are operational pressures to proceed. 'To develop and reinforce this culture employees should be praised if they stop work or do not approve modifications because there is reasonable doubt about safety implications' (p. 7) (INSAG, 2002). For further information on similar techniques and non-technical safety skills see the book website.

Environment

A detailed review of construction activities and their impact on the environment should be undertaken prior to the start of construction operations during the constructability review.

Environmental issues to be covered would include noise, spillage, waste materials, emissions and effects of earthwork, on ground water, soil erosion and drainage.

Operations which create the risk of spillage (e.g. vessel opening, removal of pipe work) require investigation before the event. Prior to such operations, the PMO in conjunction with the client should assess the risk of such occurrences. A full-risk assessment should be conducted and as a result, the necessary precautions established, that is, prescribed personal protective equipment and containment measures such as provision and placement of absorbent matting/materials and methods of disposal. This information will be included in the detailed job method statement.

All waste material designated as toxic or hazardous must be notified to the PMO. A licensed subcontractor must be used to transport and dispose of material to a licensed waste disposal site. Hazardous waste such as contaminated materials must be packaged and labelled in accordance with current legislation and good practice. Disposal of material must be done in accordance with approved method statements and comply with environmental protection legislation. The disposal contractor will be provided with a full description of the waste material.

Additional guidance is provided through the ECI environmental guide.

Emergency preparedness

Emergency preparedness or contingency development that concerns the arrangements will be instigated to reduce the consequences of accidents and environmental violations, which are related to the several activities. These arrangements include:

- Emergency management
- The control mechanism for effective handling of accidents and environmental violations
- Emergency response
- The actions to be taken to limit the consequences of accidents and environmental violations
- Response training (e.g. training in first aid, fire fighting, evacuation, management of environmental clean-up)
- Emergency drills
- Actual testing of the effectiveness of the emergency management, emergency response and response training

Section III

Key points

- The SHE-MM provides the construction team with policy, standards and proce-dures whereby all involved in the construction process may identify and address safety, health and environmental issues related to the work to be undertaken.
- The model comprises seven phases or sections.
- In this text we have concentrated on the planning phase whereby hazards may be identified and controlled. This provides a basis for all organisations that will be involved in construction to further develop the plans for safety, health and the environment within their own organisational work plans and schedules.

Section IV
Delay and Forensic Analysis

Introduction

It seems to be a fact of life that construction and engineering projects finish late. Not only high-profile and complex projects that are subject to reports in the national and specialist press but also small domestic projects – everyone has a story to tell about how their project was delayed, the reasons for delays, and the resulting problems. Despite industry efforts to improve the situation, experience from the United Kingdom shows that over the last 10 years, little has changed with only about 45% of projects (the combined time for design and construction phases) finishing on time. In 2012 the figure fell to 34%, the lowest level since 2000 when the data was first collected.[1]

This section of the book details the practicalities of the analysis of delay and disruption. We look at categories and types of delay and the concepts of prospective and retrospective analysis. Delay and disruption can only be analysed by comparison with the as-planned schedule. The information required to analyse delays is discussed including as-built and progress records. We look at how the as-planned schedule should be reviewed as and when events and progress impact

[1] UK Industry Performance Report 2012. Constructing Excellence (October 2012)

A Handbook for Construction Planning and Scheduling, First Edition. Andrew Baldwin and David Bordoli.
© 2014 John Wiley & Sons, Ltd. Published 2014 by John Wiley & Sons, Ltd.

the original intentions of the contractor. The following types of analysis are considered: As-Planned compared with As-Built, Impacted As-Planned, Collapsed As-Built, Time Impact Analysis and Windows Analysis. We consider two common protocols for delay analysis: the Society of Construction Law (SCL) Delay and Disruption Protocol (SCL,2002) and the American Association of Cost Engineering International (AACEI) Recommended Practice for Forensic Schedule Analysis, AACE International, 2008). The section is not a commentary on the law or a discussion based on the interpretation of contracts. We leave that to lawyers. We do refer to general principles of the law; for more details on legal aspects of delay analysis, two texts are recommended: *Keating on Construction Contracts* (Furst et al., 2011) and *Delay and Disruption in Construction Contracts* (Pickavance, 2010). The section gives guidance on the basis and practice for forensic delay analysis primarily, but not exclusively, from the construction contractor's perspective.

Chapter 13
Delays

Delay and disruption: Definitions

Delay and disruption are terms that are often linked almost as if they are one and the same thing. However, they are two different phenomena, and most will have some understanding of what a delay is even though there is not a precise definition in relation to construction and engineering projects.

As a verb, delay means:

- To make (someone or something) late or slow: *the completion of the project was delayed by the poor weather*.
- To be late or slow; to loiter: *the contractor delayed starting the excavation no longer*.
- To postpone or defer: *they decided to delay the excavation until the weather improved*.

As a noun, delay means:

- A period of time by which something is late or postponed: *a 5-week delay; long delays in erecting the formwork*.
- Something that results in late, slow or postponed action: *the delay event was the lack of steel fixers*.

To make sense of a delay, it has to be compared to something. The most obvious reference is the as-planned schedule, what was intended before the any delay occurred or some specific date. For instance:

- A delay to the contract completion date
- A delay to the planned start of 'brickwork'
- A delay to the expected progress of the installation of ductwork

Section IV

A Handbook for Construction Planning and Scheduling, First Edition. Andrew Baldwin and David Bordoli.
© 2014 John Wiley & Sons, Ltd. Published 2014 by John Wiley & Sons, Ltd.

Disruption in terms of construction is more readily defined (Furst et al. 2012):

> "Disruption occurs where there is disturbance of the contractor's regular and economic progress and/or delay to a non-critical activity even though, on occasion, there is no or only a small ultimate delay in completion."

Again, disruption is a comparative measure related to the planned productivity of the affected parts of the works and in relation to the cost of carrying out the work.

Delay and disruption can happen at any time in the project cycle; this section deals with delay and disruption in the construction phase. Most construction contracts require the assessment of the effects of delaying events to be made as soon as it is practicable after the delaying event has occurred and a forecast of the likely effect of the delaying event on the progress and completion of the project. Some contracts require the contractor to estimate the effects of a delaying event, for instance, a proposed change to the scope of the works, ahead of the change being made, thus assisting the employer to decide whether to invoke the change or not. For one reason or another, assessments and agreements of the effects of delaying events might not have been made during the project. When the project is eventually completed late, the parties are likely to dispute the responsibility for the overruns and any additional costs. At this stage, a forensic analysis looking back over the project to find the causes of delay to the project and the responsibility for the delays will be required. Estimates of the effects of disruption can be made prior to the disruption materialising, but generally, the effects of disruption can only be ascertained after the effects and costs are known.

Delays

Virtually all construction contracts contain extension of time clauses; these allow the contract completion date to be re-fixed following changes or delays caused by the employer. Without such clauses, the employer would not ordinarily be able to change the scope of the contract, for instance, to add extra facilities, without a further contractual agreement for an increased time to completion or 'at large' – the contractor only having to complete the works in a reasonable time. Extension of time clauses also allow the liquidated damages clauses for non-completion by the contract date to be maintained. Again, without extension of time clauses, the employer would have to prove general damages should the contractor run over time.

The contractor also benefits from the extension of time provisions. Without such clauses, the contractor would be in risk of breach of contract if it did not complete on time and could be liable to pay general damages to the employer. Extension of time clauses in conjunction with the liquidated damages provisions means the contractor always knows the liability for running late and means the contract completion date can be extended if the employer is in default, for instance, by not providing information on time.

Categories of delay

The progress and/or completion of a project may be delayed by events not taken account of in the original schedule. Delaying events can generally be divided into three categories:

- Those for which the contractor is responsible – from the point of view of the contractor, these are often described as **culpable** or **inexcusable** delays and are usually non-compensable.
- Those for which the employer is responsible – from the point of view of the contractor, these are often described as **excusable** delays and are usually compensable.
- Those for which neither the contractor nor employer is responsible – these are often described as **neutral** delays and are excusable but non-compensable.

The list of events in each category will be specific to the particular contract, and in most cases, the contractor will not be granted an extension of time for culpable delays and may become liable to pay the employer liquidated damages. These events usually include such causes as progress being slower than planned, late delivery of materials, putting right incorrect work and, where the contractor is responsible for producing the design, for the late provision of information.

Where the project is completed late because of excusable delays, the contractor will usually be granted an extension of time, will not have to pay liquidated damages to the employer and is likely to be able to claim the direct costs and losses resulting from the delay to completion. Examples of excusable delays usually include events such as late possession of the site, employer-inspired changes and variations, and late information from the design team when the employer is responsible for producing the design.

When the project is delayed by neutral events, the risks and cost are split; the contractor will usually be granted an extension of time and will not have to pay the employer liquidated damages, but the employer will not be able to claim from the employer the direct losses and expense costs resulting from the delay. Neutral events can include causes such as exceptionally adverse or infrequent weather, delays by local authorities and statutory bodies and as a result of 'force majeure'.

Types of delay

In addition to the three categories of delay (culpable, excusable and neutral), there are different types of delay. These are broad groupings, and each type of delay is modelled in a different way when using critical path analysis:

Date

This type of delay is where an activity cannot start and/or finish until a specific date irrespective of when preceding activities were planned or were to be carried out. It is modelled by the addition of a milestone with a constrained date (Figure 13.1).

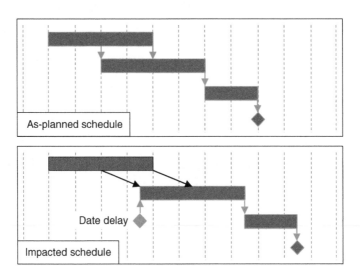

Figure 13.1   Date delay modelling.

Examples of this type of delay are:

- The delivery of plant or material scheduled for a specific date without which the activity cannot proceed
- The start of an activity determined by the availability of labour or a specialist subcontractor who is unable to start (perhaps because of previous commitments) until a specified date
- The release of information without which the activity cannot proceed
- A delay to the start of the works

Total

This type of delay is where a complete stoppage of work is caused. It is modelled by changing the calendar for the relevant activities. The new calendar will include additional 'holidays' to represent the affected periods (Figure 13.2).

Examples of this type of delay are:

- Strikes and lockouts
- Postponement of the works
- The inability to temporarily gain access or egress from the site
- The effects of weather not catered for in the original schedule

Extended

This type of delay is where the as-planned duration of an activity is increased. It is modelled by increasing the duration of the relevant activities (Figure 13.3).

Examples of this type of delay are:

- An increase in the work content of an activity as a result of a variation increasing the amount of work to be carried out

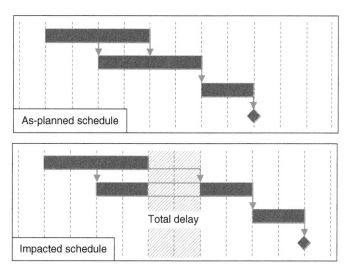

Figure 13.2   Total delay modelling.

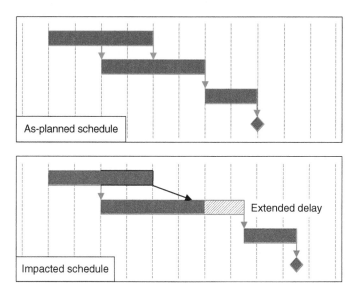

Figure 13.3   Extended delay modelling.

- A change in the circumstances in which the work is to be carried out resulting in lower productivity for operatives and/or plant – for instance, additional noise restrictions, working in confined spaces and working at a different time of year
- Restrictions in the supply of labour, plant or materials that result in reduced overall output or intermittent working

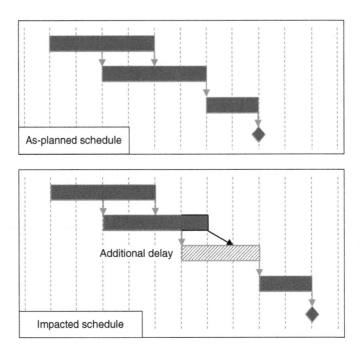

Figure 13.4   Additional delay modelling.

Additional

There may be occasions where it is relevant to insert additional activities to the schedule. To model this type of delay, it is necessary to add additional activities and link them into the as-planned network. A judgement has to be made with regard to the duration of the new activity taking into account work in progress, material deliveries, resources available, construction logic and so on (Figure 13.4).

There are two general reasons for this:

- To add activities for new or additional work that, subsequent to the production of the as-planned network, has been incorporated into the project.
- When the effect of extending an existing activity as a result of an increase in similar work is not appropriate. For instance, it might be more relevant to add additional activities to reflect the additional or delayed part of the original activity, especially when the original activity is nearing completion or is already complete.

Progress

At any point during a project's life, the progress of each activity can be assessed and the resulting data compared to its planned progress. The progress achieved may have an effect on the critical path and may change the path, the overall planned project duration or part of the project. This type of delay is modelled by the introduction of progress to the network. This is a function built into virtually all critical path analysis software and allows updating and reanalysis of the network to take account of the progress achieved to date (Figure 13.5).

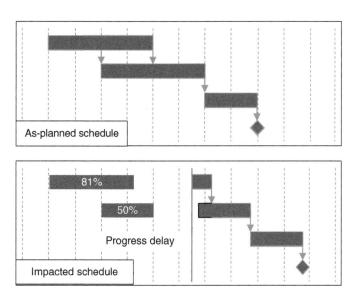

Figure 13.5   Progress delay modelling.

Generally, if the actual progress is less than planned, especially with regard to activities on the critical path or those with minimal float, the planned project completion is likely to be delayed and the overall project period increased. Similarly, if actual progress is greater than planned, especially with regard to activities on or near the critical path, the planned project completion is likely to be advanced and the overall period reduced.

Progress delay modelling is the most effective way of modelling the contractor's performance as it will reflect, if accurately assessed, culpable delays such as reduction in output due to poor management, over-optimistic schedules and reworking that often the contractor is unwilling to acknowledge. Lower productivity and hence reduced progress can result from the effects of disruption so the cause of lower productivity should always be examined and not merely accepted as being a culpable delay on the part of the contractor. When progress is greater than planned, this may reflect measures taken to accelerate the works and mitigate delays (or potential delays), enthusiastic and effective management or an indication that the schedule was overly pessimistic. Again, where productivity is unexpectedly greater than planned, the cause of the apparent increase in productivity should be examined and an attempt made to isolate the cause.

Sequence

A delaying event might affect the planned sequence of activities envisaged on the as-planned schedule. This is different to a change in approach or change mind of the preferred sequence; neither of those is strictly delaying events; rather, they are carrying out the works out of sequence. Whilst this type of delay is uncommon, it will generally result in activities that were originally planned to be carried out concurrently and to be carried out sequentially. The method of modelling this type of delay is to adjust the logic of the schedule to take account of the revised sequence (Figure 13.6).

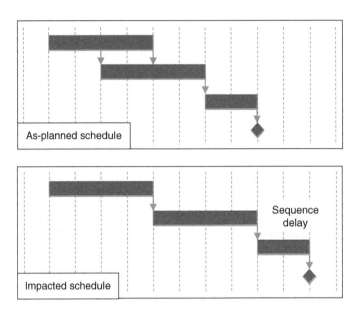

Figure 13.6   Sequence delay modelling.

The primary reason for this is:

- Changes in specification of materials, techniques or legislation that result in activities no longer being able to be carried out concurrently

Fragnets

The modelling of delay types illustrated previously shows a simple single insertion into the as-planned schedule. In some cases, the delay is a little more complex, and the understanding of the delaying event would be better illustrated by more scheduled information. In this case, a 'fragnet'[1] is produced. For instance, a 'date delay' representing the release date of some vital information can be modelled by a single date-constrained milestone representing when the information was made available or it can be expanded showing the sequence of activities leading to the release of the information (Figure 13.7).

Prospective versus retrospective delay and other concepts

Virtually all contracts require the contractor to give notice to the employer when it becomes apparent that the progress of the works is being delayed and that there is likely to be a delay to the completion of the project. The precise terms of

---

[1] A fragment network – an activity or short sequence of activities broken down into a subnetwork showing greater detail.

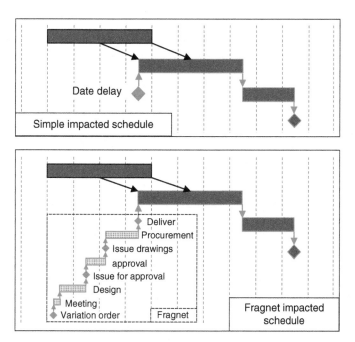

Figure 13.7   Fragnet modelling.

the extension of time provisions will be specific to the particular contract but generally include some requirements regarding the timing of the notice and what it is to contain. Some contracts have 'condition precedent' clauses which require the contractor to give notice within a specific time of the delay event happening or becoming apparent and in a specific manner. If those conditions are not met, then the contractor may be denied an extension of time and/or associated loss and expense.

When a claim for delay is made during the project in line with the requirements of the contract, it is the likely effects of the delaying event on the completion date that have to be forecast – the future completion date calculated. Forecasting ahead in this manner is known as a **prospective analysis,** and as the future is unknown, it can only be a theoretical calculation. Some contract conditions allow the estimate to be reviewed when the project is complete, but in most cases, such a review only allows an increase in the extension of time, not a reduction to that already granted. Other contracts do not allow the extension to be reviewed, and the initial estimation of the completion date is fixed no matter what eventually transpires. Whatever the contract conditions say, there is a conflict between the requirements of the contractor who wants to maximise the extension of time granted and the employer who wants to minimise the extension of time.

For whatever reason, upon completion of the project, there may be disputes between the contractor and the employer as to the extent of the extension of time awarded, as either a review of what was initially granted or potential conflicts in what each party considers to be a reasonable extension granted, not granted, or even some administrative issues which prevented either the contractor giving notice of delays or the employer properly exercising his obligations in granting an extension of time. When the project is complete, there is an opportunity to consider what happened on the project and take this into account when granting or adjusting and extension of time. Looking back in this manner is known as **retrospective analysis**.

It is possible to use prospective analysis methods when the project is complete, the analyst putting himself into the position of the project when the delaying event occurred. In carrying out a prospective analysis retrospectively, the analyst has two choices about how to treat information about the project that is now known as the project has now been completed. Using **blindsight**, the analyst takes into account information only known at the time the delaying event occurred. Using **hindsight**, the analyst takes into account information that is now apparent after the completion of the project, for instance, how long variation works actually took to complete.

Whichever approach is adopted, prospective or retrospective, if the analysis is carried out correctly and thoroughly, the outcome will be the same. The argument as to which approach is the better is generally based on the preference of the analyst. Where there have been criticisms of the approach adopted by analysts, it is almost always that they have not carried out the analysis correctly; it is not the method that is at fault but the manner in which it was undertaken.

Most methods of analysis depend in some way upon critical path analysis. It is delays to activities that are on or near the critical path, particularly those with no or minimal float to absorb delays, that will result in a delay to the completion of the project. Even though a schedule might not be logically linked or analysed to identify the critical path (the schedule may, for instance exit as a simple bar chart), there will be a critical path through the project whose activities, when delayed, will result in a delay to the completion of the project. Analysts who are experienced in the type of construction process under investigation and are knowledgeable about planning and programming techniques *may* be able to deduce where the critical path is likely to be, but the only accurate way is to calculate the critical path using a networked schedule. This was a feature of the *Mirant v Ove Arup* (2007) case:

> "[Mr H] was employed by [A] as the Consortium Project Director between January 1996 and July 1999. He wrote many of the contemporaneous documents which chart the progress (and the problems) of the movement of the boiler foundations and of the remedial works. He was, in the course of this Project, as was entirely proper, astute to protect [A's] interests. He warned [Mr S] very early of the consequences of delay in undertaking the remedial works to the boiler foundations and made [A's] approach as to how the problems should be tackled equally clear. Although the programming experts have demonstrated that his understanding that the critical path always goes through the boiler is incorrect, he was also an impressive and helpful witness."

> "The critical path can be defined as 'the sequence of activities through a project network from start to finish, the sum of whose durations determines the overall project duration'. See BS.6079 – 2.2000 Part 2, 2.41."

"In the helpful work, Delay and Disruption Contracts by Keith Pickavance (2005), the author makes the point that the Critical Path Method requires detailed and sophisticated analysis and that in complex projects it is unlikely that a critical path can be identified inductively, i.e. by assertion. 'It can only reliably be deduced from the mathematical sum of the durations on the contractor's schedule to be completed in sequence before the completion date can be achieved.' This is an important cautionary word in this case where a number of witnesses were convinced, without the benefit of any such analysis, that they knew where the critical path lay."

Key points

- Delay and disruption are two different phenomena. Delay is linked to time, and disruption occurs when there is a disturbance to a contractor's regular and economic progress.
- To make sense of delay, it has to be compared with something. The most obvious reference is the as-planned schedule. Disruption is a compatible measure related to productivity.
- Delaying events can generally be divided into three categories: those for which the contractor is responsible (culpable or inexcusable), those for which the employer is responsible (excusable) and those which are neutral (excusable but non-compensable).
- A contractor will usually be granted an extension of time for excusable delays and may become liable to pay the employer liquidated damages for culpable delays.
- There are different types of delays: date, total, extended, additional, progress, and sequence delays.
- Analysis of delays may be undertaken prospectively (looking forward from when the delaying event occurred) or retrospectively (looking backwards from the completion of the project).
- Virtually all contracts require the contractor to give notice to the employer when it becomes apparent that the progress of the works is being delayed.
- Most contracts require the contractor to submit at the start of the project a schedule for the work. (This may be called 'the contract schedule', 'the master schedule' or 'the accepted schedule'.) This is the as-planned schedule/schedule, the basis for comparison.
- The as-planned schedule is integral to determining and modelling the effect of changes, relevant events and the like.
- Experienced analysts may be able to deduce where the critical path is likely to be, but UK law courts have determined that the only accurate way is to calculate the critical path using a networked schedule.

Chapter 14
Factual Information

The As-Planned schedule

At the beginning of a project, one of the contractor's key tasks is to produce a schedule for the project. This is often an updated version of the schedule originally produced during the tender period. Most contracts require the submission of a schedule, called the contract, master schedule or accepted schedule, at the start of the project which will show how the contractor intends to carry out the works. The contract documentation will generally specify the format of the schedule and other parameters such as the software to be used, maximum duration of activities and what is to be displayed on the schedule (float, contingency, resources and so on). When considering delay analysis, this schedule is known as the as-planned schedule or baseline schedule.

All but the simplest of schedules are produced using software that uses critical path analysis to calculate the timing and sequence of activities and the overall planned duration of the project (see Chapter 4). In the vast majority of cases, the schedule, in its hard copy form, is reproduced as a bar chart, but other formats such as time chainage diagrams and line of balance charts (see Chapter 3) may be adopted.

The detail and accuracy of the as-planned schedule will be dependent on the information available for its production, the time available for production and the skill and experience of the planner putting the schedule together. Unless the design and detail of the project is complete, it is rarely possible to produce a schedule that is totally accurate from beginning to end. In most cases, the initial design will be complete but sections of the work later in the project period may not yet be fully developed. There may, for example, be provisional sums to consider or the construction schedule may require input from specialist subcontractors. The CIOB, in its *Guide to Good Practice in the Management of Time in Complex Projects* (CIOB, 2011), recommends that 'Schedule Density', the detail of activities and resources shown, varies according to the completeness of the information available. Typically, full definition (high density) applies to the immediate period of, say, 3 months' work with reducing detail at medium (4–9 months) and low density (9 months plus) which is expanded progressively as the detailed information becomes available (see Chapter 4).

The manner in which the as-planned schedule is put together, how the detail is shown and the links between activities will determine how useful it is in acting as a proactive management tool. It must represent the scope of the works and the preferred and realistic timing and sequence of carrying out the activities that comprise the project. Contemplation with respect to its use in the analysis of delays is generally not one of the considerations uppermost in the mind of those responsible for its compilation. In most cases, the emphasis is on the sequence and timing of the works. (The schedule is often developed in a relatively short period of time, without full information and working to deadlines for submission.)

Using critical path analysis and project management software allows the as-planned schedule to be used in a dynamic nature which can forecast the effects of actual progress on the sequence and timing of future activities and allow forward planning of future work. Similarly, the effect of any delays can also be determined, and by simulating various scenarios, strategies for mitigating delay can be investigated, thus giving the opportunity to take corrective action. Being able to make decisions about the future sequence and timing of the works at an early stage is likely to reduce the likelihood of delays manifesting themselves and/or their effect, thereby negating or reducing the costs of recovering delays if analysis of the impact is left until the delay actually materialises.

The as-planned schedule is also integral to determining and modelling the effects of changes, relevant events and the like. During the currency of a project, and if necessary retrospectively on completion of the project, the as-planned schedule is fundamental for preparing claims for disruption and extensions of time because it allows the planner to demonstrate the effects of events on the likely completion of the works.

However, if the schedule is not constructed correctly, whilst it may *appear* to reflect the preferred sequence and timing of the works in the static state, it will not realistically forecast changes when current progress, delays and such like are added to the schedule. In fact, it is possible that an incorrectly developed schedule will provide misleading and inaccurate results that could exacerbate discussions/negotiations in delay situations and/or lead to inappropriate action or changes being made to reduce the effects of delaying events. It is essential therefore that the schedule is formulated so that it can be of maximum benefit to the management of the project.

Correcting the As-Planned schedule

It is not uncommon that the as-planned schedule contains some anomalies and technical errors that could affect the outcome of a delay analysis. Should the as-planned schedule containing errors be used as the as-planned schedule and risk an inaccurate result, or should it be rejected as unsuitable and the analysis be abandoned? The third option exists of correcting or repairing the as-planned schedule for use as the as-planned schedule in the analysis.

Section IV

The preference should always be that the contemporaneous schedule be used where possible and the minimum corrections made to make the schedule functional. The contemporaneous schedule, in its original state, will reflect the contractor's intent and the intended way for carrying out the works. The contractor (and employer) will have made decisions based on the schedule including its inbuilt errors. Correcting or altering the schedule excessively may 'distinguish' some of the problems that resulted in the actual delays and make analysis of the causes impossible. The following are examples of corrections that should be considered as acceptable and necessary. Wherever possible, changes, corrections, additions, and the like should be made jointly between the parties so as to reduce the perceived or actual accusations of manipulation:

Key contract dates

The as-planned schedule should reflect the project start and finish dates and any sectional completion and other key dates in the contract. If any of these dates are incorrect, then works that commence at these dates will be scheduled to start either earlier or later than they should be, and similarly, where work is normally required to be finished by a completion date, it will have been unduly restrained (less time allowed for the work that was available) or already in delay when the works started.

It may not be possible to correct the key dates without affecting the timing of all the activities in the schedule. If this were the case, then the corrected schedule would not reflect what was the (albeit misconceived) intent of the parties at the start of the project. More realistically, the incorrect dates should be maintained and a note made of the inconsistencies. In extreme cases, if the error was not recognised during the project and amendments or revisions made to the schedule, then it is possible that a cause of delay would be the inaccurate scheduling.

Even though the as-planned schedule might have been 'accepted' or 'approved' by the employer, it is unlikely that a tribunal will find that the employer would be culpable for any delay caused. Generally, the accuracy of the schedule remains the responsibility of the contractor.

Missing logic links

Work activities in the schedule are logically linked by dependencies, the most common of which is the 'finish-to-start' (FS) link (see Chapter 4). This dependency link determines that the earlier activity must finish before the following activity can start. If such a link is omitted from the schedule, then any change to the sequence and timing of the earlier activity will not result in a knock-on effect to the following activity. This effect is diagrammatically shown in Figure 14.1.

An activity without a successor dependency (a dependency going out of the activity) has no effect on the sequence and timing of subsequent activities if it is delayed (or if it is ahead of schedule). Similarly, activities without predecessor dependencies (a dependency coming into the activity) cannot be affected by the sequence and timing of preceding activities if they are delayed (or ahead of schedule).

To enable a delay analysis to be carried out using critical path methods, it is necessary to ensure that every activity, except for project start and completion milestones,

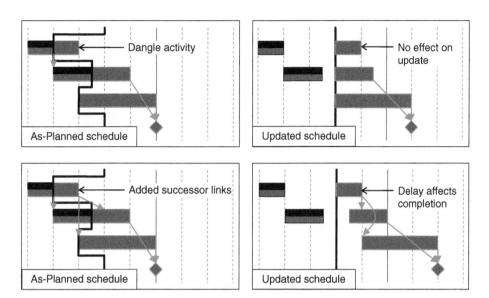

Figure 14.1   (a) Unlinked activity does not affect project completion. (b) Knock-on effect of dependency links.

should have at least one predecessor start link and at least one successor finish link – there should be no dangle activities on the schedule (see Chapter 3). These corrections should be made to the schedule and, where possible, bearing in mind the added logic should reflect the dependencies between the activities. They should be made so as not to affect the critical path of the project in its un-progressed state.

Constraints

Constraints are dates added to an activity that can override their start or finish that would normally arise from its relationship with other activities. The most common constraints are 'Start On' type constraints; these are often used to denote a fixed date when an activity or milestone can start, such as a specific date for starting the project. Similarly, 'Finish On' type constraints are often used to fix activities or milestones dates such as the Contract Completion Date.

Constraint dates are frequently used in the compilation of schedules as a short-cut method of determining the start of an activity or sequence of work. However, if the schedule relies on constrained activities to order and sequence the work rather than on logical dependency linking between the activities, the schedule's effectiveness as a management tool is severely restricted. Where possible and where logical, the constraint dates on activities should be replaced by logic links that reflect the actual dependencies between the activities. In any case, even if a constrained date is justifiable, the activity or milestone should still have at least one start predecessor link and at least one finish successor link.

As the name of the constraint suggests, 'Start On' and 'Finish On' dates are fixed with no flexibility. This can cause problems in schedules when progress is not as planned. Such constraints form fixed points in a schedule which means that if the project gets ahead or behind schedule, those dates do not change and subsequent activities (that are linked to the fixed dates) also do not move and do not dynamically react to the actual progress of the works.

'Finish On' constraints can lead to severe problems if the project falls into delay. In normal circumstances, the schedule would react dynamically, and the schedule activities would move along in time and subsequently indicate a delay to completion. However, if the completion date is fixed by a 'Finish On' constraint, it cannot move and an impasse is caused by preceding activities needing to finish later than the fixed completion date. This results in a condition known as 'negative float' – broadly a measure of how far the schedule activity is behind schedule. This also results in the schedule not reflecting realistic sequence and timing for future work.

In most cases, 'Start On' constraints should be replaced by 'Start On or After' constraints and 'Finish On' constraints by 'Finish On or After' constraints. The substitute constraints act in a similar way to the original constraints but allow some flexibility in the schedule so that it does not remain rigidly fixed. (Further discussion on constraints is included in Chapter 4.)

Activity durations

The duration of an activity is primarily a function of the quantity of work to be undertaken, the resources applied to it, the method used to carry out the work and the conditions under which the work is carried out.

The method statement accompanying the schedule should detail the resources and method to be used to carry out the work. However, in many instances, especially in relation to specialist work, those compiling the schedule may not have carried out these calculations, have used their experience or have used periods quoted by specialist subcontractors to finalise activity durations. In compiling the schedule, the planner might have adopted the 'operation orientated' approach where a time period is allowed for the activity based on broad parameters and experience and the subcontractor undertaking the work is responsible for completing the activity within the scheduled duration using the most appropriate resources and method.

It is not usually possible to say that a duration allowed for an activity is wrong – all durations are made on the presumption of an output or work rate which itself is determined by the resources and method. Changing the duration of an activity, either increasing it or decreasing it when making corrections for the as-planned schedule, is neither reasonable nor necessary. The proof of the accuracy of the activity duration will be determined once the activity is complete. Only then is it possible to show that the activity did not progress at the rate it was planned (it resulted in a delay to the progress of the works) or that it progressed faster than planned (and potentially resulted in producing additional float in the schedule) or that it progressed at the rate shown on the schedule.

Sequence of activities

In most cases, there are a number of ways in which work can be sequenced to achieve the desired outcome although some might not be contractually or legally permitted (especially in relation to health and safety legislation). When examining the as-planned schedule, the sequence of work should not be changed just because there was a better way to do it or because it was not scheduled in the sequence that the analyst would have preferred.

Similarly, it is possible to schedule work in a physically or logically impossible sequence. In all but the most simplest of cases, such as swapping the order of two activities to correct an error (such as concrete being poured before the formwork is fixed), the impossible sequence should be left in the as-planned schedule. The contractor (and employer) would, if the anomaly had not been recognised, have arranged information, subcontractors, materials and so on to match the planned sequence. However, when the work was actually carried out, the impossible sequence would have been recognised and work actually carried out in a feasible solution albeit potentially delayed by the misinformation prior to the organisation of the works. The delays that are manifested will become apparent by the actual out-of-sequence working (the As-Built Schedule; see Chapter 4), the delays to starting activities and the rate at which the activities were completed.

Missing activities

When activities have been missed out of the as-planned schedule, adding them back into the as-planned schedule could extend the overall period required to complete the works (if they are on or near the critical path) and/or increase the number of parallel work activities at a particular time if the missing activities are carried out concurrently with other work. Increasing the amount of parallel work can cause disruption to the works which may delay the completion of the overall works or result in higher costs for carrying out the works.

If the as-planned schedule is corrected and the missing activities are added, the schedule will not show the intent of the parties towards the project at the start of the works and could mask the effects of other delaying events at the start of the project. The correct time to take account of the missing activities is when it became apparent that they had been omitted. It is likely that contemporaneous records would show that amendments and revisions to the schedule were made to include the omissions.

Even if the missed activities were not added to the as-planned schedule, the delaying effect would manifest itself in the progress of other activities running parallel to the missed activities or those that are logically dependent on them. For instance, following activities would appear, from contemporaneous records, to have started later than planned on the schedule. Further investigation would be likely to show that this was because the work of the missed activities had to be completed before the follow-on, delayed activities could start.

Additional activities

It is rare that as-planned schedules contain additional activities. The most common cause of additional activities is that work was omitted from the project prior to works commencing but the as-planned schedule was not revised to reflect the changed scope of works.

Removing activities from the as-planned schedule may reduce the overall period required to complete the works (if they are on or near the critical path) and/or decrease the number of parallel work activities for a particular time period in the schedule.

Section IV

Leaving additional activities in the as-planned schedule might have the effect of masking early delays in the schedule. It is unlikely that those carrying out and managing the works will not become aware of the omitted work, and it is likely that they will just recognise that the schedule is wrong and amend or revise it sometime in the future.

If the additional activities are not recognised early or if the schedule is not amended or revised, the management team might not be able to establish accurate dates or other key data for subsequent work including information requirements and subcontractor and supplier orders. The effect of this is that those subsequent works could be delayed as a result. Therefore, additional activities should be removed from the corrected as-planned schedule. This might have the effect of initially introducing float into the schedule, but contemporaneous progress reports will show if the project management team were able to recognise this or if the subsequent activities were delayed as a result of incorrect pre-planning.

Scope change

It is not uncommon to find that following the tendering for a project and before the contract is let, changes are made to the scope of the project. Often, this is following value engineering exercises or merely through changes to the employer's requirements and/or the design development. In the rush to agree the contract sum, to sign the contract and to start work, it is possible that some of the changes will not have been incorporated into the as-planned schedule at the beginning of the project. Eventually, the scope changes ought to be recognised and the schedule amended accordingly. The necessary changes may however only be implemented further into the project.

Such scope changes that happen before the start of the project or before the contract is made differ from scope changes that happen during the project for which there will usually be a change order and the consequences of such will be taken account of by the normal contract provisions. Scope changes before the start of the project ought to have been taken into account *before* the works started, and because of that, they are different to 'missing activities'. Ultimately, the schedule could be showing one scope of works, whereas it ought to be showing a different scope that the contractor was aware of but had not incorporated into his schedule.

Whilst such a situation should not occur, in practice, they do and the preferred option is to incorporate the scope changes in the as-planned schedule. The overall effect on the scheduled completion date might be neutral, but if the works in question are on or near the critical path, it is more likely the changes will extend or reduce the planned completion date. The contractual effect of this will be dependent on the provisions of the particular contract, and but usually, any adverse effects resulting from something that was agreed prior to the signing of the contract will be the contractor's responsibility. Similarly, any positive effects (for instance, a reduction in the planned period for the works) will also accrue to the contractor. However, as positive effects are likely to produce an increase in float, dependent on the provisions of the contract, the contractor may not be able to take full benefit.

Software

Most construction and engineering businesses have planning and programming procedures that stipulate the type of software that is to be used in the compilation of schedules (Primavera, Powerproject, Microsoft Project, etc.), and the conditions relating to some projects also stipulate the type of software to be used. Most planners, and especially those carrying out retrospective analyses, also have their preferred software. However, *all* software operates in slightly different ways, and it is demonstrable that different results can be obtained using the same data with different software.

Converting the as-planned schedule for use on different software is generally not a good practice and should be resisted as conversion errors may be introduced. Each type of software has differing functions and carries out the analytical calculations in different ways. The contractor (and employer), when using the schedule to manage the project, would have accepted the manner in which the original software operates and have made decisions based on the results it produced. To change to another software package for the delay analysis may introduce discrepancies in analysis that were not apparent during the project, and although the differences might not be significant, it is not usually possible to be definitive about underlying variations in systems and how they could affect the outcome of an analysis.

There are situations when it is acceptable to use different software, for instance, where the original software is no longer available or where the as-planned schedule has had to be totally reconstructed because only a paper copy of the schedule exists.

Bar chart to network

Frequently, the as-planned schedule is only available in hard copy bar chart format or only available in an electronic format such as Portable Document Format (PDF) which does not allow the underlying dependencies to be determined. The bar chart will often just show the sequence and timing of the activities. In other cases, the logic might be shown, critical activities highlighted, activities coded and resourcing details and so on listed on a tabular section of the schedule.

When faced with such a situation, the analyst should either leave the schedule as it is and accept that there will be limited methods for analysing the delays to the projector or alternatively attempt to input the bar chart data into suitable software (perhaps the software that produced the schedule originally if it was not hand-drawn) and add logic from scratch and make other 'repairs' as detailed earlier.

Larger and more complex projects tend to be undertaken by larger organisations, having more rigorous specifications regarding the planning and scheduling of the works and a greater oversight from the employer and its consultants. The result is that very few large and complex projects rely only on simple bar charts with no network back-up.

Therefore, the projects that tend to have only bar charts are usually relatively straightforward and are less complex. In such cases, an analyst who has a background and experience of such projects and an understanding of network techniques should be able to make reasonable deductions as to the logic of the schedule. Assistance from the contractor's team responsible for the project will enhance the accuracy of the logic and whether it reflects the intent of the contractor. In adding logic to a bar chart, the analyst should be mindful of logic that is required to reflect the natural sequence of the works, for instance, the superstructure is dependent on the substructure and foundations being in place (known as 'absolute' or 'hard' logic), and logic that is added to steer the works on a particular course or to take account of other factors such as limited resources and the availability of a specialist piece of equipment that has to move from one location to another (known as 'preferential' or 'soft' logic).

After inputting the bar chart data and adding logic linking, critical path analysis of the reconstructed schedule will sequence the activities according to the logic applied. A comparison of the timing and sequences of the original bar chart and of the activities in the reconstructed network will indicate if the added logic is reflective of what would have been envisaged originally – the sequence and timing should be very similar if not identical. Adjustment of the preferential logic may be needed to ensure the two schedules match. In some cases, this may not be possible (there could be some intrinsic errors in the bar chart that require correction, as in the preceding text), or it may require the addition of constrained dates, the ultimate preferential logic, to fix the sequence and timing or the networked schedule.

The logic additions to the bar chart should be fully documented and, particularly with regard to preferential logic, the reasons for their inclusion stated. Such transparency will reduce accusations of manipulation of the schedule and will confirm the reasonableness of the approach.

Where corrections are made to the original as-planned schedule to form the as-planned schedule for use in the analysis, they should be kept to a minimum where possible so that it reflects the original intent of the contractor.

Ideally, corrections to the as-planned schedule should be made jointly and agreed between the parties but this rarely happens. Therefore, *all* corrections and changes should be comprehensively documented to demonstrate that the changes were necessary and that they have been carried out in a transparent and reasonable manner. The basis of most delay analyses is the as-planned schedule, and it is more appropriate and preferable to use a 'reasonable repaired schedule' than an 'unreasonable contemporaneous schedule'.

The AACEI Recommended Practice for Forensic Schedule Analysis (AACE International, 2011) for baseline (as-planned schedule) selection, validation and rectification says:

1. *Ensure that the baseline schedule is the earliest, conformed plan for the project. If it is not the earliest, conformed plan, be prepared to identify the significant differences and the reasons why the earliest, conformed plan is not being used as the baseline schedule.*
2. *Ensure that the work breakdown and the level of detail are sufficient for the intended analysis.*
3. *Ensure that the data date is set at notice-to-proceed (or earlier) with no progress data for any schedule activity that occurred after the data date.*
4. *Ensure that there is at least one continuous critical path, using the longest path criterion that starts at the earliest occurring schedule activity in the network (start milestone) and ends at the latest occurring schedule activity in the network (finish milestone).*
5. *Ensure that all activities have at least one predecessor, except for the start milestone, and one successor, except for the finish milestone.*
6. *Ensure that the full scope of the project/contract is represented in the schedule.*
7. *Investigate and document the basis of any milestones dates that violate the contract provisions.*

8. *Investigate and document the basis of any other aspect of the schedule that violates the contract provisions.*

9. *Document and provide the basis for each change made to the baseline for purposes of rectification.*

10. *Ensure that the calendars used for schedule calculations reflect actual working day constraints and restrictions actually existing at the time when the baseline schedule was prepared.*

11. *Document and explain the software settings used for the baseline schedule.*

Details that should be recorded are:

- A description of the inadequacy found
- Details of the correction made including activity identifications, resources and links added/removed
- Justification for the correction including a description of the work and activities involved and any supporting evidence (drawings, photographs and so on)
- Details of the effects of the changes on the remainder of the schedule (including the critical path and float)

As-built/progress records

Having produced the schedule, the contractor should use it as a tool for managing the project. A key part of managing the project is knowing how the works are progressing compared to the plan. This entails collecting progress data and analysing it to measure performance against the plan, to determine the project position and the effects of any changes and to forecast the likely completion date of the project (see Chapter 4).

Ideally, progress records should be kept continuously by all staff involved in the project (employers, designers, contractors, subcontractors); however, in the majority of cases, it is the contractor who is responsible for recording the progress, and on all but the most well-run sites, there are no comparable records kept by the employer or, preferably, by some independent third party. Even if one party is made responsible for record keeping, there should be, preferably, some independent verification of the data or agreement with the employer that the data are accurate. It is much better for records be agreed contemporaneously rather than some time later or when the project is complete and a dispute has arisen.

With respect to delay analysis, the most important records are those that record what, when and why things have gone wrong and when things do not proceed according to plan. The problem is that at such times, management tends to become stretched: measures are being employed to rectify things, and routine, mundane tasks, such as record keeping, are let slide for the more urgent task of getting things back on track. However, this is exactly when records are needed most of all and when viewing the project historically the time that they seem most elusive. It is in all parties' interest that accurate records are kept and the onus is initially on the employer and his advisors to ensure there are adequate provisions in the contract documents that stipulate the amount, type and level of detail to be recorded. Whilst the contract documentation may well be prescriptive and provide details of records to be kept and who is responsible for keeping them, it does not follow that those accurate and comprehensive records will in fact be kept in practice. The SCL 'model records clause' (SCL, 2002, pp. 71–72) does not mention any sanctions if the records are not kept and delivered in accordance with the requirements. Employers should be mindful of this and retain some onus for ensuring the contractor (in respect of the SCL clause) complies with record keeping obligations.

The SCL Protocol states, '...*The starting point for any delay analysis is to understand what work was carried out and when it was carried out*'. Therefore, the minimum progress data that ought to be recorded for each activity on the as-planned schedule is the actual start and finish date of each activity and the progress of the activity at the end of each week. The best way for this to be achieved is for the staff on the project to keep detailed daily site diaries. To ensure all data are captured, all site management staff should keep diaries – those that know each section of the work intimately are those that are able to keep the best records. If it is left to one person, say the most senior manager on the project or worse still a visiting planner, then the records are likely to be sparse, incomplete, inaccurate and seldom kept at holiday times or when something important is happening.

The SCL model records clause[1] states that for a project:

The records shall be in a form as agreed between the parties and shall include:

2.2.1   *identification of contractor/subcontractor working and their area of responsibility;*

2.2.2   *operating plant/equipment with hours worked, idle or down for repair;*

2.2.3   *work performed to date giving the location, description and by whom, and reference to the contract schedule;*

2.2.4   *test results and references to specification requirements. Lists deficiencies identified, together with the corrective action;*

2.2.5   *material received with statement as to its acceptability and storage;*

2.2.6   *information or drawings reviewed with reference to the contract specification, by whom, and actions taken;*

2.2.7   *job safety evaluations;*

2.2.8   *progress photographs;*

2.2.9   *a list of instructions given and received and any conflicts in plans and/or specifications;*

2.2.10   *weather conditions encountered;*

2.2.11   *the number of persons working on-site by trade, activity and location;*

2.2.12   *information required from and by the Employer/CA;*

2.2.13   *any delays encountered. (SCL, 2002)*

Whilst all of this information is not absolutely necessary to undertake a delay analysis, such details will provide, in addition to the basic progress data, documentation to assist in determining the cause and liability for delays to the progress of the project. Most of the data listed are produced during the normal course of

[1] This is the '*records clause for medium to high value or medium to highly complex projects*', at page 71 of the SCL Protocol.

Section IV

the management and administration of the project and are often summarised in weekly or monthly reports to the employer and are included in the minutes of various daily, weekly and monthly meetings.

If contemporaneous, verified and agreed progress data are not available, an attempt has to be made to reconstruct these data after the event. The accuracy of the progress data will be dependent upon the source documents used to compile the data.

- Daily site diary entries are likely to be accurate to be within a day as are other data such as material test certificates (cube tests, welding records, piling logs) and time sheet and labour records especially when produced by an 'electronic clocking-in system' that, for instance, depends on fingerprint or similar individualised data for operation.

  Photographs are generally reckoned to have daily accuracy if date stamped and the metadata of the electronic file will show the date and time the photograph was taken provided that the camera date and time was set up correctly initially. Whilst it is possible to change or misrepresent metafile data, comparison with other contemporaneous documents and photographs of known events is likely to expose such errors. Photographs, whilst providing impressive records, can generally only corroborate other evidence – for instance, that an activity has started, not when it did start, or that an activity is complete, not when it was completed. Unless a series of photographs are taken in set positions at regular intervals, the nature of photography tends to be that photographs are only taken when something has happened or when something has changed.

- Some documents tend to be produced or presented on a weekly basis, e.g. weekly progress reports, subcontractor reports, plant records, requests for information, 'daywork' sheets and some site diaries. Some of the data recorded can be interpreted as accurate to within a day other data may have an accuracy of 1–7 days. How these data are recorded and interpreted may affect assessments of progress.

- Monthly meeting minutes and the monthly progress reports that are often appended to them have an accuracy of, say, 1–30 days. Subcontractor and supplier invoices are generally submitted on a monthly basis and therefore have a similar accuracy.

Notwithstanding the accuracy of the data relative to when it was recorded, there are two further issues affecting the accuracy of contemporaneously recorded progress data.

Firstly, whilst it ought to be relatively to accurately assess the percentage progress of an activity given the amount complete divided by the total amount of the activity, there is an almost universal tendency to estimate, or even guess, the percentage progress. Furthermore, most seem only capable of assessing progress in multiples of 5% or 10%. These estimates go on optimistically until 90% completion is achieved when the 'ninety-ninety rule of project schedules' (Bloch, 1977) kicks in:

> "The first 90% of the task takes 90% of the time, and the last 10% takes the other 90%."

Thereafter, there is a slow progression to 100% completion.

Secondly, it is unlikely that two or more persons recording data for the same activities independently will arrive at identical estimations. In addition to variances in estimation, there is likely to be a subconscious (or otherwise) skewing of the data to suit the recording parties' ends. As suggested previously, the contractor will want to maximise the possible extension of time for delays that are not his responsibility and the employer will want to minimise delays that are not the contractor's responsibility.

Whilst the foregoing may suggest that progress records are too unreliable to be of any use in any analysis of delays, this is not the case. Agreement of the records between the parties contemporaneously should attest to their acceptability and accuracy as should validation by an independent third party. Verification and checking progress data against other contemporaneous data will confirm or contest their accuracy.

It is often the case that progress data is incomplete; it is recorded faultlessly for a couple of months, then a month or two will be missed, and then the recording resumes. In these instances, it is usual to interpolate data for the missing periods. Though activity progress tends to follow an 'S-curve', it is generally accepted that a straight line interpolation is sufficiently accurate, especially on short-duration activities. Interpolation should be carried out mathematically, taking account of the project calendar (the working and nonworking days for the activity) and not relying on 5% or 10% estimates.

Where actual start and finish dates are not recorded, examination of other contemporaneous records will have to be examined to determine such dates. The start and finish dates of preceding or succeeding activities will also provide some indication of the activity in question's start and finish dates. Where supporting records do not assist in accurately determining the actual start and finish dates, the usual convention is to assume that the activity starts at the beginning of the week (or month if that is the frequency of the progress report) when the first progress of the activity is recorded. Similarly, when the finish date cannot be accurately determined, the convention is to assume that the finish date is the date of the report at which the activity was first reported to be 100% complete. Although analysing delays using critical path methods and the interpolated or assumed start and finish dates will provide a seemingly accurate 'answer', awareness of the scope of possibilities of the accuracy of the progress data suggests that results should be stated more in terms of 'around' or 'about' rather than some precise figure.

As-built schedule

The as-built schedule illustrates what did happen on the project as opposed to what was planned to happen. In its simplest format, it is a record, usually in bar chart format, of when activities actually started and actually finished. The list of activities on the as-built schedule should be similar to those on the as-planned schedule but will also include additional activities and/or exclude as-planned activities that result from variations, scope changes and so on. The effectiveness of the as-built schedule can be enhanced. By plotting the as-built data alongside/beneath the as-planned data, the variances between the two are immediately apparent.

Although the easiest and most frequent method of compiling, an as-built schedule is to use the as-planned schedule as a guide to the selection of activities and the records, particularly the progress reports and actual start and finish dates, to compile the schedule. If for some reason an as-planned schedule does not exist, it is possible to compile an as-built schedule from first principles provided there are sufficient records available, including as-built drawings and assistance from staff that constructed the project. After the event, when the project is complete, a schedule is developed of what was constructed and the records are scrutinised to determine the actual start and finish dates. Whilst this would appear to be a somewhat daunting task, it is, theoretically, what might be envisaged in the 'collapsed as-built' method of analysis (see Chapter 4).

It is preferable that the as-built schedule is produced contemporaneously as the works proceed. Traditionally, construction progress was indicated by 'marking up' the schedule at the end of each week by drawing a bar on the as-planned schedule parallel to the corresponding activity and showing by this bar when work on the as-planned schedule had been carried out and the actual duration of the activity. These days, more often than not, the as-built progress data (along with actual start and finish dates) are entered into the computer file of the software used to produce the as-planned schedule, and an as-built schedule can then be automatically produced.

Depending on the sophistication of the software (or the diligence of the person marking up the bar chart), the as-built schedule may show only the overall as-built period (from actual start to actual finish) or alternatively show intermittent working if progress was more of a 'stop/start' nature. This is illustrated in Figure 14.2. The software can also capture the progress made on each activity during each reporting period, and this may be added to the marked-up schedule. When this is done, the as-built schedule will also give some idea of the rate of progress as well as the overall period for the works.

Whilst the actual start and actual finish dates are important, what is of equal significance is the 'effective start' and 'substantial completion'. An activity may start but not make any significant progress. (Some managers will attempt to start activities as soon as they are able to ensure some early progress or to provide an indication that work is progressing as fast as possible.) However, progress that dawdles along at 5% for a several months does not significantly contribute to the progress of the project, and an effective start can only be said to have been achieved when progress on an activity reaches around 10% and progresses at something like its planned rate. Similarly, some activities can reach 90% completion, and the remaining 10% drags on for months. A judgement has to be made when the activity is substantially complete. Usually, this is when sufficient work has been completed to allow succeeding activities to start – in some cases, the activity will have to be 100% complete to enable following activities to start.

A further sophistication of the as-built schedule is to develop the schedule from a mere bar chart into a network with the addition of logic linking. On the face of it, this is a straightforward process and is similar to the addition of logic to an as-planned bar chart as described in 'Bar chart to network'. The major difference is that the as-planned schedule is more likely to have been developed with logical sequences in mind, whereas the as-built schedule is likely to exhibit some illogical and out-of-sequence working that often results during the construction of a project.

Section IV

Section IV

As-built schedule

Line	Name	Planned start	Planned duration	Planned Finish	Actual start	Actual Duration	Actual Finish
1	**Steelwork design & procurement**						
2	Finalised data from RE	25 Jan	0w	25 Jan	25 Jan	0d	25 Jan
3	Steelwork design	25 Jan	3w	12 Feb	25 Jan	3w	12 Feb
4	Design approval period	15 Feb	2w	26 Feb	15 Feb	4w	12 Mar
5	Design approval	26 Feb	0w	26 Feb	12 Mar	0d	12 Mar
6	Manufacture & delivery	01 Mar	3w	19 Mar	15 Mar	3w	02 Apr
7	**Site works**						
8	Clear site and RL excavation	01 Feb	3w	19 Feb	01 Feb	4w	26 Feb
9	Exc & FRC bases	22 Feb	4w	19 Mar	01 Mar	4w	02 Apr
10	Steelwork erection	22 Mar	5w	23 Apr	05 Apr	6w	21 May
11	Roof cladding & rooflights	12 Apr	7w	28 May	17 May	7w	02 Jul
12	Wall cladding & loading doors	12 Apr	8w	04 Jun	19 Apr	8w	11 Jun
13	Fill & FRC ground slab	26 Apr	4w	21 May	24 May	4w	18 Jun
14	Services installations	10 May	7w	25 Jun	07 Jun	7w 2d	27 Jul
15	Racking	24 May	3w	11 Jun	07 Jun	3w	25 Jun
16	Signange, fittings & fixtures	14 Jun	1w	18 Jun	28 Jul	1w	03 Aug
17	Handover key date 1	25 Jun	0w	25 Jun	03 Aug	0d	03 Aug

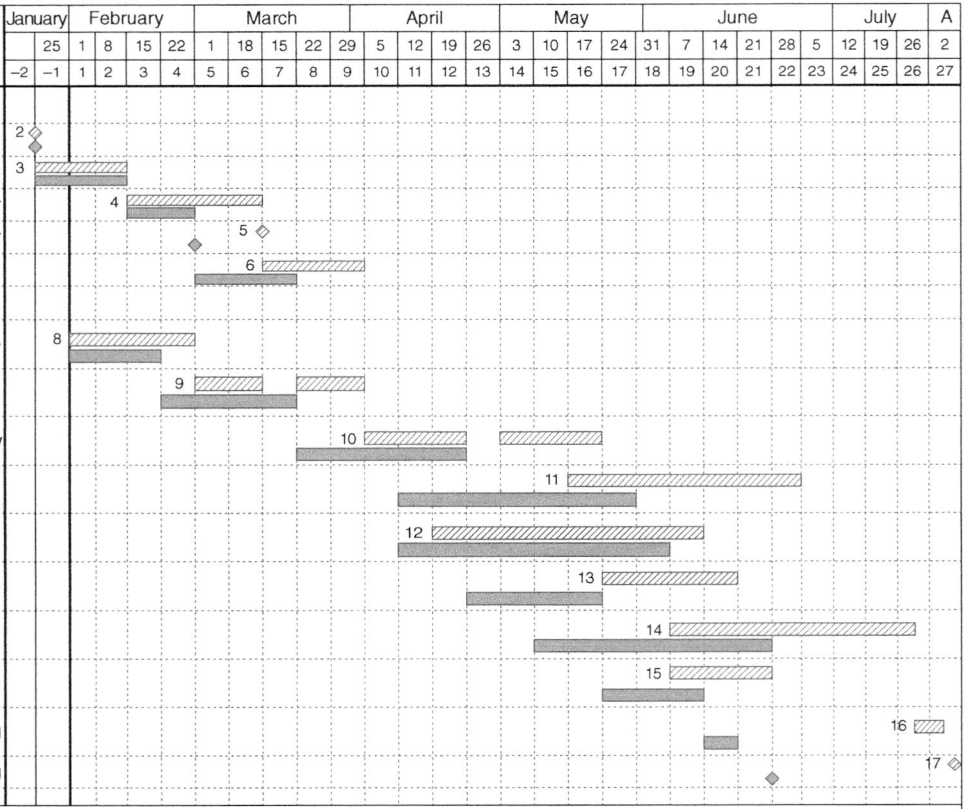

Legend

As-built (Actual)

Baselines

As-planned baseline

Figure 14.2 As-built bar chart schedule.

Furthermore, delays, errors and mishaps will have affected the timing and sequence of operations and these would never have been envisaged when the project was originally planned.

A competent planner will be able to add logic with a selection of leads, lags and constraint dates to any selection and sequence of activities such that, when rescheduled, the activities are 'logically' fixed in the sequence and with the timing extracted from the records. This does not mean the logic inserted is correct and a totally different set of logic linking could also have the same 'correct' effect. In most cases, there is little certainty in the absolute or hard logic, and most added dependencies are likely to be soft or preferential. Finding out what actually happened between two dependent activities will take thorough examination of the project records and assistance from site staff that were present and managed the construction of the project. The amount of work involved in rationalising each logic link in a schedule with thousands of activities is virtually impossible to imagine, and even so, there can be no absolute certainty that the linking accurately reflects reality. Only in very simple projects where most activities can only be carried out sequentially, one after the other, can there be any confidence in the as-built logic – and those types of projects do not usually require sophisticated forensic delay analysis.

Transparent and detailed documentation of the decisions made for each of the added logic links must be made to enable other parties to have any measure of confidence that the critical path has not been purposely or unwittingly manipulated or steered in a specific direction. Networking and critical path analysis is such that if it can be shown that a few links are incorrect, the whole of the schedule can take on a different 'shape' that would move the activity dates and sequences such that the linking is shown to be unreliable.

Key points

- The detail and accuracy of the as-planned schedule will be dependent on the information available at the time of production.
- It may be necessary to correct the as-planned schedule to reflect changes in the original schedule before its use as the as-planned schedule in the delay analysis.
- Accurate as-built, progress records are imperative for comparing how the works are progressing compared with the plan.
- The as-built schedule illustrates what did happen on the project as opposed to what was planned to happen.
- The type of data available and the scheduling techniques adopted will determine the type of analysis that may be undertaken.

Section IV

Chapter 15
Protocols and Methods of Analysis

There are numerous texts and papers written about delay and disruption in the construction industry. Many of these are written by one or two academics or practitioners; the subject and the guidance offered distilled from their experience, view and interpretation of the law. However, there are two publications that have been compiled by learned groups and attempt to provide a consensus, and more balanced and independent view of the subject of delay and disruption and how resolving issues may be approached.

The Society of Construction Law Delay and Disruption Protocol

Following a presentation to the Society of Construction Law (SCL) in the United Kingdom in 2000 about the way a dispute concerning a delay to a project had been resolved and subsequent discussion of the matter, the Society decided to produce a protocol for delay and disruption. A drafting subcommittee was formed by a group of members of the Society, and a draft publication was issued for consultation in November 2001. Following consultation, a protocol was formally published in October 2002 (SCL, 2002). Because of the make-up of the drafting subcommittee and the UK base of the SCL, the Protocol is somewhat UK-centric, and whilst most of the recommendations represent good practice at the time of the publication, the legal principles that support some of the guidance may not be applicable to all jurisdictions.

The first paragraph of the introduction says:

"The object of the Protocol is to provide useful guidance on some of the common issues that arise on construction contracts, where one party wishes to recover from the other an extension of time and/or compensation for the additional time spent and the resources used to complete the project. The purpose of the Protocol is to provide a means by which the parties can resolve these matters and avoid unnecessary disputes."

The Protocol has 21 'core principles relating to delay and compensation' (pp. 5–9) regarding:

1. Schedule and records,
2. Purpose and extension of time,
3. Entitlement to extension of time,
4. Procedure for granting extension of time,
5. Effect of delay,
6. Incremental review of extension of time,
7. Float as it relates to time,
8. Float as it relates to compensation,
9. Concurrent delay – its effect on entitlement to extension of time,
10. Concurrent delay – its effect on entitlement to compensation for prolongation,
11. Identification of float and concurrency,
12. After the event delay analysis,
13. Mitigation of delay and mitigation of loss,
14. Link between extension of time and compensation,
15. Valuation of variations,
16. Basis of calculation of compensation for prolongation,
17. Relevance of tended allowances,
18. Period for evaluation of compensation,
19. Global claims,
20. Acceleration, and
21. Disruption.

The core principles form the basis of four 'guidance sections' (Section 1 at pp. 10–34, Section 2 at pp. 35–41, Section 3 at pp. 42–45, and Section 4 at pp. 46–49):

1. Guidelines on the Protocol's position on core principles and on other matters relating to delay compensation,
2. Guidelines on preparing and maintaining schedules and records,
3. Guidelines on dealing with extensions of time during the course of the project, and
4. Guidelines on dealing with disputed extension of time issues after completion of the project – retrospective delay analysis.

The Protocol concludes with four appendices (Appendix A at pp. 52–62, Appendix B at pp. 63–70, Appendix C at pp. 71–72, and Appendix at pp. 73–82):

A. Definitions and glossary,
B. Model specification clause,
C. Model records clauses, and
D. Graphics illustrating points in this Protocol.

The Protocol was initially generally well received although there were some criticisms from some commentators (presumably those that did not agree with its guidance). Although much of the Protocol relates to avoiding disputes by good practice in preparing and maintaining schedules and records and in dealing with extensions of time during the course of a project, the majority of the focus from most commentators of the Protocol was on the final section of guidance, the subject of retrospective delay analysis. Since its publication, the Protocol has generated discussion and elevated the issues involved in, in particular, retrospective delay analysis. This discussion has resulted in more discussion, writing, research, training and understanding of the problems involved. As a result, 10 or more years after its publication, the Protocol seems somewhat scant in its content (for instance, it details only four methods of retrospective delay analysis), and its overt support of time impact analysis as the best and preferred technique (paragraph 4.8 at p. 47) is now viewed as injudicious:

> "Time impact analysis is based on the effect of Delay Events on the Contractor's intentions for the future conduct of the work in the light of progress actually achieved at the time of the Delay Event and can also be used to assist in resolving more complex delay scenarios involving concurrent delays, acceleration and disruption. It is also the best technique for determining the amount of EOT that a Contractor should have been granted at the time an Employer risk event occurred. In this situation, the amount of EOT may not precisely reflect the actual delay suffered by the Contractor. That does not mean time impact analysis generates hypothetical results – it generates results showing entitlement. This technique is the preferred technique to resolve complex disputes related to delay and compensation for that delay."

Table 15.1 (adapted from paragraph 4.13 of the Protocol (p. 48) illustrates the factual material required for each type of analysis for each of the four

Table 15.1 Methods of analysis and the required factual material 1.

Type of analysis	As-planned schedule without network	Networked as-planned schedule	Updated as-planned networked schedule	As-built records
As-planned versus as-built	X	or X	and X	or X
Impacted as-planned		X		
Collapsed as-built				X
Time impact analysis		X	or X	and X

Adapted from SCL Protocol table 4.13.

methods of analysis covered by the Protocol. In comparison to more recent texts and technical papers, the range of methods described is rather limited. What seems surprising is that the 'windows' method of analysis is not included – it was in the draft consultation copy (SCL, 2001) but was dropped in the final publication. Perhaps the reason for the omission is that 'windows' and 'watersheds' are not methods of analysis in themselves; they are merely aspects of conducting a given method of analysis...' (Pickavance, 2005, at paragraph 14.327, p. 572).[1] Nevertheless 'windows analysis' is the most accepted method of critical path analysis (Mirant v Ove Arup [2007], at paragraph 131 and 132).

AACEI recommended practice no. 29R-03 – Forensic schedule analysis

In 2003, the Claims & Dispute Resolution Committee of the Association for the Advancement of Cost Engineering International (AACEI) launched the Recommended Practice (RP)/Forensic Schedule Analysis project. The purpose of the project was to '*provide a unifying, standard reference for the forensic application of CPM scheduling...in order to alleviate, if not eliminate, the confusion among practitioners regarding terminology, definitions and techniques of forensic scheduling*' (D'Onofrio and Hoshino, 2010, at p. 2), and the desired result being to '*decrease the number of unnecessary disagreements concerning technical implementation and allow the practitioners to concentrate their skills on resolving disputes over substantive issues*'. The RP was first published in June 2007 and subsequently revised in June 2009 and April 2011 (Hoshino et al., 2011). It is likely that the SCL Protocol had some influence on the RP if only as a reference document. The RP states (footnote 3 at p. 11):

> "The only other similar protocol known at this time is the 'Delay & Disruption Protocol' issued in October 2002 by the Society of Construction Law of the United Kingdom. The DDP has a wider scope than this RP."

Again, reflecting the make-up of the drafting subcommittee and the U.S. base of the AACEI, the Protocol is somewhat U.S.-centric, and whilst most of the document represents good practice at the time of the publication, the legal principles that support some of the guidance may not be applicable to all jurisdictions.

Although the SCL Protocol has a wider scope than the AACEI RP, it is a much more substantial document – 134 pages of the RP compared to 82 pages of the Protocol. The AACEI RP has five major sections (Section 1 at pp. 10–18, Section 2

Section IV

---

[1]   This was the edition first published after the publication of the SCL Protocol.

at pp. 18–37, Section 3 at pp. 38–98, Section 4 at pp. 98–125, and Section 5 at pp. 125–131):

1. Organisation and scope,
2. Source validation,
3. Method implementation,
4. Analysis evaluation, and
5. Choosing a method.

The Protocol concludes with two appendices (Appendix A at p. 133 and Appendix B at p. 134):

A. Nomenclature correspondence figure, and
B. Taxonomy of forensic schedule analysis.

The AACEI RP has also received criticism not least that although it is titled 'Recommended Practice', there are no recommendations as to the 'best' method of analysis, and it treats all methods as neutral (the opposite is a criticism of the SCL Protocol due to its support for time impact analysis). The AACEI RP focuses on the technical aspects of forensic delay analysis, and the protocols can be used as a step-by-step guide to undertaking forensic delay analysis.

The AACEI RP details some 17 specific retrospective analysis methods under nine basic implementation types (the AACEI RP is only concerned with retrospective analysis). The common names for the techniques used by the SCL Protocol are insufficient to describe the methods so the AACEI RP has developed a Taxonomy and Nomenclature (pp. 11–16). Although initially this appears confusing, the logical hierarchy allows for accurate description of the method without resorting to non-standardised common names. The taxonomy and nomenclature hierarchy are shown in Table 15.2; the numbers in Layer 4 relate to the chapter headings of section 3 of the AACEI RP, 'Method Implementation'.

Given the comprehensive nature of the methodologies described in the AACEI RP, it would be surprising if something the equivalent of 'windows analysis' was not included. Along with the strict taxonomy, the AACEI RP also lists the common names of the methods; these are collected in Table 15.3, and five of the nine basic implementation methods are commonly called 'windows analysis'!

The table at section 5.3 of the ACCEI RP (p. 128) illustrates the factual material required for each of the nine basic methods described in ACCEI RP (see Table 15.4).

Methods of analysis

There are many types of analysis – the SCL Protocol lists 4, and the AACEI RP has at least 17. Delay analysts say that it is virtually impossible to use a single technique 'straight out of the box' and that all delay and disruption disputes have their idiosyncrasies that require the analyst to be flexible in his approach and to

Table 15.2 AACEI RP taxonomy and nomenclature hierarchy.

Layer 1	Layer 2	Layer 3		Layer 4	Layer 5
Timing	Basic methods	Specific methods		Basic implementation	Specific implementation
Retrospective	Observational	Static logic	3.1	Gross	
		Dynamic logic	3.2	Periodic	Fixed periods
					Variable periods or grouped
			3.3	Contemporaneous as-is	All fixed periods
					Variable periods or grouped
			3.4	Contemporaneous split	All fixed periods
					Variable periods or grouped
			3.5	Modified / recreated updates	All fixed periods
					Variable periods or grouped
	Modelled	Additive modelling	3.6	Single base model	Global insertion
					Stepped insertion
			3.7	Multiple base model	Fixed periods
					Variable periods or grouped
		Subtractive modelling	3.8	Single simulation model	Global extract
					Stepped extract
			3.9	Multiple simulation models	Periodic modelling
					Cumulative modelling

Section IV

Table 15.3 Common names for methods of analysis.

Taxonomic description	Common name
3.1 Observational : Static logic : Gross	As-planned vs. as-built AP vs. AB Planned vs. actual As-planned vs. update
3.2 Observational : Static logic : Periodic	As-planned vs. as-built AP vs. AB Planned vs. actual As-planned vs. update Window analysis Windows analysis
3.3 Observational : Dynamic logic : Contemporaneous updates : All Periods	Contemporaneous period analysis Contemporaneous project analysis Observational CPA Update analysis Month-to-month Window analysis Windows analysis
3.4 Observational : Dynamic logic : Contemporaneous updates : Grouped periods	Contemporaneous period analysis Contemporaneous project analysis Contemporaneous schedule analysis Bifurcated CPA Half-stepped update analysis Two-stepped update analysis Month-to-month Window analysis Windows analysis
3.5 Observational : Dynamic logic : Modified/recreated updates	Update analysis Reconstructed update analysis Month-to-month Window analysis Windows analysis
3.6 Modelled : Additive : Single base	Impacted as-planned (IAP) Impacted baseline (IB) Plan plus delay Impacted update analysis Time impact analysis (TIA) Time impact evaluation (TIE) Fragnet insertion Fragnet analysis
3.7 Modelled : Additive : Multi base	Window analysis Windows analysis Impacted update analysis Time impact analysis (TIA) Time impact evaluation (TIE) Fragnet insertion Fragnet analysis

Table 15.3 *(Continued)*

Taxonomic description	Common name

3.8 Modelled
    : Subtractive
    : Single simulation
	Collapsed as-built (CAB)
	But-for analysis
	As-built less delay
	Modified as-built

3.9 Modelled
    : Subtractive
    : Multi simulation
	Collapsed as-built (CAB)
	Windows collapsed as-built
	But-for analysis
	Windows as-built but-for
	As-built less delay
	Modified as-built
	Look-back window

Table 15.4 Methods of analysis and the required factual material 2.

Source schedules or data	Method								
	3.1	3.2	3.3	3.4	3.5	3.6	3.7	3.8	3.9
Baseline schedule	X	X				X	X		
Schedule updates			X	X			X		X
As-built records	X	X			X			X	X

Adapted from AACEI RP figure 18.

adapt methodologies to suit the facts of the matter, the contractual requirements, the information and records available, and the time that is available to carry out the analysis.

Although most analysts favour a particular type of analysis, there are a number of factors that will steer the analyst to one method or another:

- **The contract requirements.**
  Although standard forms of contract tend not to be prescriptive about the way delays are analysed, there is no reason why such clauses cannot be written into the specification and so on. Some contracts (such as the ECC[2] suite of contracts) tend towards time impact analysis as the method of assessing compensation events during the project, so if the premise (promulgated in the SCL Protocol) is, as far as is practicable, to put oneself in the position of the contract administrator at the time the delaying event occurred, then it follows that the same techniques ought to be used retrospectively.

[2] NEC Engineering and Construction Contract (ECC) (previously the New Engineering Contract).

- **The schedule, information and records available.**
  Notwithstanding what the contract might say or the preference of the analyst the records available will determine what methods of analysis can be undertaken. For instance, if as-built records are not available or are insufficient then the methods available are very limited, similarly if there is no as-planned/baseline schedule.
- **The amount of time available, the cost of the analysis and the amount in dispute.**
  If the amount in dispute is high, then the cost of the analysis will generally not be significant in comparison. The cost of the analysis is generally directly related to the total time required to undertake it (including the time/cost of ancillary staff including that of the contractor/employer). Simplistically, the quickest (and therefore the cheapest) methods of analysis are likely to be impacted as-planned and as-planned versus as-planned type analyses. The most time-consuming (and therefore the most expensive) methods of analysis are likely to be time impact analysis and collapsed as-built type analyses with the latter probably being the most time consuming due to the amount of work involved in satisfactorily constructing an as-built network. However, impacted as-planned methods are only acceptable in limited circumstances and simple as-planned versus as-built analysis can be overly subjective. There is therefore a trade-off between accuracy and proportionality of cost, and it is difficult to determine what is most acceptable, a cheap but less accurate assessment or an expensive and more accurate assessment. In some instances, there is no alternative but to undertake one of the faster methods of analysis, for instance, in statutory adjudication in the United Kingdom, the responding party has 7 days (which may be extended to 14 days) to reply to the Referral Notice; the response might include a delay analysis as part of the defence or counterclaim.
- **The nature and complexity of the dispute and the delaying events.**
  Projects with few and obvious delaying events, for instance, the failure of the employer to provide access to various areas of the works or the failure of a key subcontractor to start work as planned will not require complex analysis whereas those that involve multiple causes of delay simultaneously, and/or the responsibility of the contractor and the employer over an extended period of time on a complex project, will require a more in-depth and methodical analysis.
- **The expertise of the analyst and familiarity with the techniques.**
  Whilst this ought not to be a consideration as the person responsible for carrying out the delay analysis should be selected for his competency to investigate the matter. In reality, a lack of resources may mean that less experienced and proficient staff are allocated to matters; in such cases, a simple technique applied competently is better than a complex technique applied inadequately.
- **Agreement of technique**.
  Where possible, in a dispute, the parties should agree on which method is appropriate to use, and in formal tribunal (such as arbitration or litigation), if possible, the decision of the arbitrator or judge should be sought so as to reduce

the likelihood of future disagreement on the appropriateness of the approach. Notwithstanding the fact that the experts in *Costain v Haswell* (2009) (at paragraphs 178–180) agreed the method to be used (time impact analysis), their understanding of what that meant was different:

"Issue 1: The way the agreed methodology has been applied

[The Defendant] submits that [Mr C] has not correctly applied the time impact analysis method in that, rather than applying the impact of the delaying event himself and assessing its consequences, he has used [the Claimant's] monthly updated progress schedules for that purpose and, after the correction of certain anomalies, has accepted those schedules as correctly showing the impact of the delaying event. [Mr P], on the other hand, has carried out what he considers to be the more correct approach, namely to consider the state of the progress of the works prior to the inception of the delaying event and then impacting that event on to that schedule in order to see what the software produces as the impact of that event. It is not clear to me what difference to the actual results these alternative approaches lead to but, if it becomes material to decide, I prefer the approach of [Mr P].

The reason for that preference is that it eliminates any subjective distortion or manipulation (either advertent or inadvertent) in the production of the monthly progress schedules by [the Claimant]. [Mr P's] approach seems to me to be more rigorous and to be more in accordance with the accepted understanding of a time impact analysis approach, as agreed by the experts."

Descriptions of how to undertake or carry out a particular method of analysis have to be simple enough to understand and able to illustrate the basis of the technique. In doing so, the techniques may appear easy to apply. Examples (including those here) have a small number of activities and obvious delaying events, and the schedules are perfectly created. Reality is generally the opposite that is likely to be one of the reasons the project has suffered delays and there is a dispute. In practice,

- Schedules are complex with thousands of activities and are often poorly developed with many technical errors (one experienced analyst has even gone as far as saying every schedule he has examined in a dispute is not fit for purpose!).
- Information and records are limited and inconsistent with sections missing for crucial periods. There is an increasing tendency for narrative progress reports rather than hard tabulated data related to the contract schedule; something like 'the brickwork to the flank wall is progressing well' rather than it being tabulated as 47% complete against the relevant schedule activity.
- Delaying events tend not to be simple; otherwise, they could have been sorted out earlier. Instead of large discrete events, projects are usually delayed by multiple variations and changes over a long period of time, inconsistent provision

of information, access not being granted when required, poor management and bad workmanship, subcontractors and suppliers going out of business and unexpected and prolonged bad weather.

■ Contractors and/or employers not complying with their contractual obligations with regard to sorting out delays as the work progresses making each party unsure of what they should be doing accelerating or mitigating. There may also be queries as to whether the delays are concurrent or is the contractor merely pacing his works.

Nevertheless, the analyst armed with a broad knowledge of the techniques available, experience of the construction industry, competent in planning and programming and the use of software with an understanding of the contractual obligations of the parties and a good deal of common sense *should* be able to forensically analyse what the causes of delay to a project were.

Global claims

Global time claims (or rolled-up claims) are generally characterised by lack of analysis and a lack of particularisation between events and the delay they cause to the project. In short, there will be a list of delaying events and a statement that the combination of those events resulted in the total delay to the completion of the project.

It is possible to have some sympathy for this approach on projects that have been subject to many, many small delaying events perhaps constant minor revisions to drawings, small variations of continual late issue of information. The effect of each individual delaying event is almost immeasurable and it would be impractical if not impossible to attempt to model the delays, but the combined effect has a disruptive (see Chapter 13) and delaying effect – like death by a thousand cuts.

Historically, 'unparticularised' claims of this nature tended to be more popular prior to analytical techniques that used critical path analysis and depended on readily available computers to carry out the time-consuming analysis. Those responsible for certifying delay claims would weigh up the evidence and perhaps with some negotiation or 'horse-trading' between the parties would make a judgement and tended to make an apportionment so that each of the parties bore some of the responsibility, but an inordinate amount would usually be ascribed to 'neutral' event so that neither party could be said to be to blame. At the time, it was said that '*this sort of approach among certifiers is not merely widespread, but some would say that it is universal*' (Fenwick, E.R., 1993).

The rejection of global claims in *Wharf Properties v Eric Cumine* (1991) was made clear in comment on the case in the Building Law Reports (52 BLR 1, p. 6):

"It will no longer be possible to call in an outsider who will simply list all the possible causes of complaint and then by use of a series of chosen weasel words try to avoid having to give details of the consequences of those events before proceeding to show how great the hole was in the pocket of the claimant."

In the United States, there is some retention of jury trails for civil claims (such as construction disputes), and as such, the jury can make an impressionistic evaluation of the claim which could have been presented in nothing more than a global manner. The term 'impressionistic' was also used in *John Barker v Portman Hotel*, (1996) where the Recorder stated the principles required to make a fair and reasonable assessment of delay:

> "...calculate, rather than make an impressionistic assessment of the time taken up by events."

There has recently been a softening of the approach to global claims that they will not be automatically thrown out but there are strict guidelines as to when they are applicable. (See for instance *John Doyle v Laing Management* (2004) and *London Underground v Citylink* (2007).) In principle, a global claim may only succeed if it can be shown that there is no other way to bring the claim (that it is impossible to attribute a specific loss to a specific delaying event) and that there is no material causative factor causing the delay, disruption or cost for which the Defendant is not liable. Even if there are some delaying events that are not the employer's responsibility, the remainder of the global claim might survive, and the parties be invited to make further submissions on the surviving portion.

Global claims remain a risky proposition, and unless there is no other alternative, the recommendation of the SCL Protocol (at paragraph 1.14.1, p. 26.) is best followed:

> "The not uncommon practise of contractors making composite or global claims without substantiating cause and effect is discouraged by the protocol and rarely accepted by the courts."

One of the techniques of illustrating global claims is the scatter diagram or measles chart. The scatter diagram indicates the timing of delaying events during the project's life and generally uses the as-planned schedule as its basis. On the barchart or other timeline are plotted the incidences of the events said to be a delaying factor on the project. Usually, this comprises items such as drawing issues and revisions, responses to request for information, handovers of sections of land and so on. All instructions and the like are plotted adjacent to the relevant activity bar using milestone markers or arrows. Whilst the scatter diagram provides little analytical information in that it is unable to show directly the effect of any of the events plotted, it nevertheless shows general trends, and inferences may be drawn from the diagram – though these are not likely to be conclusive without further analysis.

Although scatter diagrams have little evidential values, they can have enormous visual impact. The impact presented by the scatter chart can provide a powerful visual message that may produce a persuasive initial impression of the merits of the case. Figure 15.1 shows an example of a scatter diagram.

Section IV

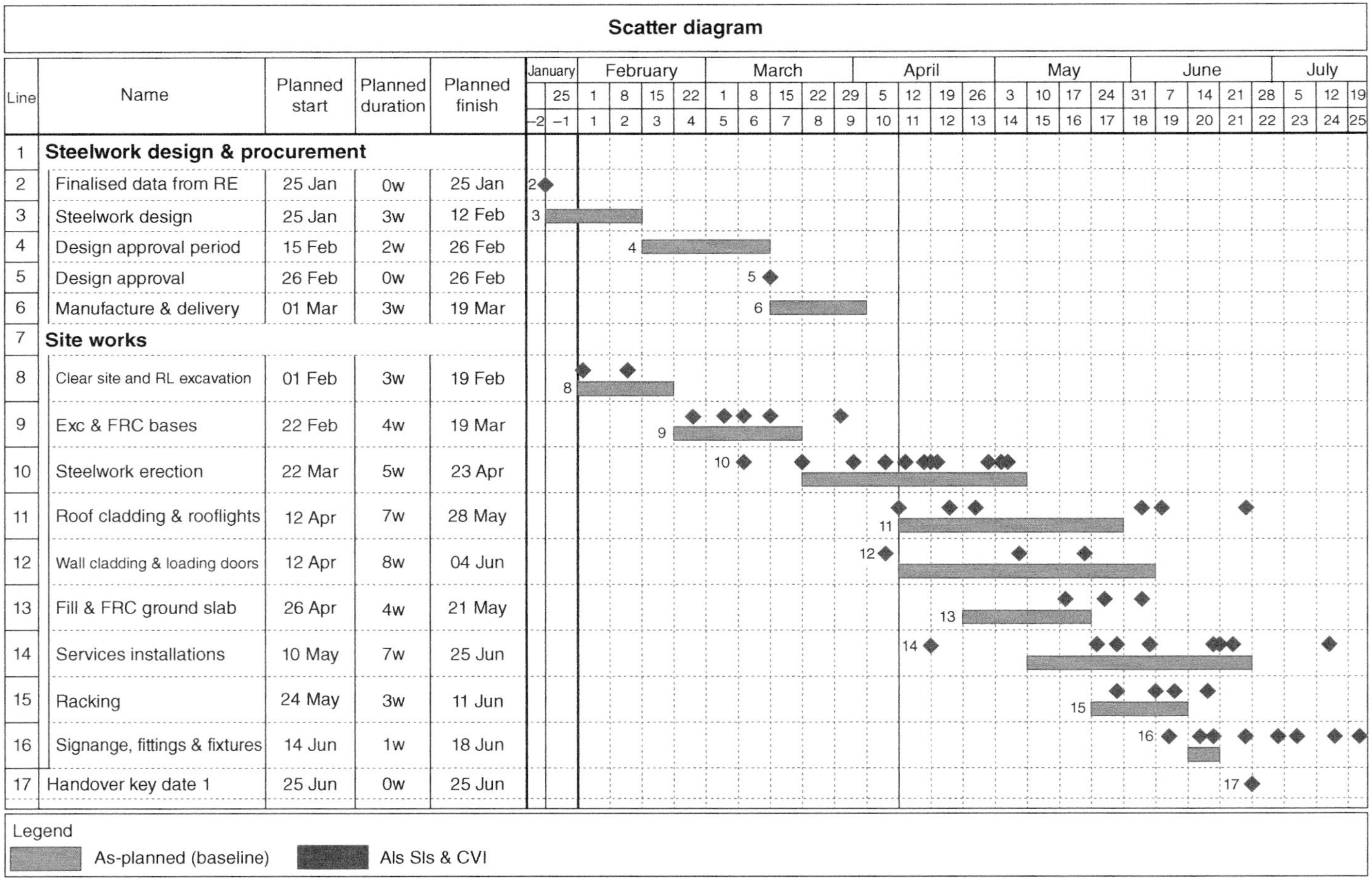
Scatter diagram
Line
Name
Planned start
Planned duration
Planned finish
January
February
March
April
May
June
July
25 1 8 15 22 1 8 15 22 29 5 12 19 26 3 10 17 24 31 7 14 21 28 5 12 19
–2 –1 1 2 3 4 5 6 7 8 9 10 11 12 13 14 15 16 17 18 19 20 21 22 23 24 25
1 Steelwork design & procurement
2 Finalised data from RE 25 Jan 0w 25 Jan
3 Steelwork design 25 Jan 3w 12 Feb
4 Design approval period 15 Feb 2w 26 Feb
5 Design approval 26 Feb 0w 26 Feb
6 Manufacture & delivery 01 Mar 3w 19 Mar
7 Site works
8 Clear site and RL excavation 01 Feb 3w 19 Feb
9 Exc & FRC bases 22 Feb 4w 19 Mar
10 Steelwork erection 22 Mar 5w 23 Apr
11 Roof cladding & rooflights 12 Apr 7w 28 May
12 Wall cladding & loading doors 12 Apr 8w 04 Jun
13 Fill & FRC ground slab 26 Apr 4w 21 May
14 Services installations 10 May 7w 25 Jun
15 Racking 24 May 3w 11 Jun
16 Signange, fittings & fixtures 14 Jun 1w 18 Jun
17 Handover key date 1 25 Jun 0w 25 Jun
Legend
As-planned (baseline)
AIs SIs & CVI

Figure 15.1 An example of a scatter diagram.

As-planned versus as-built

The as-planned versus as-built method, as it does not rely upon critical path analysis of a schedule, was the method primarily used to demonstrate the cause and effect of delaying events prior to the widespread availability of computers and software. The method suffered a decline in popularity when more dynamic scheduling methods were developed but retained a following amongst those who were not familiar with critical path analysis and similar planning techniques. Subsequently, it has had a revival in popularity as a result of some high-profile cases where critical path–based methods were erroneously applied, (see for instance *Skanska v Egger* (2004), *Great Eastern Hotel Company v John Laing* (2005), and *City Inn v Shepherd Construction* (2007)), following adverse comment about prospective methods of analysis (Barry, 2009) (such as time impact analysis) used retrospectively (when the project was complete) and the concerns of the 'black box syndrome'.[3] This is the only method that does not explicitly rely on a networked as-planned schedule and/or a networked as-built schedule so it is ideal for projects that have only been planned using bar charts or when a networked as-planned schedule is not available.

In its simplest form, the as-planned versus as-built analysis is merely a comparison of the as-planned schedule and the as-built schedule, almost always set out in simple barchart format with the as-planned and the as-built plotted adjacent to each other (see Figure 4.8). Where the timing and or duration of as-planned and as-built differs, in particular when the durations are longer than planned and/or the activities start or finish later than planned, these are the areas where delays are likely to have happened and initial examination of the causes of delay should be focused in these areas. On simple projects or where there have been few delays, it is possible to produce a narrative that is adequate to show the cause and effect of the delays. Such a narrative should be produced by someone who is experienced in the construction industry and experienced with the type of project under investigation, someone who understands planning and programming of work and workflow and was preferably involved in the construction of the particular project and has first-hand knowledge of what affected the progress of the project.

In making his assessment, the analyst will have in mind that only delayed activities on or close to the critical path of the project will have resorted to a delay to completion. Even though, in the simple scenario, the as-planned schedule and the as-built schedule are not logic linked and have not been subject to critical path analysis, it is implicit that the analyst needs to take into account the critical path of the project. Without calculation of the critical path (which entails networking of the schedule), the envisaged work flow and critical path of what was built and what delayed the project may not be correct.

---

[3]   One of the criticisms of computer/software-based analysis is that it depends solely on the input of data, the workings of the software and the reliance on the output without any knowledge of its internal workings; its implementation is opaque. The output is totally dependent on the data that are input. In most cases, those not familiar with the workings of the software will have no way of scrutinising the output data for veracity.

Section IV

The analysis in the simple form is purely opinion based albeit with a narrative that describes the causal links between delaying events and delays to the completion of the project. However, if the tacit assumptions about the critical path are incorrect, the conclusions drawn from that will also be incorrect. Many analysts draw onto the as-built schedule their envisaged critical path to give the impression of legitimacy. Alternatively, the as-built barchart can be logic linked *ex post facto* to provide a more dependable critical path (subject to the difficulties of retrospective logic linking).

Whatever method of analysis is used, the production of a simple as-planned versus as-built comparison is a valuable first step as it will provide a readily accessible overview of the project and the potential delay hotspots.

The as-planned versus as-built methods are described in the AACEI RP as 'observational' techniques although it tends to rely more on a networked and analysed as-built schedule so that the critical path is more readily visible.

There are a number of refinements to the basic as-planned versus as-built analysis:

- The schedule is split into sections or 'windows'; these can represent regular periods of time (say, monthly or quarterly) or more likely into periods representing key points in the projects; the key points are usually referred to as 'watersheds' (such as, completion of the substructures, completion of the frame, completion of the envelope, completion of the fitting out and so on). Splitting of the analysis in this way allows closer scrutiny of the delaying events and focuses on delays happening within a discrete section of the work.
- The as-built schedule is based on contemporaneous updates of the schedule carried out during the project. This requires that the as-built schedule was fully networked and updated and analysed at intervals during the project and does not just rely on the as-built start and finish dates of the activities but gives some indication of the progress of activities between their start and finish dates and takes into account any contemporaneous changes and amendments made to the planned schedule during the project. At each update point (a window or watershed), a more accurate analysis of what was delaying the project at the time can be established.
- If contemporaneous updates to the schedule were not made, these can be recreated retrospectively, the major difference being that the recreated updates will be based only on the as-planned schedule (which must be networked) and will not have been amended during the project as contemporaneous updates might have been.

Where analysis is based purely on observation, visual comparison between planned and as-built activities, then the as-planned versus as-built methods,is most useful when the schedule is relatively small in size so that it can be visually assimilated. It is not possible to take account of the start and finish dates of thousands of activities over numerous pieces of paper and to envisage where the critical path is likely to be at any given point.

The major advantage of the as-planned versus as-built methods is that actual as-built data are used so the analysis ought is based on what actually happened. However, that is only true in simple applications when the project is looked at as a whole. In all enhancements, the method depends on contemporaneous or *ex post facto* updates of the as-planned schedule at each period the delay observed is a prospective projection based on the as-built effects on future work. In simple applications of the method, the fact that networked as-planned or networked as-built schedules are not required is often cited as an advantage; however, delays to the project completion can only result from delays to activities on or near the critical path so the critical path will have to be envisaged or deduced by the analyst. This may not be accurate. By comparing the activity variances, it is possible to say when the delays occurred. This is of assistance to those responsible for calculating the cost of the delays for the periods they actually occurred. The method is also dependent on accurate as-built information. The quality of that information will influence the accuracy of the results.

Some of the more sophisticated as-planned versus as-built methods, in particular those that use updates of the as-planned schedule, are often not thought of as as-planned versus as-built methods but are considered to be 'windows' methods even though they are a comparison of what was planned against what was built.

Impacted as-planned

The impacted as-planned method of analysis is the simplest form of dynamic modelled analysis – that is, it depends on critical path analysis and insertion of delaying events into the schedule.

The only information required for this method is a compliant networked as-planned schedule (modified, repaired or reconstructed as required – see Chapter 14) and a list of delaying events. The SCL Protocol suggests that the method 'is based on the effects of Employer Risk Events on the planned schedule of work' (SCL, 2002, paragraph 4.6, p. 45). This is *not* the case, it can be used to model the effects of delaying events that are the responsibility of the employer only, the contractor only, neutral events only or any other combination.

After identifying and securing the as-planned schedule, the next step is to list and categorise the delaying events. The following data need to be assembled:

- Reference number
- Nature of the delaying event
- Date the delaying event occurred
- Responsibility: employer, contractor or neutral
- Type of delay: date, total, extended or addition (progress delays are not modelled in the impacted as-planned method)

There are then two options for impacting the delays: single insertion and individual insertion.

Single insertion

All of the delaying events are modelled and inserted into the as-planned schedule in a *single* step, and the schedule is then rescheduled.

Following rescheduling, it is likely that the critical path will have changed under the influence of the impacted events, and the planned completion date of the

Section IV

schedule is likely to have moved indicating the delay to the planned completion of the project. (It is more probable than not that, if the analysis is carried out retrospectively, the change to the planned completion will not match the actual completion of the project.)

Examination of the critical path will indicate which of the delaying events are on the critical path and hence contributed to the delay and the late completion of the project.

By inserting the delaying events category by category, a very approximate measure of concurrent delay can be established. If only the excusable events are inserted into the schedule, the resultant delay will be the maximum for which the contractor can be compensated subject to any concurrent delay. If only the culpable events are inserted into the schedule, the contractor will be responsible for the resultant delay. The amount of compensable delay is the delay resulting from the excusable events (employer responsible) minus the amount of delay resulting from the culpable delays (contractor responsible). If the culpable delays are equal to or greater than the excusable delays, then there is no compensation due for the employer delays.[4]

By inserting both the excusable delay events and the neutral delay events into the schedule, the resultant delay will be the maximum entitlement to extension of time for the contractor. If the culpable delays are greater than the maximum entitlement to delay, the difference is that for which the contractor may be liable to pay liquidated damages.

The indications of concurrent delay and the like are only approximate; the effects of concurrency can only be properly determined using methods that rely on as-built and progress records.

Individual insertion

In this option, the delaying events are inserted into the as-planned schedule one by one in chronological order and the network rescheduled after each individual insertion.

The change to the planned completion date after each insertion and rescheduling is recorded (on the delay event list); this represents the delay resulting from each delaying event – some delaying events might not result in a further delay to the planned completion date, their effect being subsumed by earlier delaying events.

When all the delaying events have been modelled, the final resultant impacted schedule will be identical to an impacted schedule where all the delaying events were impacted in a single insertion. The individual insertion option provides cause and effect detail for the individual delaying events.

The impacted as-planned method is illustrated in Figures 15.2a–15.2g.

Although the impacted as-planned method is straight-forward and can be carried out without the need to compile detailed as-built records, its use and

[4]   The general rule for concurrent delays, subject to the terms of the particular contract, is that where there is concurrent delay and an extension of time is granted, liquidated damages are not levied but direct loss and expense resulting from the delay is not payable.

(a)

As-planned schedule

Line	Name	Start	Duration	Finish
1	**Steelwork design & procurement**			
2	Finalised data from RE	25 Jan	0w	25 Jan
3	Steelwork design	25 Jan	3w	12 Feb
4	Design approval period	15 Feb	2w	26 Feb
5	Design approval	26 Feb	0w	26 Feb
6	Manufacture & delivery	01 Mar	3w	19 Mar
7	**Site works**			
8	Clear site and RL excavation	01 Feb	3w	19 Feb
9	Exc & FRC bases	22 Feb	4w	19 Mar
10	Steelwork erection	22 Mar	5w	23 Apr
11	Fill & FRC ground slab	26 Apr	4w	21 May
12	Roof cladding & rooflights	12 Apr	7w	28 May
13	Wall cladding & loading doors	12 Apr	8w	04 Jun
14	Services installations	10 May	7w	25 Jun
15	Racking	24 May	3w	11 Jun
16	Signange, fittings & fixtures	14 Jun	1w	18 Jun
17	Handover key date 1	25 Jun	0w	25 Jun

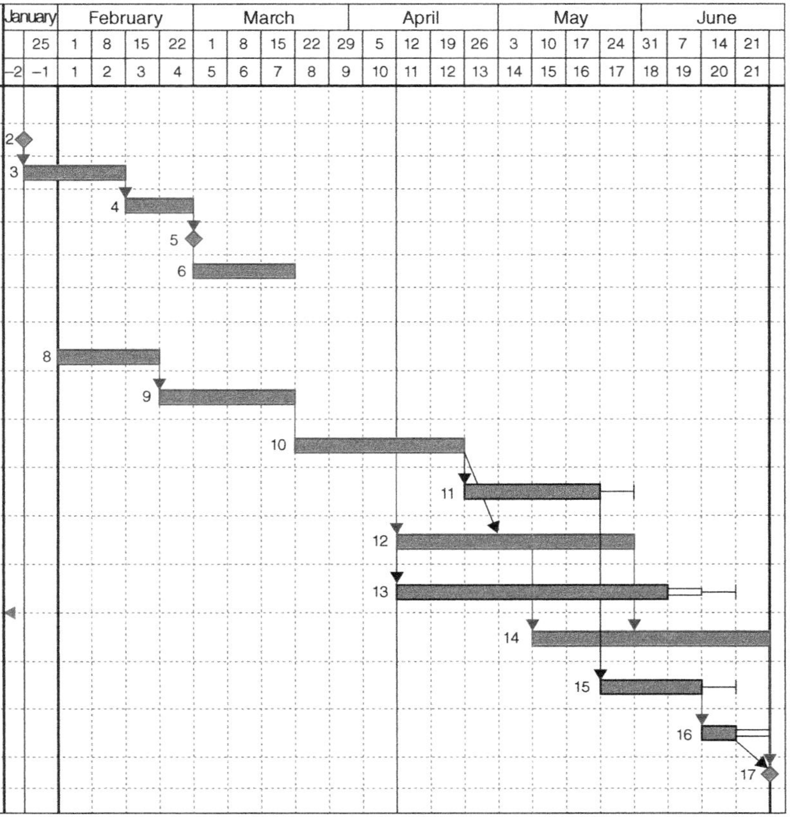

Figure 15.2   The impacted as-planned method.

Section IV

(b)

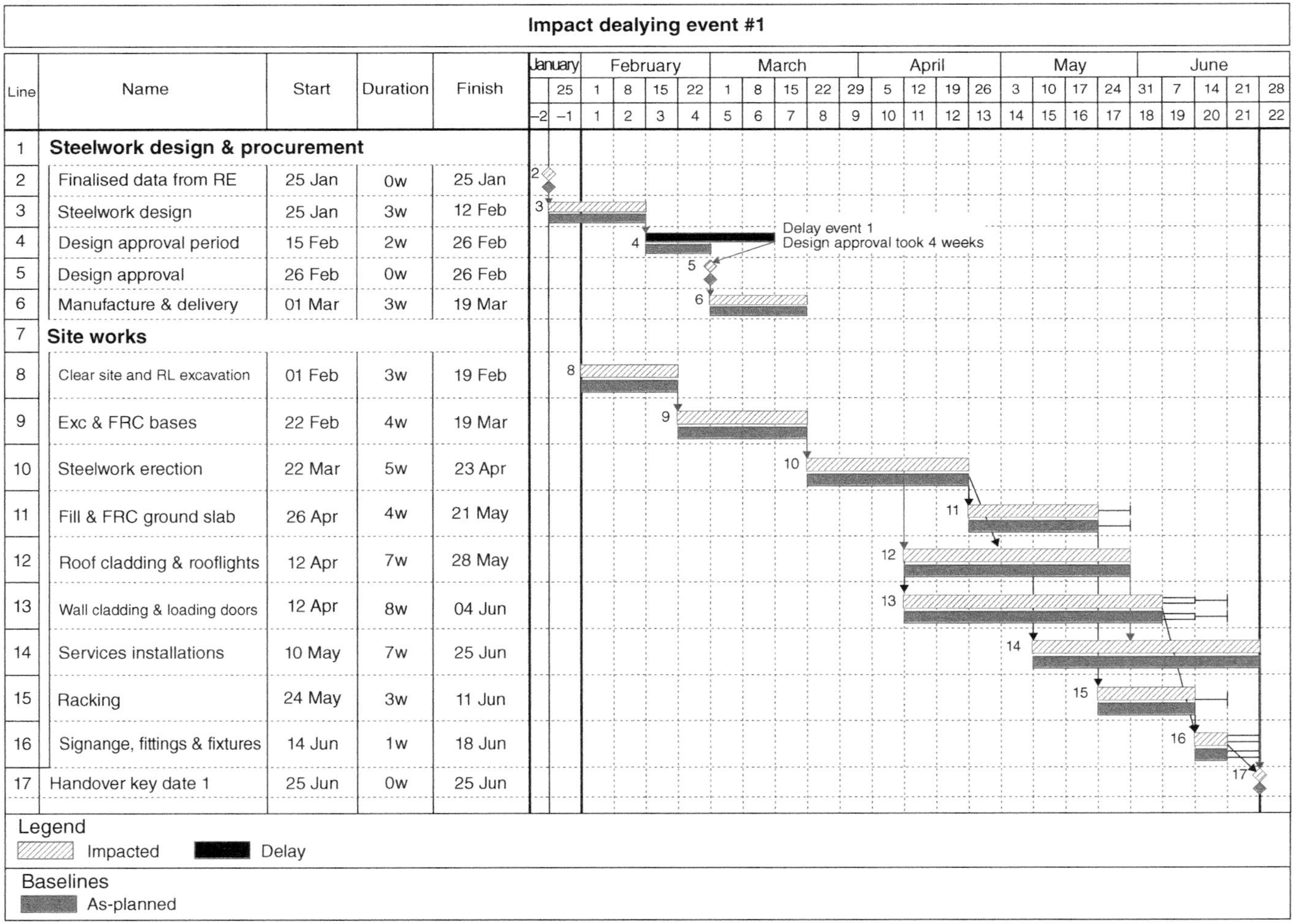
Impact dealying event #1
Line
Name
Start
Duration
Finish
January
February
March
April
May
June
25 1 8 15 22 1 8 15 22 29 5 12 19 26 3 10 17 24 31 7 14 21 28
–2 –1 1 2 3 4 5 6 7 8 9 10 11 12 13 14 15 16 17 18 19 20 21 22
1 Steelwork design & procurement
2 Finalised data from RE 25 Jan 0w 25 Jan
3 Steelwork design 25 Jan 3w 12 Feb
4 Design approval period 15 Feb 2w 26 Feb
5 Design approval 26 Feb 0w 26 Feb
6 Manufacture & delivery 01 Mar 3w 19 Mar
7 Site works
8 Clear site and RL excavation 01 Feb 3w 19 Feb
9 Exc & FRC bases 22 Feb 4w 19 Mar
10 Steelwork erection 22 Mar 5w 23 Apr
11 Fill & FRC ground slab 26 Apr 4w 21 May
12 Roof cladding & rooflights 12 Apr 7w 28 May
13 Wall cladding & loading doors 12 Apr 8w 04 Jun
14 Services installations 10 May 7w 25 Jun
15 Racking 24 May 3w 11 Jun
16 Signange, fittings & fixtures 14 Jun 1w 18 Jun
17 Handover key date 1 25 Jun 0w 25 Jun
Delay event 1
Design approval took 4 weeks
Legend
Impacted
Delay
Baselines
As-planned

Figure 15.2 *(Continued)*

(c)

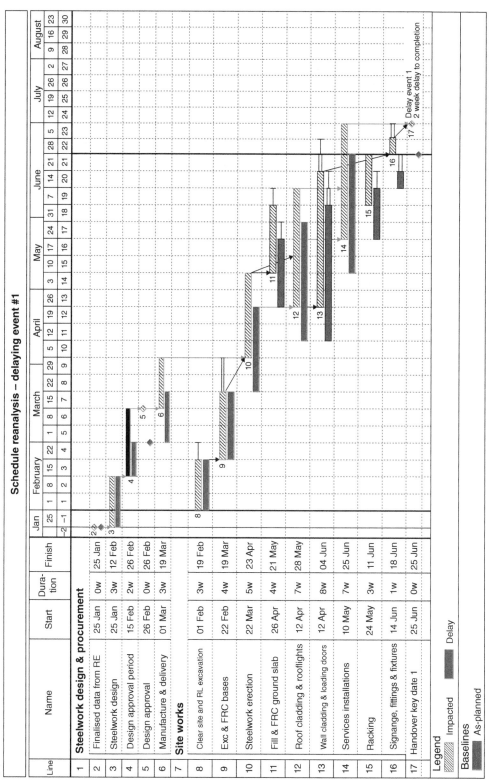

Figure 15.2 (Continued)

Section IV

Impact delaying event #2

Line	Name	Start	Dura-tion	Finish																					

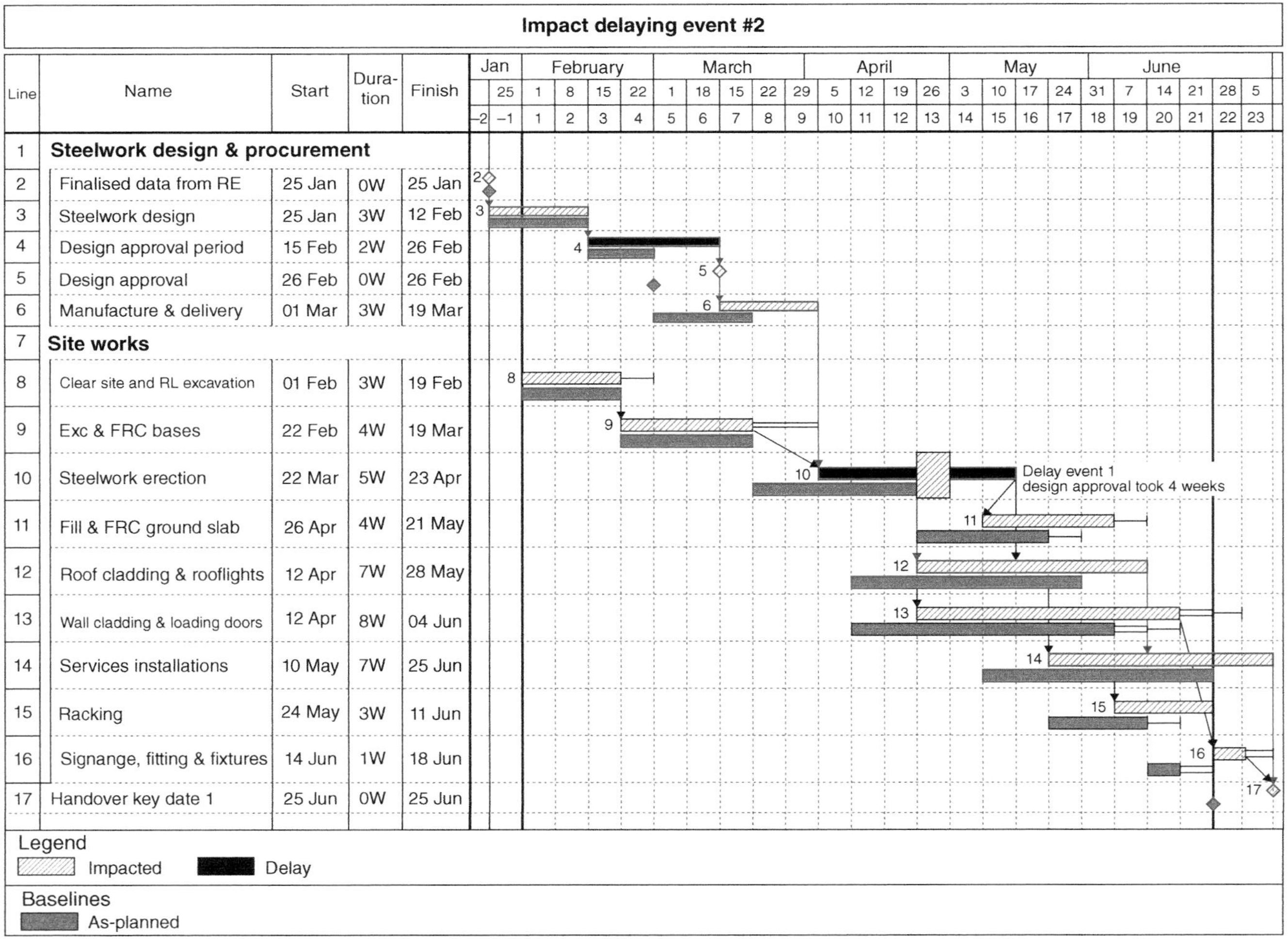

Figure 15.2 (*Continued*)

(e)

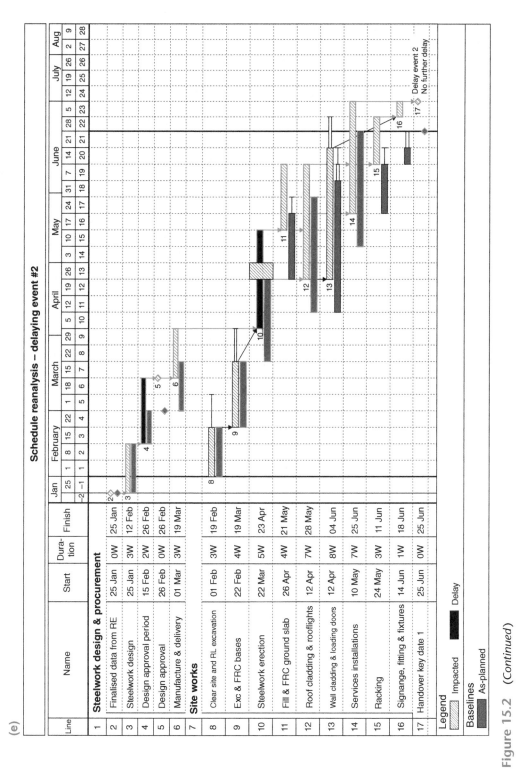

Figure 15.2 (*Continued*)

Section IV

(f)

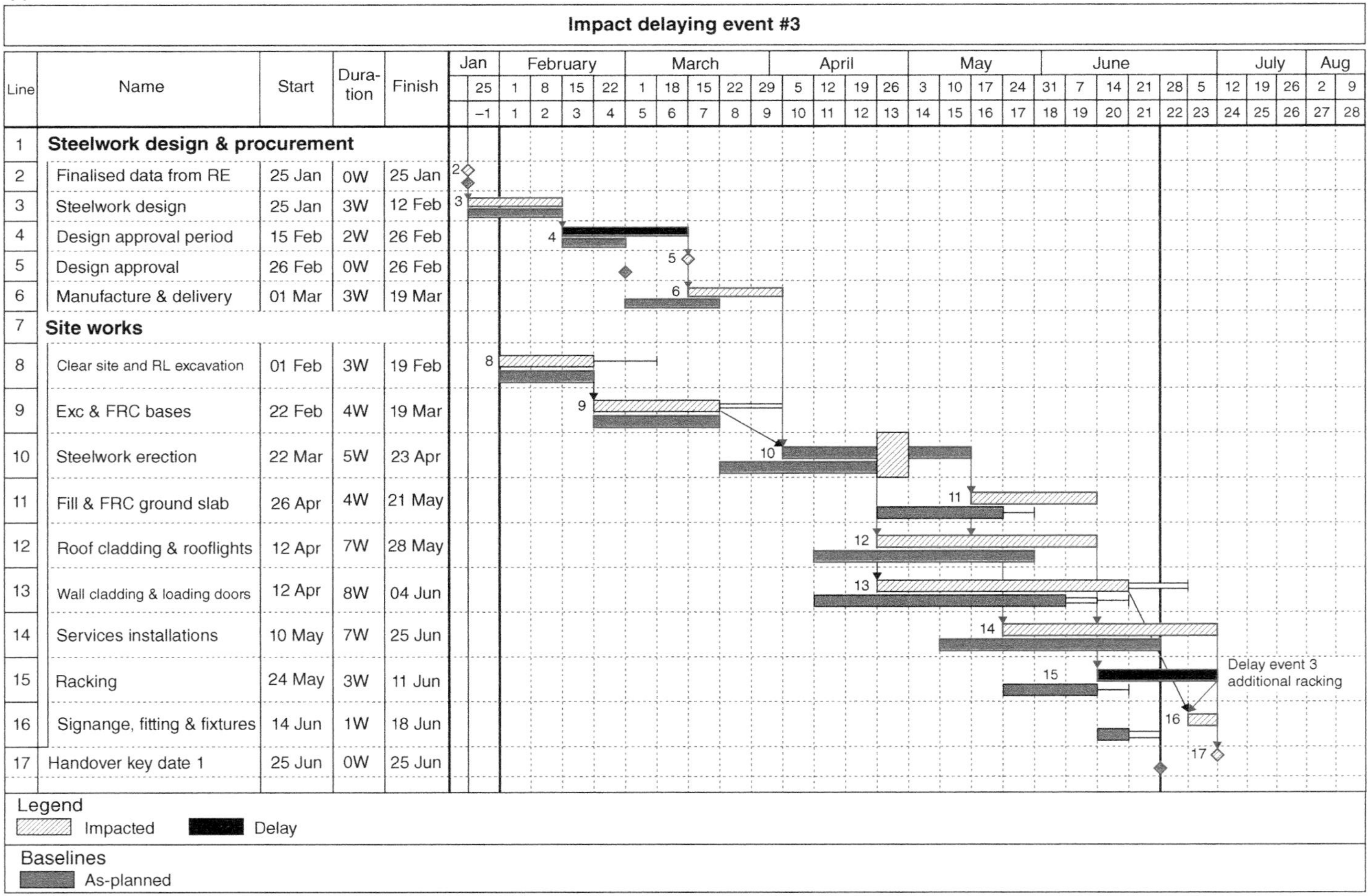
Impact delaying event #3
Line
Name
Start
Dura-tion
Finish
Jan
February
March
April
May
June
July
Aug
25 1 8 15 22 1 18 15 22 29 5 12 19 26 3 10 17 24 31 7 14 21 28 5 12 19 26 2 9
–1 1 2 3 4 5 6 7 8 9 10 11 12 13 14 15 16 17 18 19 20 21 22 23 24 25 26 27 28
1 Steelwork design & procurement
2 Finalised data from RE 25 Jan 0W 25 Jan
3 Steelwork design 25 Jan 3W 12 Feb
4 Design approval period 15 Feb 2W 26 Feb
5 Design approval 26 Feb 0W 26 Feb
6 Manufacture & delivery 01 Mar 3W 19 Mar
7 Site works
8 Clear site and RL excavation 01 Feb 3W 19 Feb
9 Exc & FRC bases 22 Feb 4W 19 Mar
10 Steelwork erection 22 Mar 5W 23 Apr
11 Fill & FRC ground slab 26 Apr 4W 21 May
12 Roof cladding & rooflights 12 Apr 7W 28 May
13 Wall cladding & loading doors 12 Apr 8W 04 Jun
14 Services installations 10 May 7W 25 Jun
15 Racking 24 May 3W 11 Jun
16 Signange, fitting & fixtures 14 Jun 1W 18 Jun
17 Handover key date 1 25 Jun 0W 25 Jun
Delay event 3
additional racking
Legend
Impacted
Delay
Baselines
As-planned

Figure 15.2 *(Continued)*

(g)

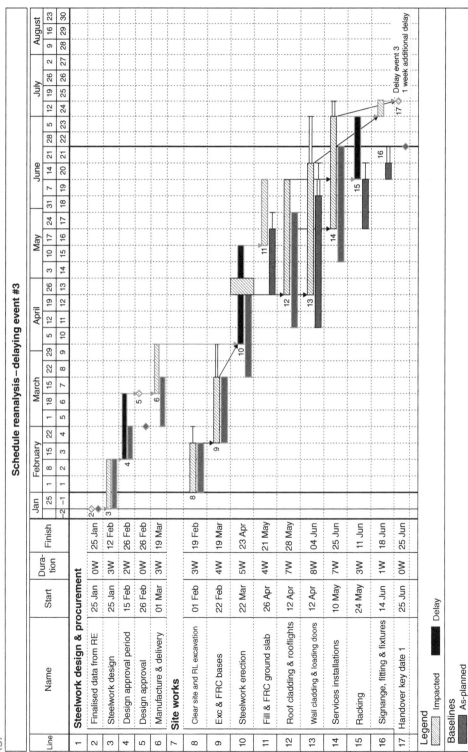

Figure 15.2 (Continued)

Section IV

acceptability are very limited as it assumes that, apart from the delaying events modelled, everything went according to plan. This is seldom the case. In almost every project, the actual progress differs from that planned and the failure of methods to take account of the status of the project at the time the delaying event happened – as a result of not modelling 'progress delays'– means that the results must be treated with the utmost scepticism.

The method relies only on the logic of the as-planned schedule and does not take account of any changes to sequences or steps taken to accelerate the works or to mitigate delays. At best, the results are hypothetical and do not reflect reality; at no time does the impacted schedule represent the actual status of the project, and the completed final impacted schedule bears little resemblance to the as-built schedule, what actually happened.

There is one situation where the impacted as-planned method *may* be acceptable. If there are very few delaying events and they all happen in the very early stages of the project, then the results might reflect a contractual entitlement to extension of time only. In the early stages of a project, the schedule is generally more accurate and the works less affected by change than in later stages, and the influence of actual progress is likely to be less.

The AACEI RP describes the method as Modelled/Additive/Single Base. (AACEI, 2011, paragraph 3.6, p. 70)

Time impact analysis

Time impact analysis is similar to the impacted as-planned method; the major difference is that time impact analysis uses multiple baseline schedules rather than just the single baseline of the original as-planned schedule. In the basic method of implementation, the multiple successive schedules are updates of the original as-planned schedule. The use of updated schedules means that the schedules reflect the status of the project and progress at the time the delaying events occur.

The method is a prospective method in that it uses the networked schedule to project the likely effect of the delaying event and is generally the way in which the likely effects of delaying events are determined contemporaneously during the project. The method is very similar to that detailed in the ECC suite of contracts for the determination of compensation events during the project. By updating the schedule with actual dates and progress, the data to the left of 'time now' reflect the as-built situation and the schedule to the right of 'time now' reflects the intended schedule through to completion taking account of progress and delaying events up to that point.

The basis for the method is the as-planned schedule. It must be networked, and as progressing and updating are involved, it must be fully networked and, if necessary, modified, repaired or reconstructed as required to ensure that the logic is comprehensive and there are no dangles. When the analysis is carried out during the project, the method uses estimates of the delays (durations, sequences and so on) to impact the schedule. When the method is used retrospectively, the SCL proposes that (SCL, 2002, paragraph 4.19, p. 49.)

"The Protocol recommends that, in deciding entitlement to EOT, the adjudicator, judge or arbitrator should as far as is practicable put him/herself in the position of the CA at the time the Employer Risk Event occurred. He/she should use the Updated Schedule to establish the status of the works at that time. He/she should then determine what (if any) EOT entitlement could or should have been recognised by the CA at the time."

This application seems to suggest that the 'blindsight' approach should be adopted and that the analyst should only take into account, when modelling the delaying event, what was known at the time. For instance, where a work activity is extended due to an increase in quantity, a reasonable assumption, at the time, would be that the activity duration should be extended in proportion to the original scope of work. Contemporaneous records may provide evidence of what, at the time, the management team thought the effects of the delaying event might be. This approach has led to criticism when dealing with delays retrospectively 'why look into the crystal ball, when you can read the book'.[5] When dealing with delays retrospectively, the analyst will have data and evidence of the actual duration and sequences that resulted from the delaying event. Using this information, the 'hindsight' approach will lead to a more realistic and accurate response of the schedule and reduces the speculative and subjective assumptions.

The data required when undertaking a basic time impact analysis are:

- The networked as-planned schedule;
- The details of the delaying events (listed and categorised as previously described);
- The progress records; and
- The as-built details for the project.

Progress records and as-built data whether collected from diaries, weekly reports or progress meeting minutes should be cross-checked and verified against other records such as photographs, quality records, time sheets and so on to verify their accuracy. The method depends on updating the schedule to the point just prior to the delaying event occurring. Most probably, progress and as-built data for a particular date, if it is not on the date of a specific progress report, will not be available; therefore, it will be necessary to interpolate the specific data between the records for two known dates.

The stages in undertaking time impact analysis are:

1. Collect all the relevant data including
    - The as-planned network (repaired as necessary),
    - The list of delays, and
    - The as-built and progress data.
2. Identify the first delaying event and the date it happened.
3. Identify the progress records for the date, or just prior to the date, that the delaying event occurred. If actual progress records for the date do not exist for the date in question, they must be synthesised by interpolation between

⁵ Attributed to Aneurin Bevan (1897–1960) but used more recently in *Beware the Dark Arts*, p. 5.

Section IV

known progress dates. To be completely accurate, the progress records should be that of the actual date of the delaying event. Inspection of the progress records will determine how significant a time difference is. On short duration and fast-moving projects, the significance is liable to be greater.

4. Using the progress data, update and reanalyse the network and recalculate the critical path. This represents an updated network taking account of the progress of the project (and any variances in progress, faster or slower than planned) at the delay event date. At this stage, the network logic around the progress date should be examined to ensure it is still robust; out-of-sequence progress may result in illogical sequences and dangles (see Chapter 13 and 17).

    The updated network will show whether, as a result of progress thus far, the project is likely to be completed ahead, on or behind schedule. If the update indicates that the project is likely to finish ahead of schedule, the contractor has, by his efforts, created additional float in the project (terminal float). If the update indicates that the project is likely to finish behind schedule, absent of any unidentified delaying events, the contractor has created a likely 'culpable' delay to the project for which it may wish to take action to reduce.

    A record of the static or changed forecast completion date is made. At this stage and all subsequent intermediate stages, it is good practice to save and rename a copy of the updated network and use the copied version for the next stage of the analysis.

5. Using the updated network, model the first delaying event. The method of modelling will be dependent on the type of delay. In some cases, it might be necessary to use more than one of the methods. For instance, an additional activity may also require a date constraint. When the delay has been modelled, the network is reanalysed. If the delaying event was on or near the critical path, then the end date of the schedule (calculated by the previous progress update) will be extended. The difference between the previous finish date and the new date is the delay attributed to the delaying event, the 'cause and effect' link having been established.

    If the prior progress update was *ahead* of schedule *and* the delaying event now modelled on or near the critical path, the effect of the delay will be to consume part or all of the newly generated terminal float thus reducing or negating the potential delay to the project (subject to the contract conditions regarding the ownership of float).

    A record of the new forecast completion date is made and the impacted network saved and renamed for use in the next stage of the analysis.

6. The procedure for modelling subsequent delaying events follows the sequence described in stages 2–5.

7. Following modelling of the last delaying event, recording the resulted and copying and renaming the network, the impacted network should be progress updated to the actual end of the project. This could be a number of updates representing the, say, monthly progress reports or a single update at the end of the project, when all activities should be complete. Continuing progress updates to the actual completion date will confirm if any further time was gained or lost during the period. The final updated schedule will represent, and should coincide with, the as-built schedule for the project.

The time impact analysis method is illustrated in the following example:

- The starting point for the analysis is the networked (linked) as-planned programme as shown previously in Figure 15.2.
- The first delaying event is identified and the date that it took place. Progress at that date is identified. In this case, the design approval took longer than expected so progress is measured up to the end of week 2 as shown in Figure 15.3a. The dropline shows that activity 8, 'Clear site and RL excavation', is 1 week behind schedule.
- When the schedule is reanalysed (the critical path recalculated), the position or status of the project at the time of the delaying event is established. The slow progress of activity 8 results in the critical path being extended, and the milestone 17 'Handover Key Date 1' is delayed until 2 July, 1 week later than the original as-planned date of 25 June, as shown in Figure 15.3b. This is considered to be the effect (or likely effect) on completion due to the thus far slow progress of clearing the site and reduced level excavation; this would usually be considered to be a culpable delay on behalf of the contractor.
- The first delaying event is then modelled in the progress updated schedule. As previously, the design approval took 4 weeks as shown in Figure 15.3c.
- When the schedule is reanalysed, the critical path is further extended and the handover date is delayed until 9 July, 1 week more than the previous date of 2 July and 2 weeks later than the original as-planned date of 25 June(see Figure 15.3d). The additional 1 week is said to be the effect (or likely effect) on completion of the delay to the drawing approval period beyond the delay already apparent due to the slow progress of activity 8.
- Progress just prior to the second delaying event is input to the delay schedule, and the schedule reanalysed. Figure 15.3e shows the effect of progress and the likely effect on the completion of the project. The analysis shows that a further week has been lost; the additional lost week appears to be due to the extended period taken to manufacture and deliver the steelwork (activity 6). Usually, this would be considered a culpable delay, the responsibility of the contractor.
- The next delaying event, the stoppage of the steelwork erection whilst the wind-bracing calculations are checked, is modelled on the previously impacted schedule as shown in Figure 15.3f.
- When the schedule is reanalysed, the critical path is not extended any further, and the delay to completion remains at 3 weeks, as shown in Figure 15.3g. Thereis no effect (or likely effect) on the completion date as a result of the steelwork erection.
- Progress just prior to the third delaying event is input to the delay schedule and the schedule reanalysed. Figure 15.3h shows the effect of progress and the likely effect on the completion of the project; there is no effect (or likely effect) on the completion date.
- The third delaying event, the additional racking, is then modelled on the schedule previously impacted by the second delaying event as shown in Figure 15.3i.
- When the schedule is reanalysed, the critical path is not extended any further and the handover date remains at 16 July(see Figure 15.3j). There is no effect (or likely effect) on the completion date as a result of the installation of the additional racking.

Section IV

(a)

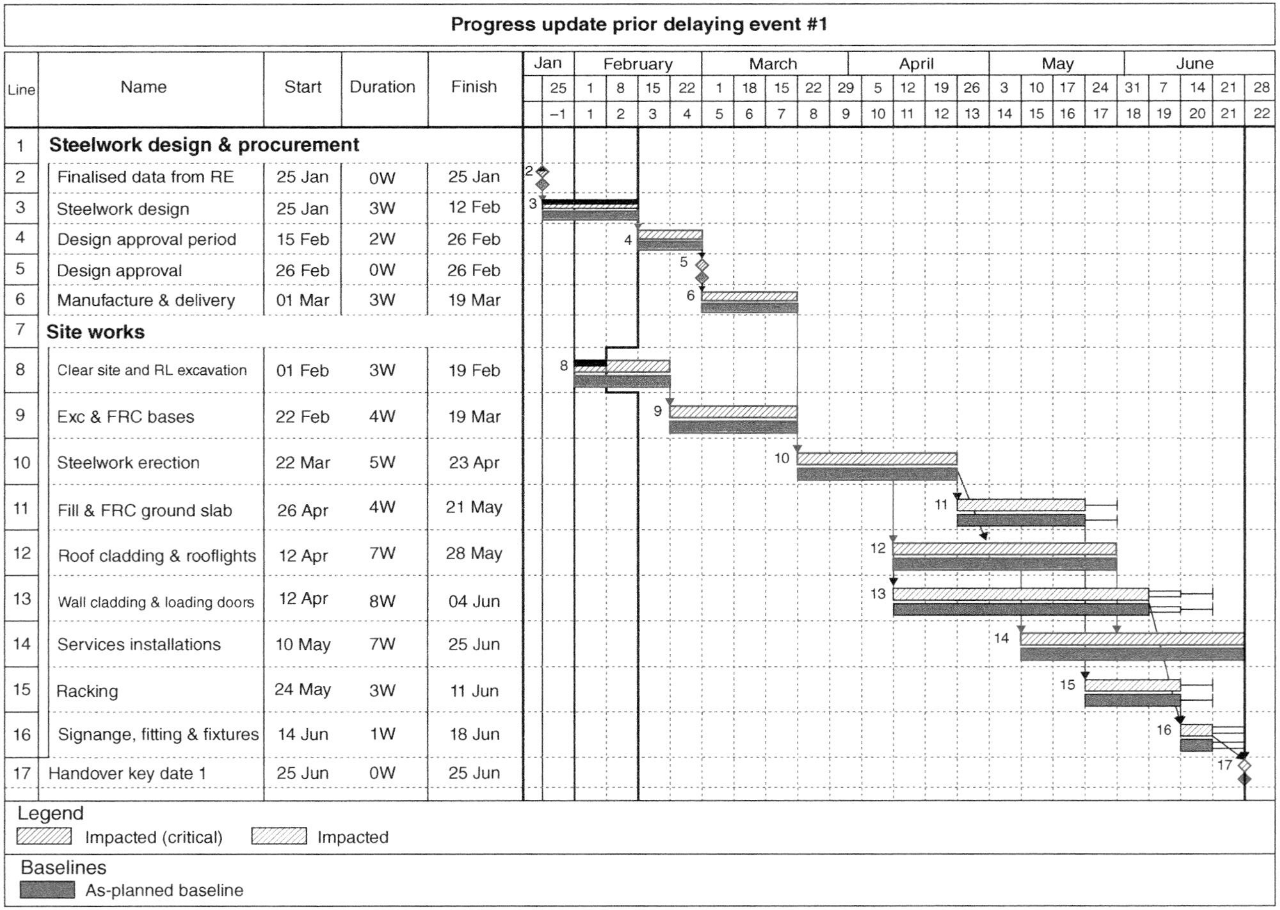
Progress update prior delaying event #1
Line
Name
Start
Duration
Finish
Jan
February
March
April
May
June
25 1 8 15 22 1 18 15 22 29 5 12 19 26 3 10 17 24 31 7 14 21 28
−1 1 2 3 4 5 6 7 8 9 10 11 12 13 14 15 16 17 18 19 20 21 22
1 Steelwork design & procurement
2 Finalised data from RE 25 Jan 0W 25 Jan
3 Steelwork design 25 Jan 3W 12 Feb
4 Design approval period 15 Feb 2W 26 Feb
5 Design approval 26 Feb 0W 26 Feb
6 Manufacture & delivery 01 Mar 3W 19 Mar
7 Site works
8 Clear site and RL excavation 01 Feb 3W 19 Feb
9 Exc & FRC bases 22 Feb 4W 19 Mar
10 Steelwork erection 22 Mar 5W 23 Apr
11 Fill & FRC ground slab 26 Apr 4W 21 May
12 Roof cladding & rooflights 12 Apr 7W 28 May
13 Wall cladding & loading doors 12 Apr 8W 04 Jun
14 Services installations 10 May 7W 25 Jun
15 Racking 24 May 3W 11 Jun
16 Signange, fitting & fixtures 14 Jun 1W 18 Jun
17 Handover key date 1 25 Jun 0W 25 Jun
Legend
Impacted (critical)
Impacted
Baselines
As-planned baseline

Figure 15.3 The time impact analysis method.

(b)

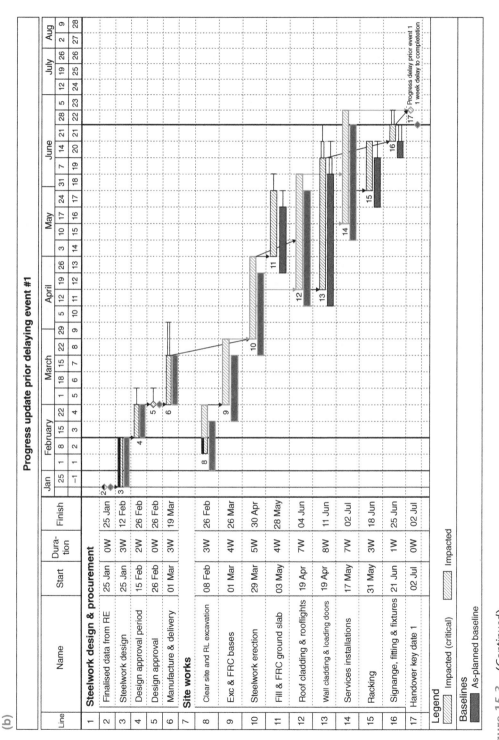

Figure 15.3 (Continued)

Section IV

(c)

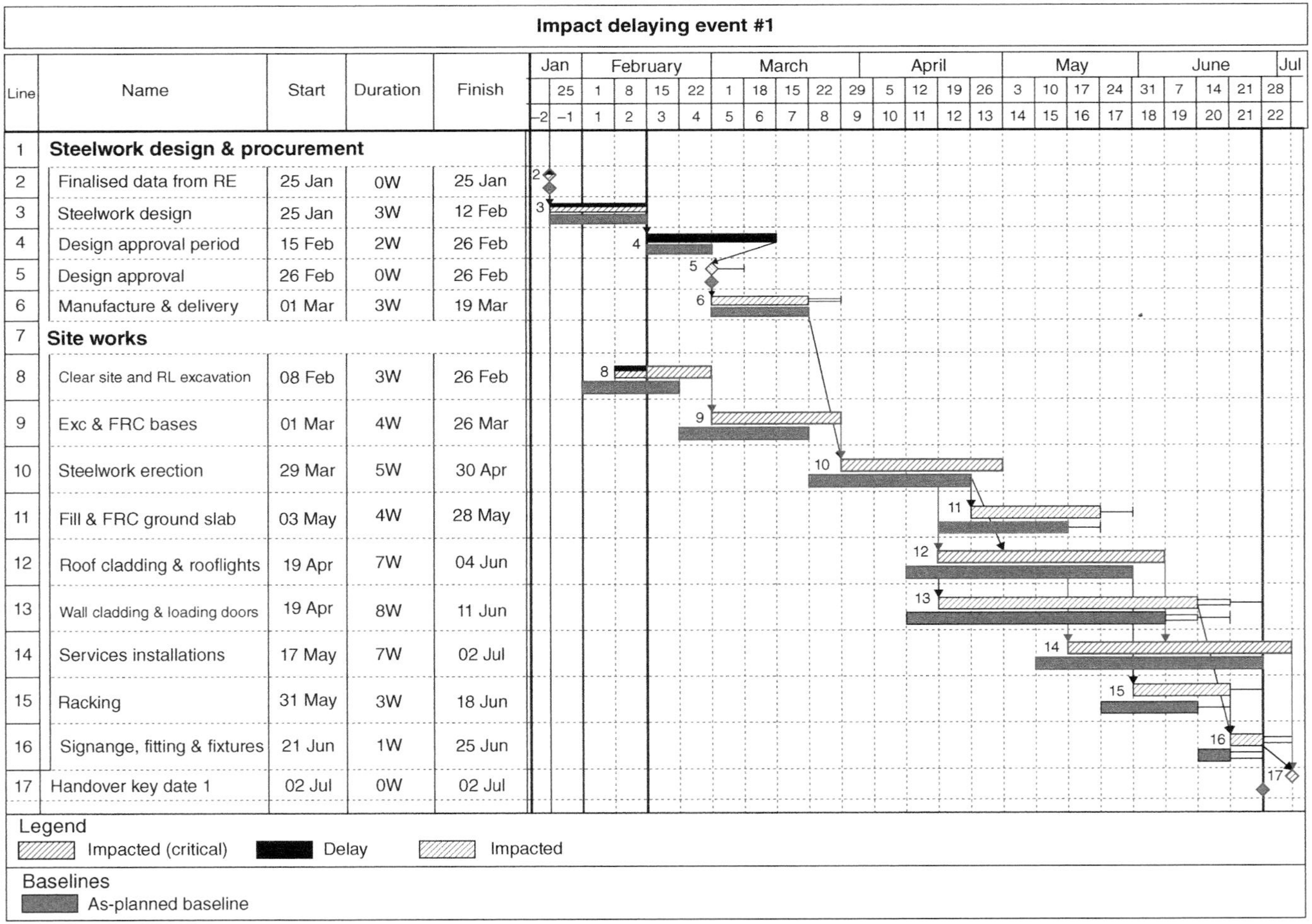
Impact delaying event #1
Line
Name
Start
Duration
Finish
Jan
February
March
April
May
June
Jul
25 1 8 15 22 1 18 15 22 29 5 12 19 26 3 10 17 24 31 7 14 21 28
–2 –1 1 2 3 4 5 6 7 8 9 10 11 12 13 14 15 16 17 18 19 20 21 22
1 Steelwork design & procurement
2 Finalised data from RE 25 Jan 0W 25 Jan
3 Steelwork design 25 Jan 3W 12 Feb
4 Design approval period 15 Feb 2W 26 Feb
5 Design approval 26 Feb 0W 26 Feb
6 Manufacture & delivery 01 Mar 3W 19 Mar
7 Site works
8 Clear site and RL excavation 08 Feb 3W 26 Feb
9 Exc & FRC bases 01 Mar 4W 26 Mar
10 Steelwork erection 29 Mar 5W 30 Apr
11 Fill & FRC ground slab 03 May 4W 28 May
12 Roof cladding & rooflights 19 Apr 7W 04 Jun
13 Wall cladding & loading doors 19 Apr 8W 11 Jun
14 Services installations 17 May 7W 02 Jul
15 Racking 31 May 3W 18 Jun
16 Signange, fitting & fixtures 21 Jun 1W 25 Jun
17 Handover key date 1 02 Jul 0W 02 Jul
Legend
Impacted (critical)
Delay
Impacted
Baselines
As-planned baseline

Figure 15.3 (*Continued*)

(d)

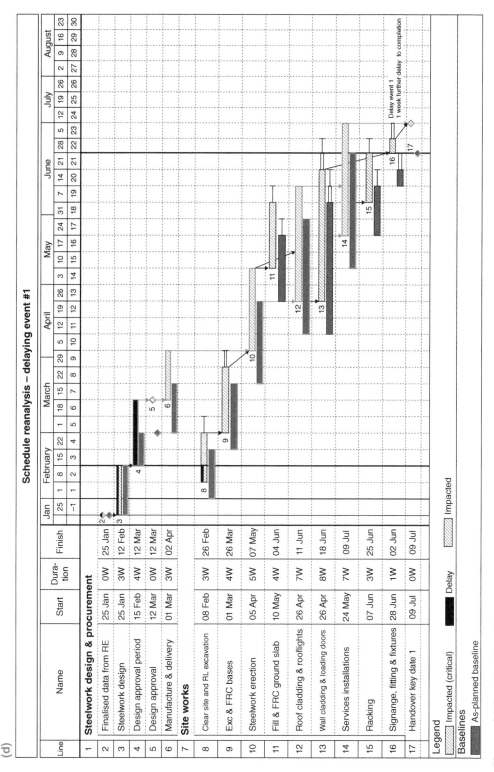

Schedule reanalysis – delaying event #1

Line	Name	Start	Dura-tion	Finish
1	**Steelwork design & procurement**			
2	Finalised data from RE	25 Jan	0W	25 Jan
3	Steelwork design	25 Jan	3W	12 Feb
4	Design approval period	15 Feb	4W	12 Mar
5	Design approval	12 Mar	0W	12 Mar
6	Manufacture & delivery	01 Mar	3W	02 Apr
7	**Site works**			
8	Clear site and RL excavation	08 Feb	3W	26 Feb
9	Exc & FRC bases	01 Mar	4W	26 Mar
10	Steelwork erection	05 Apr	5W	07 May
11	Fill & FRC ground slab	10 May	4W	04 Jun
12	Roof cladding & rooflights	26 Apr	7W	11 Jun
13	Wall cladding & loading doors	26 Apr	8W	18 Jun
14	Services installations	24 May	7W	09 Jul
15	Racking	07 Jun	3W	25 Jun
16	Signnange, fitting & fixtures	28 Jun	1W	02 Jul
17	Handover key date 1	09 Jul	0W	09 Jul

Legend

Impacted (critical)   Delay   Impacted

Baselines

As-planned baseline

Figure 15.3 *(Continued)*

Section IV

(e)

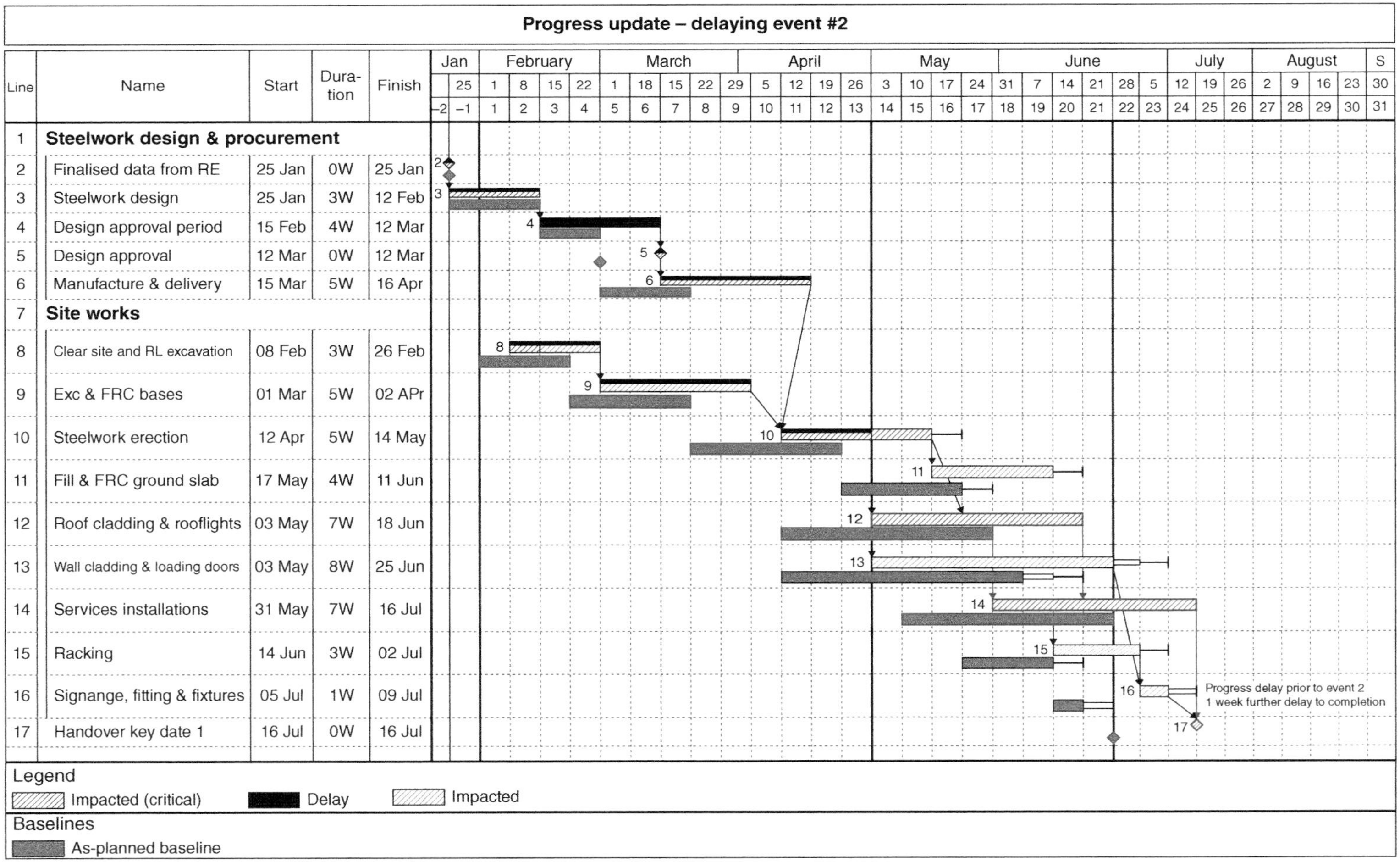
Progress update – delaying event #2
Line
Name
Start
Dura-tion
Finish
Jan
February
March
April
May
June
July
August
S
25 1 8 15 22 1 18 15 22 29 5 12 19 26 3 10 17 24 31 7 14 21 28 5 12 19 26 2 9 16 23 30
–2 –1 1 2 3 4 5 6 7 8 9 10 11 12 13 14 15 16 17 18 19 20 21 22 23 24 25 26 27 28 29 30 31
1 Steelwork design & procurement
2 Finalised data from RE 25 Jan 0W 25 Jan
3 Steelwork design 25 Jan 3W 12 Feb
4 Design approval period 15 Feb 4W 12 Mar
5 Design approval 12 Mar 0W 12 Mar
6 Manufacture & delivery 15 Mar 5W 16 Apr
7 Site works
8 Clear site and RL excavation 08 Feb 3W 26 Feb
9 Exc & FRC bases 01 Mar 5W 02 APr
10 Steelwork erection 12 Apr 5W 14 May
11 Fill & FRC ground slab 17 May 4W 11 Jun
12 Roof cladding & rooflights 03 May 7W 18 Jun
13 Wall cladding & loading doors 03 May 8W 25 Jun
14 Services installations 31 May 7W 16 Jul
15 Racking 14 Jun 3W 02 Jul
16 Signange, fitting & fixtures 05 Jul 1W 09 Jul
17 Handover key date 1 16 Jul 0W 16 Jul
Progress delay prior to event 2
1 week further delay to completion
Legend
Impacted (critical)
Delay
Impacted
Baselines
As-planned baseline

Figure 15.3 *(Continued)*

(f)

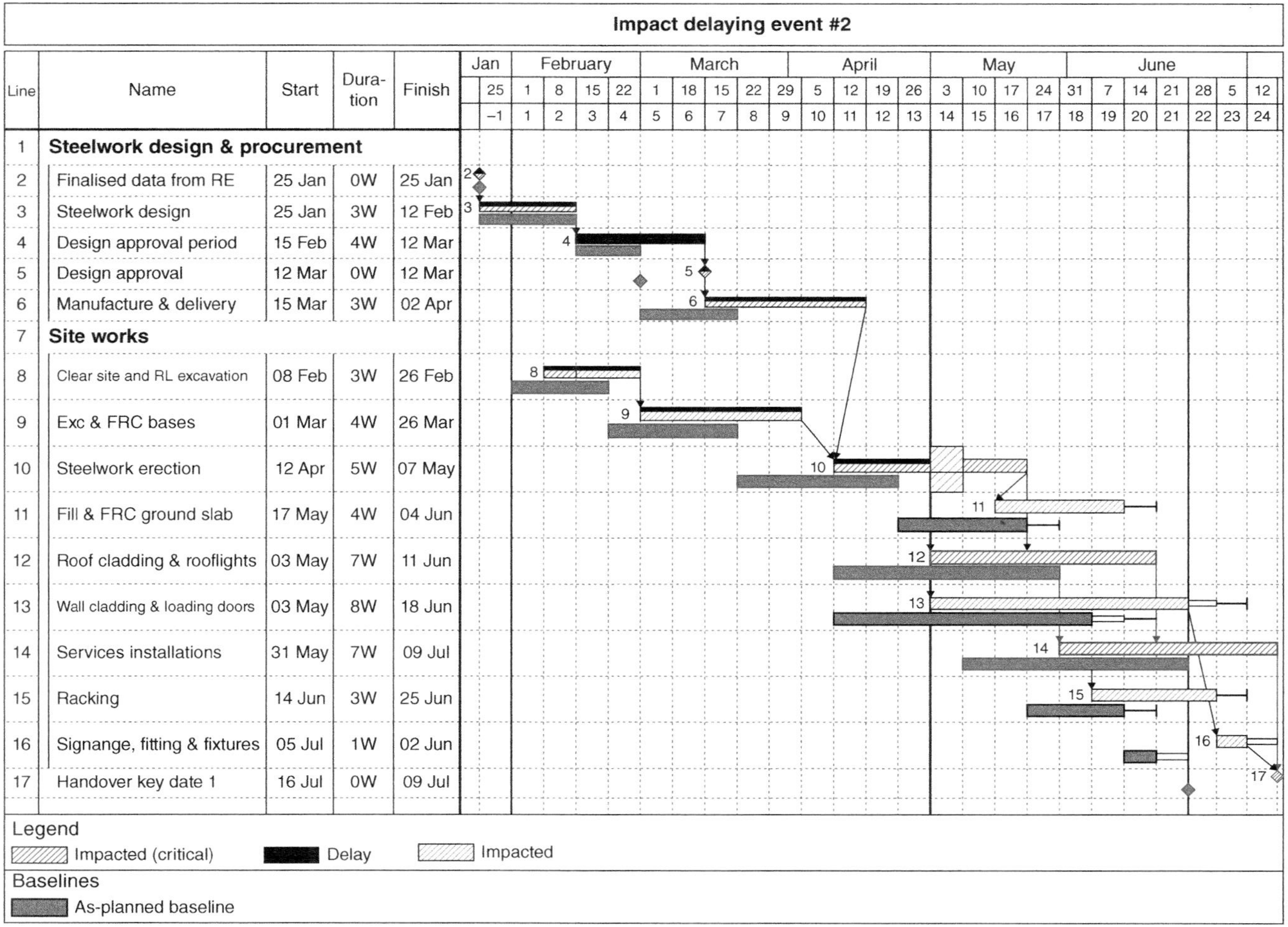

Impact delaying event #2

Line	Name	Start	Duration	Finish
1	**Steelwork design & procurement**			
2	Finalised data from RE	25 Jan	0W	25 Jan
3	Steelwork design	25 Jan	3W	12 Feb
4	Design approval period	15 Feb	4W	12 Mar
5	Design approval	12 Mar	0W	12 Mar
6	Manufacture & delivery	15 Mar	3W	02 Apr
7	**Site works**			
8	Clear site and RL excavation	08 Feb	3W	26 Feb
9	Exc & FRC bases	01 Mar	4W	26 Mar
10	Steelwork erection	12 Apr	5W	07 May
11	Fill & FRC ground slab	17 May	4W	04 Jun
12	Roof cladding & rooflights	03 May	7W	11 Jun
13	Wall cladding & loading doors	03 May	8W	18 Jun
14	Services installations	31 May	7W	09 Jul
15	Racking	14 Jun	3W	25 Jun
16	Signange, fitting & fixtures	05 Jul	1W	02 Jun
17	Handover key date 1	16 Jul	0W	09 Jul

Legend

Impacted (critical)   Delay   Impacted

Baselines

As-planned baseline

Figure 15.3   (*Continued*)

Section IV

(g)

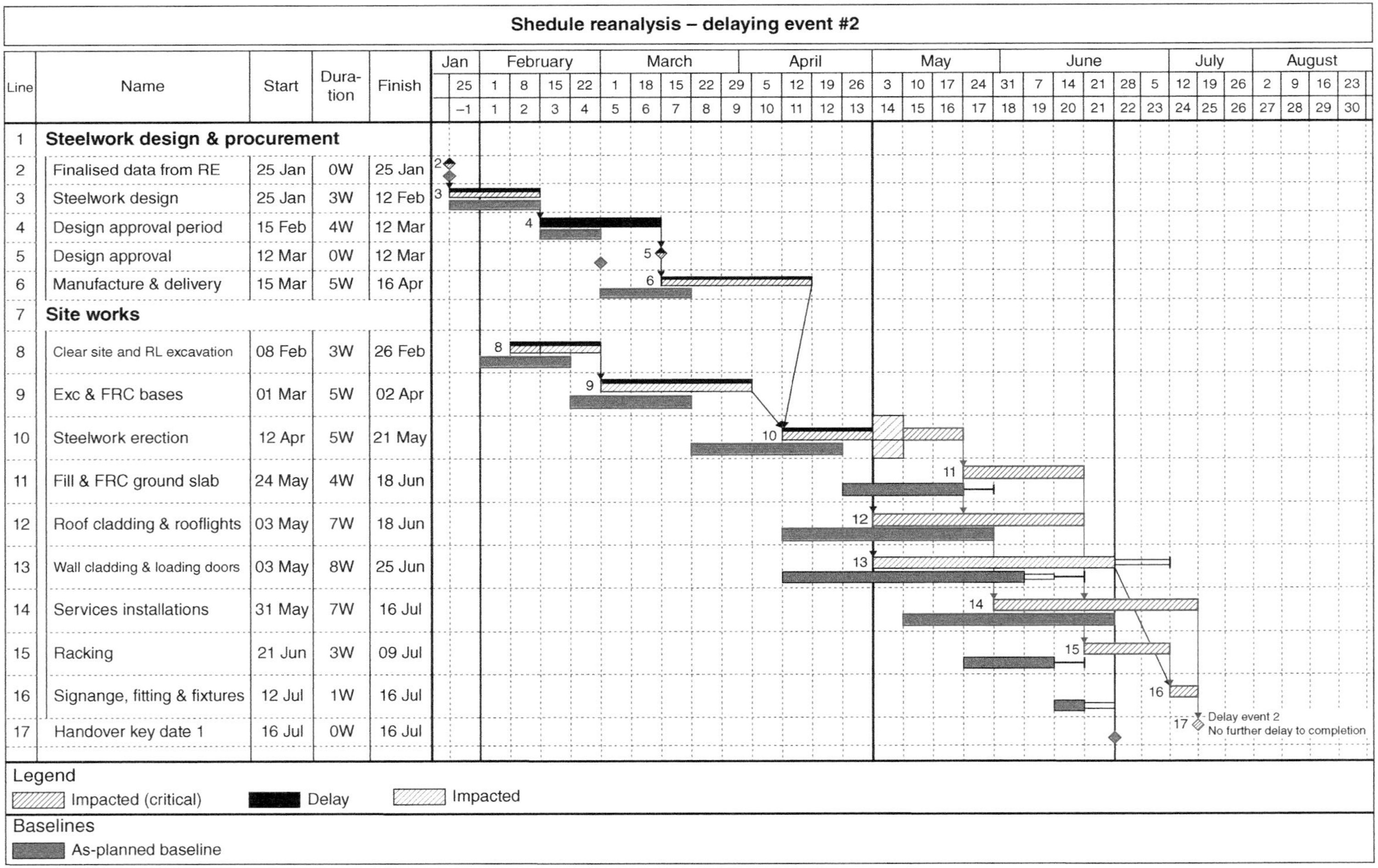
Shedule reanalysis – delaying event #2
Line | Name | Start | Duration | Finish
1 Steelwork design & procurement
2 Finalised data from RE | 25 Jan | 0W | 25 Jan
3 Steelwork design | 25 Jan | 3W | 12 Feb
4 Design approval period | 15 Feb | 4W | 12 Mar
5 Design approval | 12 Mar | 0W | 12 Mar
6 Manufacture & delivery | 15 Mar | 5W | 16 Apr
7 Site works
8 Clear site and RL excavation | 08 Feb | 3W | 26 Feb
9 Exc & FRC bases | 01 Mar | 5W | 02 Apr
10 Steelwork erection | 12 Apr | 5W | 21 May
11 Fill & FRC ground slab | 24 May | 4W | 18 Jun
12 Roof cladding & rooflights | 03 May | 7W | 18 Jun
13 Wall cladding & loading doors | 03 May | 8W | 25 Jun
14 Services installations | 31 May | 7W | 16 Jul
15 Racking | 21 Jun | 3W | 09 Jul
16 Signange, fitting & fixtures | 12 Jul | 1W | 16 Jul
17 Handover key date 1 | 16 Jul | 0W | 16 Jul
Jan February March April May June July August
Delay event 2
No further delay to completion
Legend
Impacted (critical)
Delay
Impacted
Baselines
As-planned baseline

Figure 15.3 *(Continued)*

(h)

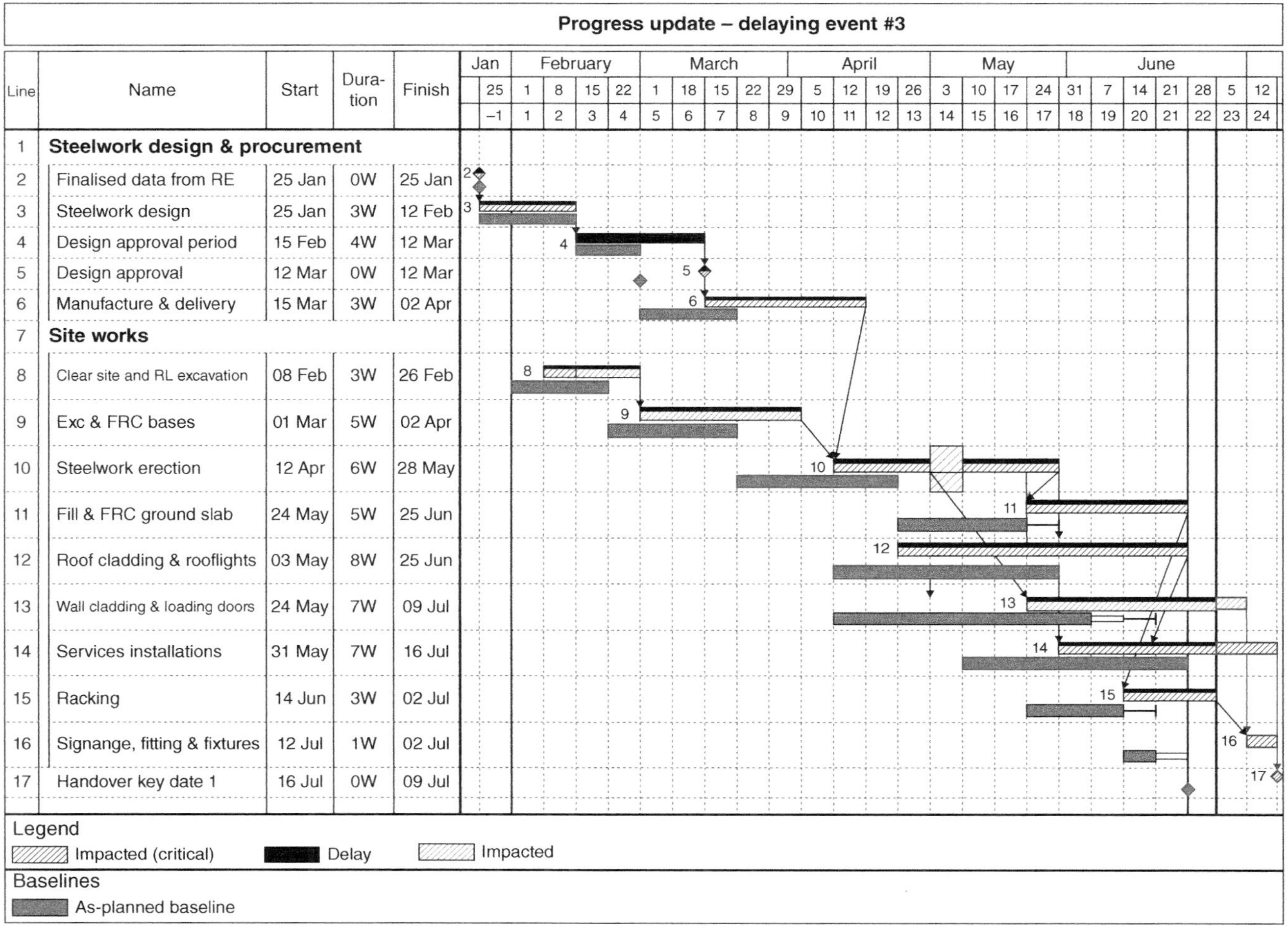

Progress update – delaying event #3

Line	Name	Start	Dura-tion	Finish	Jan		February				March				April				May					June				
					25	1	1	8	15	22	1	8	15	22	29	5	12	19	26	3	10	17	24	31	7	14	21	28	5	12
					−1	1	2	3	4	5	6	7	8	9	10	11	12	13	14	15	16	17	18	19	20	21	22	23	24
1	**Steelwork design & procurement**																													
2	Finalised data from RE	25 Jan	0W	25 Jan																										
3	Steelwork design	25 Jan	3W	12 Feb																										
4	Design approval period	15 Feb	4W	12 Mar																										
5	Design approval	12 Mar	0W	12 Mar																										
6	Manufacture & delivery	15 Mar	3W	02 Apr																										
7	**Site works**																													
8	Clear site and RL excavation	08 Feb	3W	26 Feb																										
9	Exc & FRC bases	01 Mar	5W	02 Apr																										
10	Steelwork erection	12 Apr	6W	28 May																										
11	Fill & FRC ground slab	24 May	5W	25 Jun																										
12	Roof cladding & rooflights	03 May	8W	25 Jun																										
13	Wall cladding & loading doors	24 May	7W	09 Jul																										
14	Services installations	31 May	7W	16 Jul																										
15	Racking	14 Jun	3W	02 Jul																										
16	Signange, fitting & fixtures	12 Jul	1W	02 Jul																										
17	Handover key date 1	16 Jul	0W	09 Jul																										

Legend

Impacted (critical)   Delay   Impacted

Baselines

As-planned baseline

Figure 15.3 (Continued)

Section IV

Impact delaying event #3

Line	Name	Start	Duration	Finish
1	**Steelwork design & procurement**			
2	Finalised data from RE	25 Jan	0W	25 Jan
3	Steelwork design	25 Jan	3W	12 Feb
4	Design approval period	15 Feb	4W	12 Mar
5	Design approval	12 Mar	0W	12 Mar
6	Manufacture & delivery	15 Mar	5W	16 Apr
7	**Site works**			
8	Clear site and RL excavation	08 Feb	3W	26 Feb
9	Exc & FRC bases	01 Mar	5W	02 Apr
10	Steelwork erection	12 Apr	6W	28 May
11	Fill & FRC ground slab	24 May	5W	25 Jun
12	Roof cladding & rooflights	03 May	8W	25 Jun
13	Wall cladding & loading doors	24 May	7W	09 Jul
14	Services installations	31 May	7W	16 Jul
15	Racking	14 Jun	3W	02 Jul
16	Additional racking	05 Jul	1W	09 Jul
17	Signange, fitting & fixtures	12 Jul	1W	16 Jul
18	Handover key date 1	16 Jul	0W	16 Jul

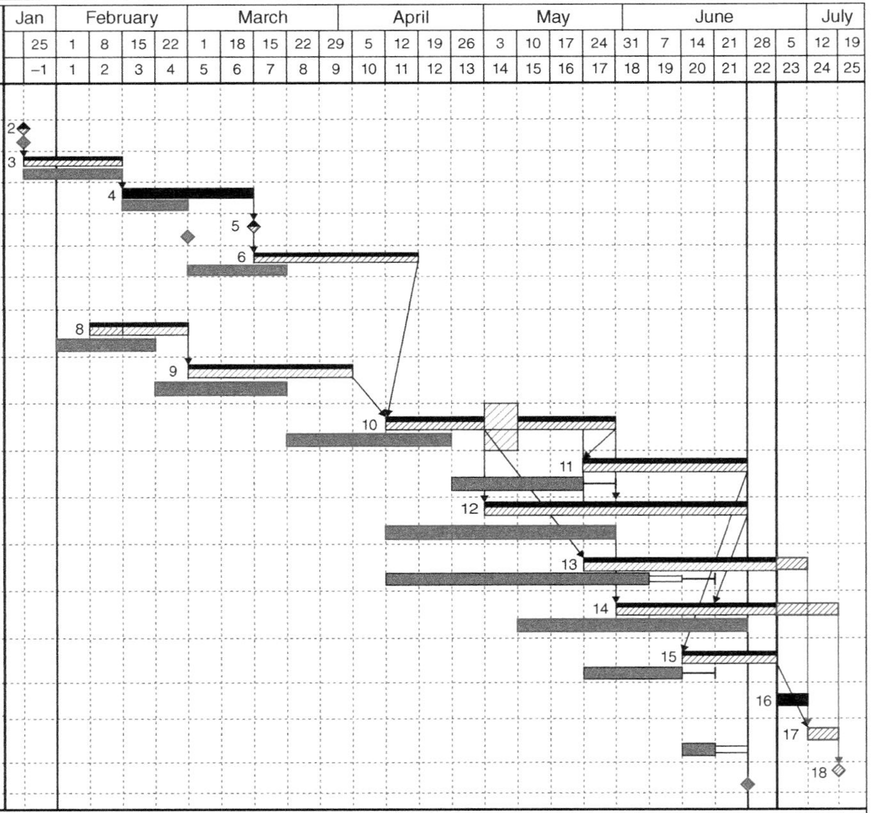

Legend

Impacted (critical)   Delay   Impacted

Baselines

As-planned baseline

Figure 15.3 (Continued)

(j)

Shedule reanalysis – delaying event #2

Line	Name	Start	Duration	Finish
1	**Steelwork design & procurement**			
2	Finalised data from RE	25 Jan	0W	25 Jan
3	Steelwork design	25 Jan	3W	12 Feb
4	Design approval period	15 Feb	4W	12 Mar
5	Design approval	12 Mar	0W	12 Mar
6	Manufacture & delivery	15 Mar	5W	16 Apr
7	**Site works**			
8	Clear site and RL excavation	08 Feb	3W	26 Feb
9	Exc & FRC bases	01 Mar	5W	02 Apr
10	Steelwork erection	12 Apr	5W	21 May
11	Fill & FRC ground slab	24 May	4W	18 Jun
12	Roof cladding & rooflights	03 May	7W	18 Jun
13	Wall cladding & loading doors	03 May	8W	25 Jun
14	Services installations	31 May	7W	16 Jul
15	Racking	21 Jun	3W	09 Jul
16	Aditional racking	05 Jul	1W	09 Jul
17	Signange, fitting & fixtures	12 Jul	1W	16 Jul
18	Handover key date 1	16 Jul	0W	16 Jul

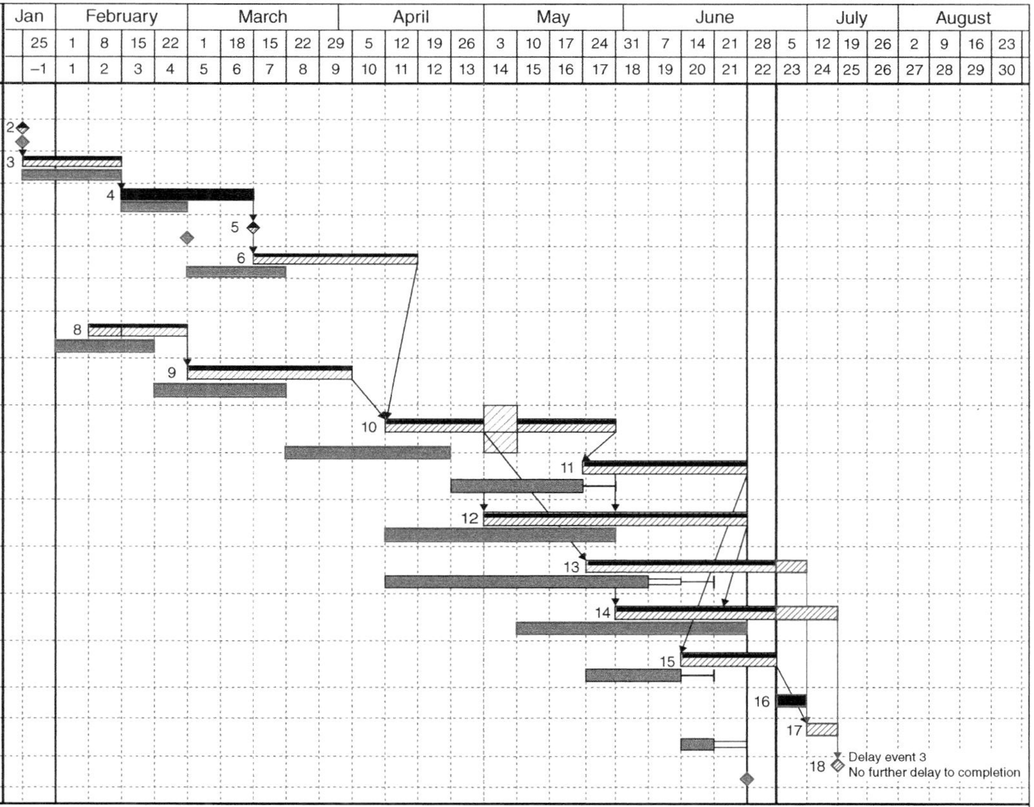

Legend

Impacted (critical)   Delay   Impacted

Baselines

As-planned baseline

Figure 15.3 (Continued)

Section IV

(k)

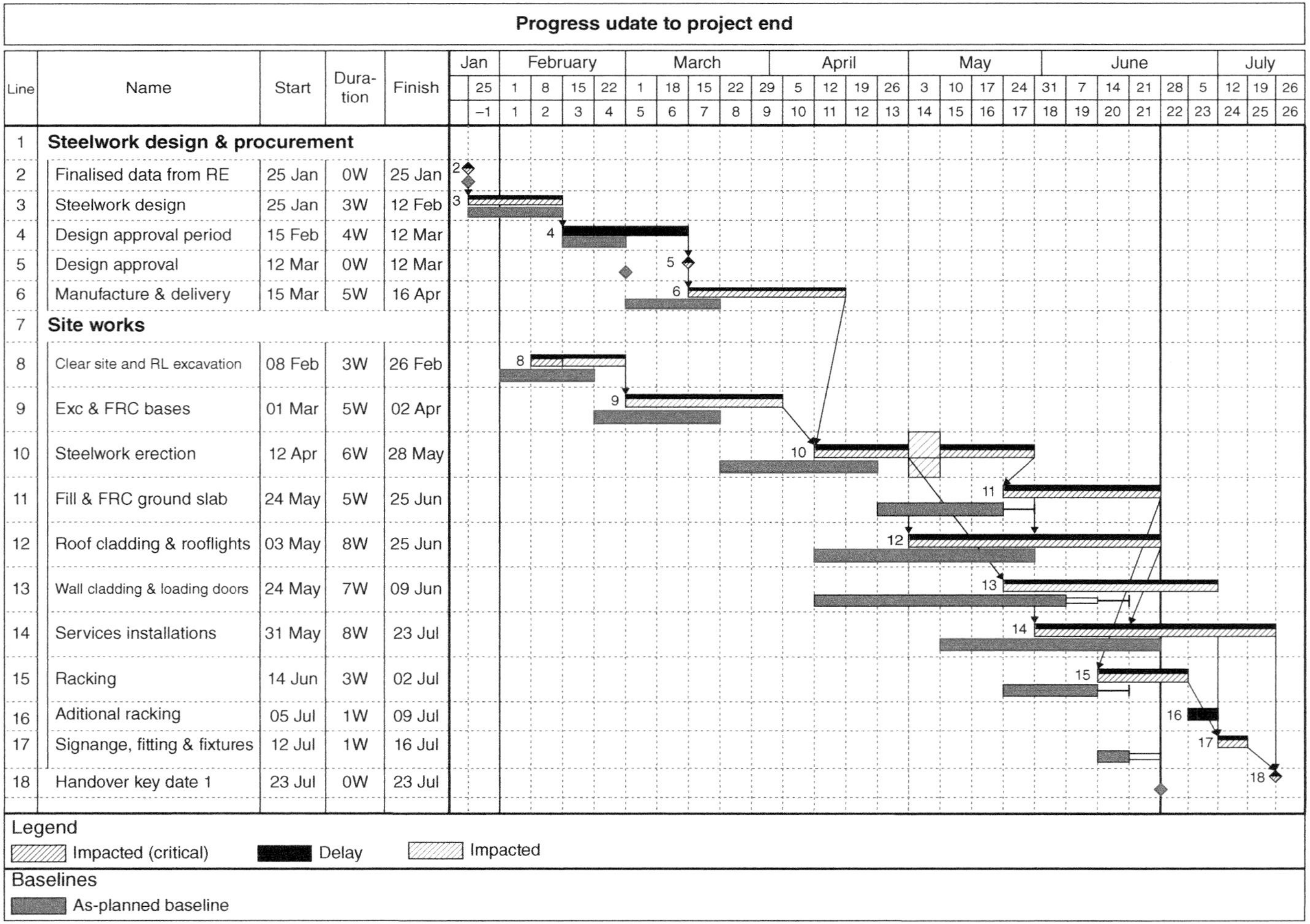
Progress udate to project end
Line | Name | Start | Duration | Finish
1 Steelwork design & procurement
2 Finalised data from RE 25 Jan 0W 25 Jan
3 Steelwork design 25 Jan 3W 12 Feb
4 Design approval period 15 Feb 4W 12 Mar
5 Design approval 12 Mar 0W 12 Mar
6 Manufacture & delivery 15 Mar 5W 16 Apr
7 Site works
8 Clear site and RL excavation 08 Feb 3W 26 Feb
9 Exc & FRC bases 01 Mar 5W 02 Apr
10 Steelwork erection 12 Apr 6W 28 May
11 Fill & FRC ground slab 24 May 5W 25 Jun
12 Roof cladding & rooflights 03 May 8W 25 Jun
13 Wall cladding & loading doors 24 May 7W 09 Jun
14 Services installations 31 May 8W 23 Jul
15 Racking 14 Jun 3W 02 Jul
16 Aditional racking 05 Jul 1W 09 Jul
17 Signange, fitting & fixtures 12 Jul 1W 16 Jul
18 Handover key date 1 23 Jul 0W 23 Jul
Jan February March April May June July
25 1 8 15 22 1 18 15 22 29 5 12 19 26 3 10 17 24 31 7 14 21 28 5 12 19 26
–1 1 2 3 4 5 6 7 8 9 10 11 12 13 14 15 16 17 18 19 20 21 22 23 24 25 26
Legend
Impacted (critical)
Delay
Impacted
Baselines
As-planned baseline

Figure 15.3 *(Continued)*

- The final progress update of the schedule, recording progress and actual dates to the end of the project are shown in Figure 15.3k. A further week has been lost in this period, to 23 July, which appears to be as a result of the delay to the completion of the service installations.

There are a number of variations to the basic procedure, and where possible, the precise method to be adopted should be documented. If possible, the precise method should be agreed between the parties in dispute especially since there is no exact definition of time impact analysis (or any other method for that matter). How the following are to be addressed should be considered:

- **Sequence of modelling**: What is the date that the delaying event should be modelled? Should it be when the parties first became aware of the delaying event or when the event first started to have an effect?
- **Concurrent delay**: Time impact analysis reflects the 'first-in-line' method of determining the liability in concurrent delay situations. True concurrency occurs when two or more delaying events occur at the same time and their effects are felt at the same time. Where two or more events happen at the same time, they should be both modelled simultaneously in step 5 and subsequent iterations. Inaddition, the two or more simultaneously occurring events should also be modelled separately to determine the individual effect of each delaying event. This analysis will demonstrate the combined effect and the individual effects of the concurrent events. Unfortunately, without precise obligations in the particular contract, there are no clear guidelines to determine the liability for concurrent delays each matter will rest on its merits ultimately to be determined by the tribunal in a dispute.
- **Mitigation**: In most jurisdictions, and often as a contract obligation, the parties are required to mitigate the effects of a delaying event.[6] How then should the extent, if any, of mitigating action be assessed? There are two approaches.
  a) At stage 5 (and subsequently after the modelling of the delaying event), consideration should be given as to what mitigating action ought reasonably to have been taken to reduce or minimise the effect of the delay. This is often a subjective assessment and should be viewed from the position of the parties at the time of the event; in these circumstances, the benefit of hindsight is *not* justifiable, 'if only we knew then what we know now, we would have done something different'. For instance, it might be reasonable to expect the contractor to use idle resources on the project but not to engage additional staff and plant.
  b) On completion of the analysis, a review of the delays and any mitigation measures taken throughout the project to determine if they were reasonable taking account of the combined effect of the various delaying events and effects. Again, the review should be taken and seen from the position of the parties at the time of the delays. For instance, an individual delaying

---

[6]   Each contract should be examined individually to determine the extent of the party's obligations to mitigate delay, for instance, 'to do what is reasonable', 'to take all reasonable steps' or 'to take all measures'.

event might not merit exclusive action, but the combined effect of a number of delaying events might have made it obvious and viable that more wide-ranging action should have been taken.

- **Progress**: The progress made on the project is integral to the results of the analysis. Where it differs from that planned, that is generally attributed to the actions of the contractor either by good fortune or design going faster than planned or through miscellaneous inadequate actions going slower than planned. In fact, some analysts contend that it is not necessary to model culpable category delays as these will be reflected in the progress of the contractor. However, contractors' progress can also be affected by:

a) Mitigating action taken to reduce the effects of delay.

b) Acceleration measures taken to reduce delays or to bring forward the completion date.

c) Pacing – the concept of adjusting productivity to match the current progress of the works in general. For instance, if the works are in delay as a result of excusable events, it may be more economical for the contractor to reduce his workforce so that its progress keeps pace with the overall progress of the project, 'why run now to stand still later?'

d) The effects of multiple and overlapping delays often have the effect of reducing overall productivity beyond the direct effects of the individual delaying event and is likely to manifest itself in progress less than that planned.

e) The effects of unidentified delaying events.

Where there are differences between planned progress and actual progress, this should not be assumed to be the 'fault' of the contractor. Occurrences should be investigated to determine the cause of progress variance.

Following long and sustained delay on the part of the contractor, there is often seen towards the end of a project (when analysed using time impact analysis and similar techniques) that there is a sudden remarkable saving of time, and the eventual delay to completion is not what was forecast. In most cases, the contractor will say that the dramatic recovery of time is a result of mitigating action it has taken; this is almost always not the case. Examination of contemporaneous documents rarely shows any plans by the contractor to accelerate or undertake mitigation measures to recover lost time. In most cases, the apparent recovery is the result of finishing the last few per cent of many activities, an expected phenomenon towards the end of a project, which, as far as the analysis results show, is enhanced by the logic linking between the remaining parts of the activities, in particular finish-to-finish with lag type dependencies. After each progress updating of the schedule, the logic of out-of-sequence and part-completed activities should be examined to establish if it is still relevant.

The application of the time impact analysis method described uses retrospectively updated schedules as the multiple bases for the analysis. During the project, the contractor might have updated and amended the schedule to take account of not only current progress and the effects of delaying events but also re-sequencing, activity duration changes and other refinements to the schedule (perhaps to take account of the CIOB's recommendations for expanding schedule density of the schedule as the project proceeds). In these cases, it would be more accurate to substitute retrospectively produced updates with the contemporaneously updated schedules. This will enhance the reliability of the forecasts of likely completion of the project. When the as-planned schedule is substituted, it is likely that some adjustment and tweaking of progress data will be required, if it does not track the changes to the up-to-date schedule.

One of the major drawbacks levied against time impact analysis is that it is costly and timeconsuming to undertake. The reasons given for this are that it requires a robust as-planned networked schedule, detailed progress and as-built records, details of the delaying events, and as it models delaying events individually, it takes a long time to perform. All of the delay analysis techniques require one or more of those components; without them, all methods become less reliable. The general requirement that causation should be demonstrated, the link between cause and effect, is in fact a positive attribute of time impact analysis in that it can model delaying events on an individual basis.

The AACEI RP describes the method as Modelled/Additive/Multiple Base (AACEI, 2011, p. 75).

Collapsed as-built

The basis for the collapsed as-built method is a networked as-built schedule, and the method is the only broad classification of methods that do not require an as-planned schedule of any type. However, an as-planned schedule is beneficial in providing pointers to the logic that might be incorporated into the as-built schedule and to provide clues as to when some activities ought to have happened (information release dates, for example).

The as-built schedule for use in collapsed as-built analysis must have the effects of delaying events included in the schedule, and where possible, these should be shown as individual activities. It is particularly difficult to produce an accurate as-built network. For instance, the start and finish dates of an activity may be known, and it is also known that the activity was subject to delay in that the quantity of the work was increased – the question is, how much of the activity is the delayed part and how much was the original part without having an as-planned schedule for reference.

The compilation of a detailed as-built schedule depends on detailed and comprehensive as-built records.

It is difficult to comprehend a situation where such detailed records exist but there is no as-planned schedule. Given the problems with constructing an as-built network and the subjective decisions and opportunity to manipulate the critical path (subconsciously or intentionally), the method is really only suitable for simple projects, where the majority of relationships between activities are sequential and the delaying events and effects are obvious and readily isolated.

Where the works were subject to multiple overlapping delaying events, acceleration, mitigation, pacing and concurrency, it is unlikely that a collapsed as-built analysis will be able to unravel the causes and liability for the delays. If the works were subject to changes in sequence from those originally envisaged (for instance, if possession of the site was given in a different order to that

Section IV

originally specified), then the as-built schedule will only collapse back to the changed sequence.

The collapsed as-built method is the opposite of the impacted as-planned method. Instead of adding delays to the schedule, the analyst subtracts the delays. After developing the as-built schedule and listing and categorising the delaying events built into the schedule, there are then two options for impacting the delays: Single Extraction and Individual Extraction.

Single Extraction

All of the delaying events are removed from the as-built schedule in a single step and the schedule is then rescheduled.

There are a number of ways in which the delaying events can be removed:

i)   By simply deleting the delay activity. (However, this is likely to result in dangles in the as-built schedule will need to be corrected by the addition of logic – with the same problems of accuracy, suitability and subjectivity.)

ii) By changing the duration of the delay modelled to zero although this will have no effect on date type delays that are modelled using 'milestones'.

iii) By deleting the activity but maintaining delay event's the logic by linking its successor activities to its predecessor activities using the same dependencies (including any lead and lag durations).

In the case of date type delays, which are usually date constrained, the method of removal is to remove the constraint date so that the milestone can collapse back and the delayed activities be unaffected by the delay event. However, this may not reflect the reality of the delay. For instance, if the date type delay represented the late issue of information without which the relevant activity could not proceed, the collapsed date should represent the date the information was originally obliged to be issued. This will require changing the constraint date to the intended issue date.

Individual extraction

In this option, the delaying events are removed from the as-built schedule one by one in reverse chronological order and the network rescheduled after each individual removal.

The change to the completion date after each removal and rescheduling is recorded on the delay event list. This represents the delay resulting from each delaying event – some delaying events might not result in a further reduction to the completion date, their effect being subsumed by earlier removed delaying events.

When all the delaying events have been removed, the final resultant collapsed schedule will be identical to a collapsed schedule where all the delaying events were removed in a single extraction. The individual extraction option provides cause and effect detail for the individual delaying events.

The collapsed as-built method is illustrated in the following example.

- The starting point for the analysis is the compilation of an as-built barchart as shown in Figure 15.4a. The data for the schedule are derived from the project documentation, progress reports, meeting minutes, photographs and so on. The as-built barchart shows that the project completed on 23 July, 4 weeks later than the contractual completion date of 25 June.

- The activities in the as-built barchart are then linked to form the as-built network(see Figure 15.4b). The as-built barchart is shown as a baseline for comparison.

  The first delaying event to be 'extracted' from the as-built network is activity 16, the additional racking. This is simulated by reducing the activity duration to zero. The schedule is then reanalysed, and the result is shown in Figure 15.4c.
- The handover date remains at 23 July so the additional racking is said not to have contributed to the delay to the completion of the project.
- The second delaying event to be extracted is the break in the erection of the structural steelwork; this is simulated by deleting the finish to start with lag link between the two segments of activity 10 as shown in Figure 15.4d.
- Again, the handover date remains at 23 July so the 1 week stop work on the structural steelwork is said not to have contributed to the delay to the completion of the project.
- The third delaying event to be extracted is the extended drawing approval period, which was, according to the project documentation, meant to be a maximum of 2 weeks. This is simulated by reducing the duration of activity 4 from 4 weeks to 2 weeks as shown in Figure 15.4e.
- The handover date has moved to 16 July, a reduction of 1 week resulting from the reduced design approval period. The increase in the design approval period is said to have caused1 week delay to the project.

Following rescheduling, it is likely that the critical path will have changed and the planned completion date of the schedule is likely to have moved to an earlier date indicating what would have happened if the delaying events had not occurred or the position 'but for' the delaying events.

It is more probable than not that the change to the date completion will not match the original contract period. In most cases, the as-built schedule will not collapse back as far as the contract date; this is an indication that if all delaying events had been correctly modelled, that there was some residual delay probably caused by various inactions of the contractor or unidentified delays by either party. If the as-built schedule collapses to a date earlier than the original contract date, this indicates that the project could have been completed in a shorter period than envisaged or that acceleration measures were employed somewhere during the project.

The AACEI RP describes the method as Modelled/Subtractive/Single Base (AACEI, 2011, paragraph 3.8, pp. 82–89).

Windows

Despite the SCL Protocol making no reference to the windows method of analysis and the AACEI RP showing that six of its nine broad categories of analysis

(a)

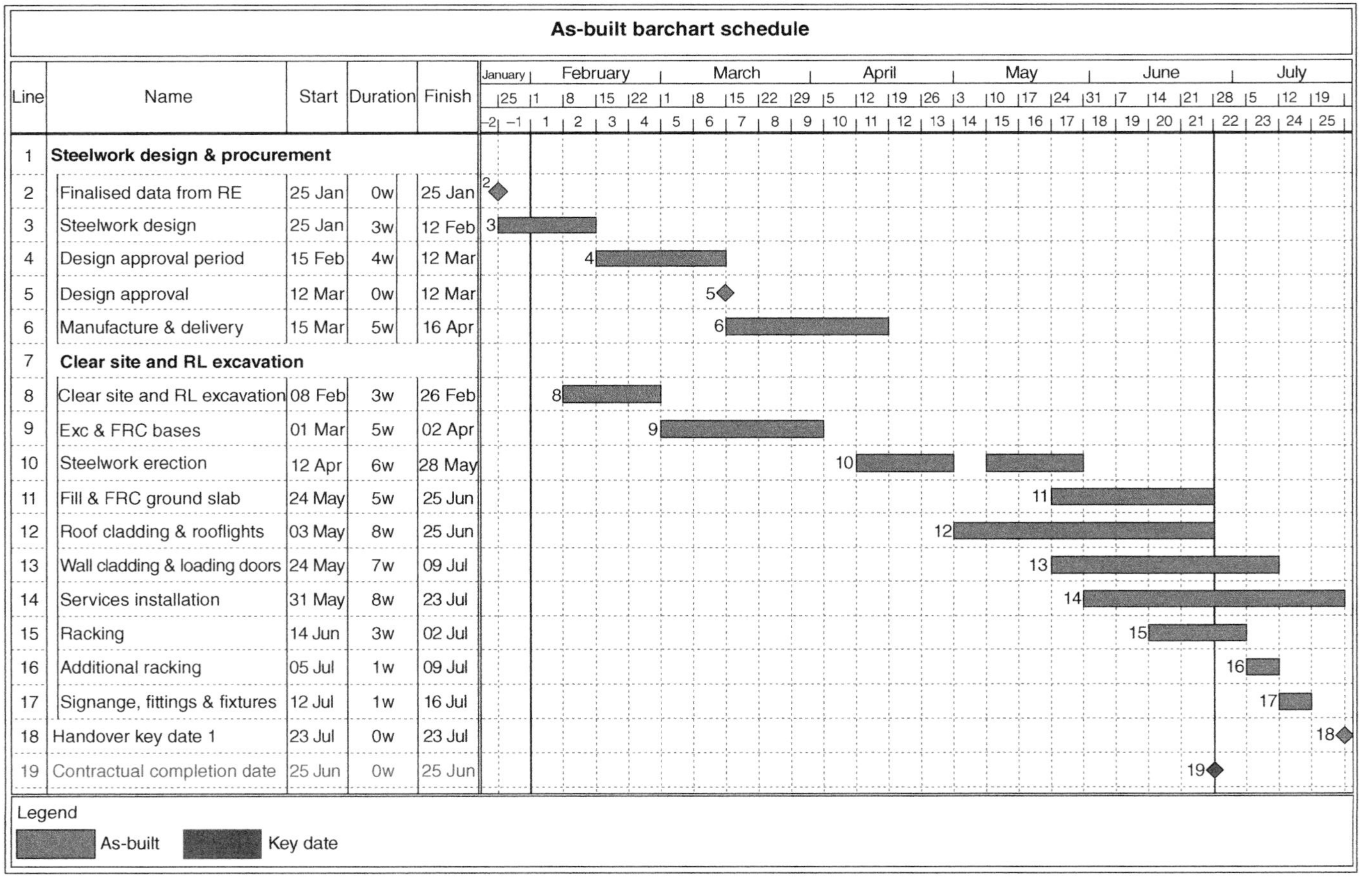
As-built barchart schedule
Line
Name
Start
Duration
Finish
January
February
March
April
May
June
July
1 Steelwork design & procurement
2 Finalised data from RE 25 Jan 0w 25 Jan
3 Steelwork design 25 Jan 3w 12 Feb
4 Design approval period 15 Feb 4w 12 Mar
5 Design approval 12 Mar 0w 12 Mar
6 Manufacture & delivery 15 Mar 5w 16 Apr
7 Clear site and RL excavation
8 Clear site and RL excavation 08 Feb 3w 26 Feb
9 Exc & FRC bases 01 Mar 5w 02 Apr
10 Steelwork erection 12 Apr 6w 28 May
11 Fill & FRC ground slab 24 May 5w 25 Jun
12 Roof cladding & rooflights 03 May 8w 25 Jun
13 Wall cladding & loading doors 24 May 7w 09 Jul
14 Services installation 31 May 8w 23 Jul
15 Racking 14 Jun 3w 02 Jul
16 Additional racking 05 Jul 1w 09 Jul
17 Signange, fittings & fixtures 12 Jul 1w 16 Jul
18 Handover key date 1 23 Jul 0w 23 Jul
19 Contractual completion date 25 Jun 0w 25 Jun
Legend
As-built
Key date

Figure 15.4 The collapsedas-built method.

(b)

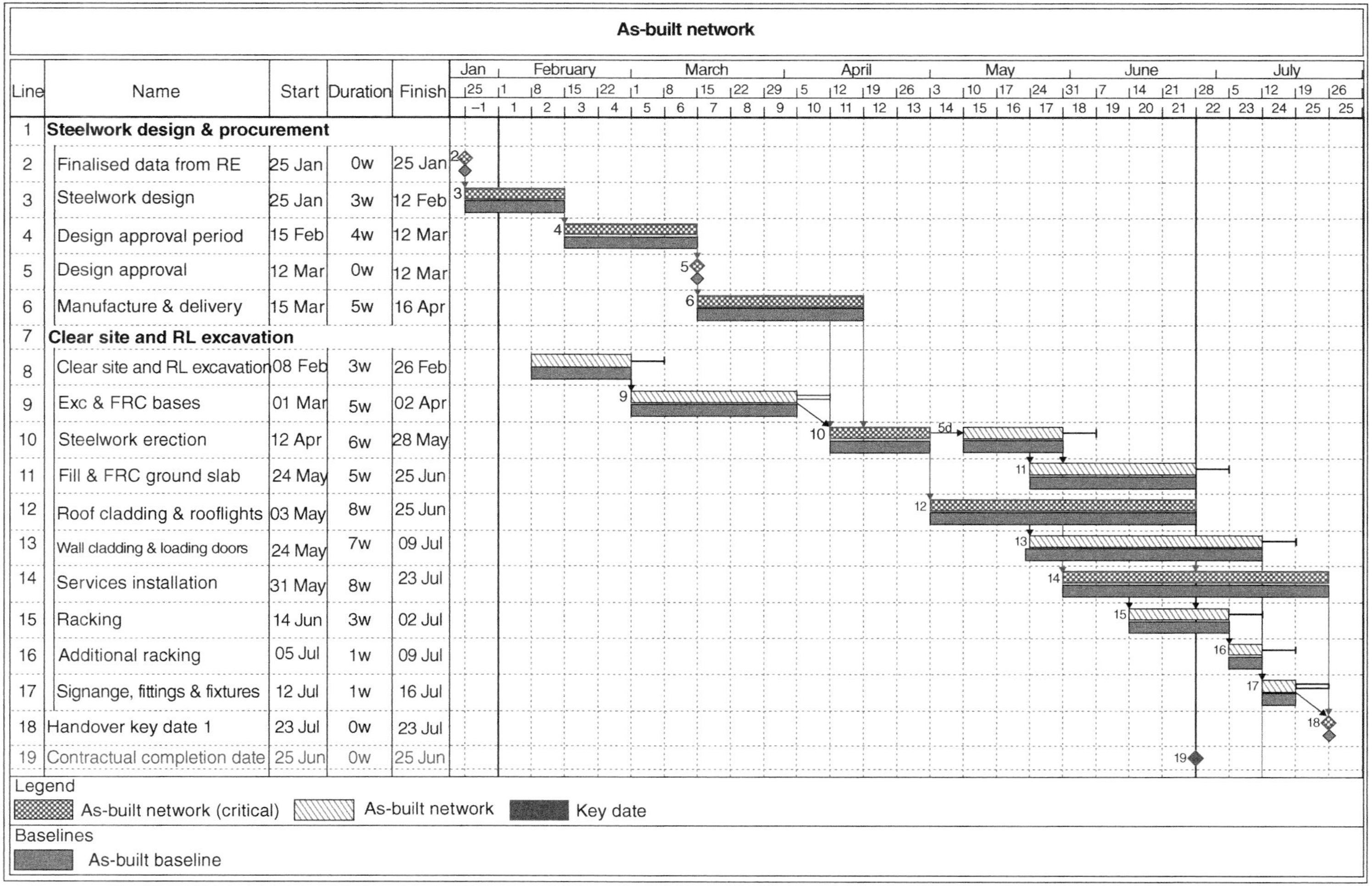

As-built network

Line	Name	Start	Duration	Finish																																					

Figure 15.4 (Continued)

Section IV

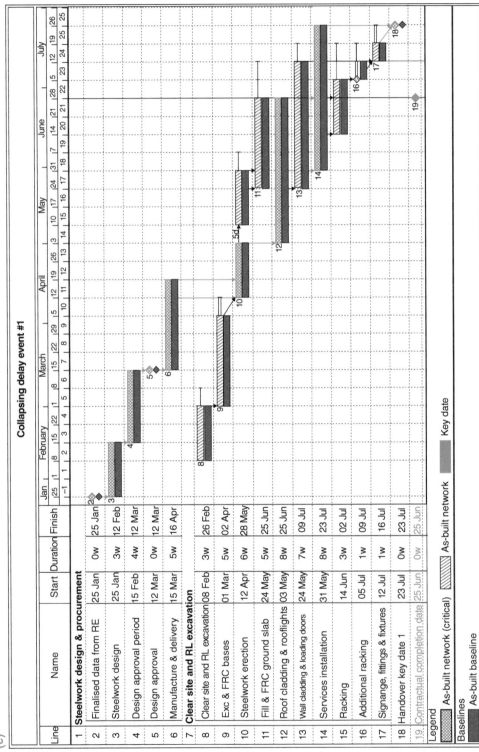

Collapsing delay event #1

Line	Name	Start	Duration	Finish	Jan	February	March	April	May	June	July
1	**Steelwork design & procurement**										
2	Finalised data from RE	25 Jan	0w	25 Jan							
3	Steelwork design	25 Jan	3w	12 Feb							
4	Design approval period	15 Feb	4w	12 Mar							
5	Design approval	12 Mar	0w	12 Mar							
6	Manufacture & delivery	15 Mar	5w	16 Apr							
7	**Clear site and RL excavation**										
8	Clear site and RL excavation	08 Feb	3w	26 Feb							
9	Exc & FRC bases	01 Mar	5w	02 Apr							
10	Steelwork erection	12 Apr	6w	28 May							
11	Fill & FRC ground slab	24 May	5w	25 Jun							
12	Roof cladding & rooflights	03 May	8w	25 Jun							
13	Wall cladding & loading doors	24 May	7w	09 Jul							
14	Services installation	31 May	8w	23 Jul							
15	Racking	14 Jun	3w	02 Jul							
16	Additional racking	05 Jul	1w	09 Jul							
17	Signange, fittings & fixtures	12 Jul	1w	16 Jul							
18	Handover key date 1	23 Jul	0w	23 Jul							
19	Contractual completion date	25 Jun	0w	25 Jun							

Legend

As-built network (critical)	As-built network	Key date

Baselines

As-built baseline

Figure 15.4 *(Continued)*

(c)

(d)

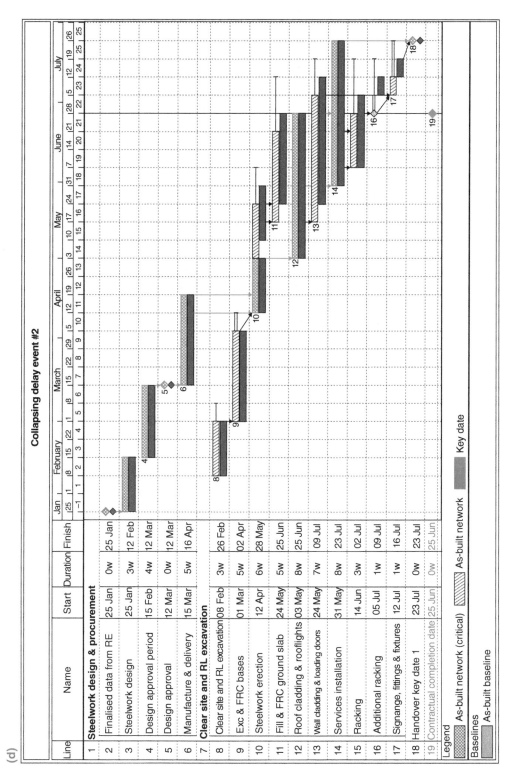

Line	Name	Start	Duration	Finish
1	**Steelwork design & procurement**			
2	Finalised data from RE	25 Jan	0w	25 Jan
3	Steelwork design	25 Jan	3w	12 Feb
4	Design approval period	15 Feb	4w	12 Mar
5	Design approval	12 Mar	0w	12 Mar
6	Manufacture & delivery	15 Mar	5w	16 Apr
7	**Clear site and RL excavation**			
8	Clear site and RL excavation	08 Feb	3w	26 Feb
9	Exc & FRC bases	01 Mar	5w	02 Apr
10	Steelwork erection	12 Apr	6w	28 May
11	Fill & FRC ground slab	24 May	5w	25 Jun
12	Roof cladding & rooflights	03 May	8w	25 Jun
13	Wall cladding & loading doors	24 May	7w	09 Jul
14	Services installation	31 May	8w	23 Jul
15	Racking	14 Jun	3w	02 Jul
16	Additional racking	05 Jul	1w	09 Jul
17	Signange, fittings & fixtures	12 Jul	1w	16 Jul
18	Handover key date 1	23 Jul	0w	23 Jul
19	Contractual completion date	25 Jun	0w	25 Jun

Legend

▨ As-built network (critical)   ▨ As-built network   ◆ Key date

Baselines
▨ As-built baseline

Figure 15.4 (*Continued*)

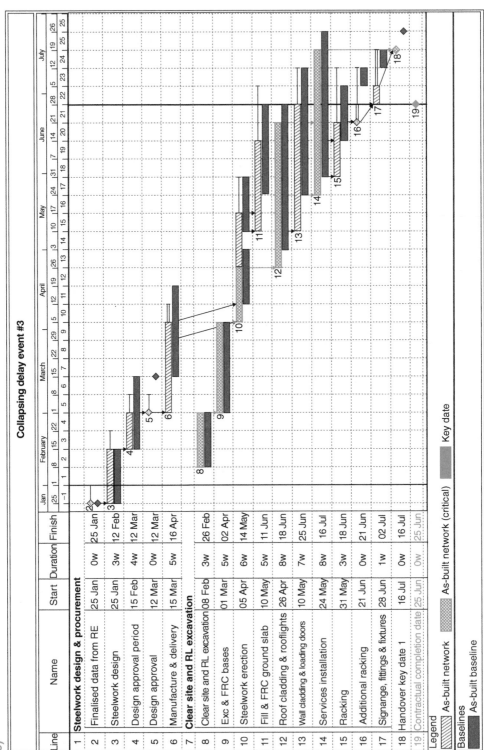

Figure 15.4 (*Continued*)

techniques are commonly called 'windows', windows analysis is the most accepted method of analysis even though it appears there is no precise definition even amongst experts (see *Costain v Charles Haswell* (2009)).

Windows generally involves dividing the project into regular or key segments (known as 'watersheds') to make the analysis more manageable or to provide focus to a specific area. Windows are also known as 'time slices', giving rise to a description 'time slice analysis'. Any of the previously described methods can be analysed using windows divisions.

As-Planned versus As-Built

The as-planned versus as-built schedule is most often divided into key date segments based on the as-built completion of those segments. The variance between the as-planned and as-built schedules in each section is analysed in detail such that the delay during each section is attributed to specific delaying events during that section. The incremental delay from each section is carried over to the next section so that as well as the delay in each section, the overall delay to the project is calculated. In *Cleveland Bridge v Severfield-Rowen* (2012):

> "…Although the two experts disagreed as to which Schedule should be used against which to measure delay, they did agree that 'at the most appropriate method for delay analysis is As-Planned versus As-Built Windows Analysis' as 'this method allows us to identify the actual critical path' and it 'is the longest path that the actual delays to completion will be located.'"

Time impact analysis

Windows analysis using time impact analysis as the basis can be undertaken using watersheds, but in the majority of cases, it uses regular intervals, usually based on the progress/as-built data presented at monthly site progress meetings.

In its original format, time impact analysis is delay event orientated in that it examines the delays individually. When the windows option is chosen, the analysis becomes time orientated such that delays within the time period are assessed as a group.

In this option, the as-planned schedule is updated at the beginning of the window period; this represents the status of the project at the beginning of the window. The delaying events occurring within the subsequent window period are then modelled at the same time and the schedule reanalysed; the resulting delay is attributed to the group of events. Further analysis of the resulting critical path within the window period will indicate the particular activities that are responsible for extending the schedule.

The size of the window has an influence on the accuracy of the analysis. The smallest window possible is one that contains only one delaying event; that equates to time impact analysis without windows. The largest possible window is one that encompasses the whole project; that equates to impacted-as-planned analysis. It is possible,by selecting differing time period windows, that the effect of some delaying events will be lost and be subsumed by subsequent delay events. Care must therefore be taken, especially when selecting watershed periods, that the time periods do not adversely affect the accuracy of the analysis.

As-Planned versus As-Built #2 Or Time Slice Analysis

The AACEI RP taxonomy allows accurate grouping of delay analysis types, and a method commonly referred to as windows analysis or time slice analysis is, in fact, an Observational/Dynamic Logic/Recreated Updates method. As it does not require the insertion of delaying events into the schedule, it is essentially an observation of what was planned compared to what was built by way of reference to updated schedules.

The method in implementation is similar to time impact analysis but without the modelling of delay events:

1. Collect all the relevant data including
    - The as-planned network (repaired as necessary), and
    - The as-built and progress data.
2. Determine the 'window' dates and the activity progress at the relevant dates; this might necessitate interpolation of data if the progress data and window dates do not coincide.
3. Update the as-planned schedule with the as-built/progress data and reanalyse the network and recalculate the critical path. This represents an updated network taking account of the progress of the project. The progress of activities is influenced by any delaying events so any change in the forecast completion date takes that into account and any other variances in progress due to mitigation, acceleration and disruption. At this stage, the network logic around the progress date should be examined to ensure it is still robust; out-of-sequence progress may result in illogical sequences and dangles.
4. Activities on the critical path within the window period are those that are driving the forecast; likely completion date and attention should be focussed there to determine what was responsible for the delays. This usually entails the examination of all the records, correspondence and so on to find anything relevant.
5. At the end of each window, the schedule is saved and copied, and the new copy renamed; the new copy of the updated network is used for the next stage of the analysis.

This method has the advantage that the critical path is calculated by the analysis rather than deduced. However, in comparison to time impact analysis, the cause and effect relationship to delays is not as immediately apparent as the delaying events themselves do not appear on the schedule.

As, in effect, the method is a comparison between planned and as-built, it is not possible to determine precisely what has contributed to the activity progress. For instance, a combination of better than expected productivity countered by a delay to the works might cancel each other out so there is no evident delay to the schedule and hence no need to examine the critical path for evidence of delays.

As less data are required to undertake this method of analysis, it tends to be less expensive and quicker to carry out than time impact analysis. It appears to be a method favoured by analysts perhaps because it is a blend of the strains of analysis, observational and dynamic.

Key points

- In the United Kingdom, a protocol for the analysis of delay and disruption has been established by the SCL. In the United States, the Claims & Dispute Resolution Committee of the Association has established a similar protocol for the AACEI. These protocols provide a basis for forensic schedule analysis as agreed by industry and legal practitioners.
- The type of data available and the scheduling techniques adopted will determine the type of analysis that may be undertaken.
- There are many types of analysis. (The SCL protocol lists 4; the AACEI lists at least 17.)
- All delay and disruption disputes have their idiosyncrasies that require an analyst to be flexible in their approach.
- The most common approaches to delay analysis areas-planned versus as-built; impacted as-planned; collapsed as-built; time impact analysis; and windows analysis.
- The windows method of analysis generally involves dividing the project into regular or key segments to make the analysis more manageable or provide focus to a specific area.
- The windows analysis the most accepted method of analysis even though there is no precise definition of the technique.

Section IV

Chapter 16
Disruption

Definitions and background

The terms 'delay' and 'disruption' go hand-in-hand, whilst disruption may cause delay to the completion of the project that is not always the case. Disruption is a result of a loss of efficiency and a reduction of productivity. Often increasing resources to counter the loss of productivity masks the effects of disruption. Similarly, taking other mitigating or accelerating action means the true impact of disruption is not known. Activities subject to the effects of disruption might not be on the critical path of the project and therefore not result in a delay to the project completion. However, there is usually an additional direct cost associated with disrupted activities due to loss of efficiency and reduced production.

The SCL Protocol defines disruption (SCL, 2002, paragraph 1.19.1, p. 31):

> "Disruption (as distinct from delay) is disturbance, hindrance or interruption to a Contractor's normal working methods, resulting in lower efficiency. If caused by the employer, it may give rise to a right to compensation either under the contract or as a breach of contract."

And the AACEI RP (AACEI, 2011, p. 122):

> "(1) An interference (action or event) to the orderly progress of a project or activity(ies). Disruption has been described as the effect of change on unchanged work which manifests itself primarily as adverse labor productivity impacts. (2) Schedule disruption is any unfavorable change to the schedule that may, but does not necessarily, involve delays to the critical path or delayed project completion. Disruption may include, but is not limited to, duration compression,

A Handbook for Construction Planning and Scheduling, First Edition. Andrew Baldwin and David Bordoli.
© 2014 John Wiley & Sons, Ltd. Published 2014 by John Wiley & Sons, Ltd.

out-of-sequence work, concurrent operations, stacking of trades, and other acceleration measures."

As with delay, the ability to demonstrate disruption relies heavily on comprehensive contemporaneous records. Again, when site management is occupied or diverted from their normal duties attempting to combat the effects of disruption, their management of the project is likely to become less efficient and some duties are left undone; one of the consequences is that when detailed records are required, most of all the site management staff are least able to spend time producing and collecting them.

Productivity is a measure of output, for instance, the planned productivity of a bricklaying gang of four could be 5 m²/h. When work is disrupted (or for other reasons), productivity might drop to, say 3 m²/h. Efficiency is the comparison of what is actually produced or performed with what can be achieved with the same consumption of resources (money, time and labour). In terms of the bricklaying gang, its efficiency is 3 m²/5m² = 0.6 = 60% efficiency. For a 120 m² piece of brickwork, the gang would usually be expected to complete the work in 120 m²/5 m² = 24 h. However, due to reduced efficiency, the work will take 120 m²/3 m² = 40 h, a loss of 16 h due to the effects of the disruption.

One way of countering the effects of the loss of efficiency due to disruption is to increase the resources undertaking the work. At 60% efficiency to complete the 120 m² of brickwork in the planned time of 24 h, the gang size will need to be increased to 100%/60% = 167% or 8.33 bricklayers. Eight bricklayers at 60% efficiency would take 25 h or nine bricklayers just over 22 h. However, one of the effects of increasing gang sizes is that to maintain productivity, a commensurate rise in supervision is also required. Furthermore, increasing gang sizes in confined areas or on small sections of work can further reduce efficiency thus creating a never-ending requirement to increase the gang size to maintain overall productivity.

Crew overloading/crowding and congestion at work faces are two of many possible causes of disruption. Keane and Caletka (2008) list 41 causes are shown in Table 16.1.

Whilst establishing the cause of the disruption, it is also necessary to determine the responsibility or categorisation of the disrupting cause: excusable, neutral or culpable. Comprehensive records and documentation are required to isolate the disrupting causes, and often there will not be a single cause of disruption but more like multiple causes with mixed responsibility. Disruption of activities on non-critical paths may not result in a delay to the completion of the project. Nevertheless, there will be an additional cost to the project as a result of inefficient working.

Methods of analysis

There are a number of methods for analysing the effects of disruption.

Measured mile

The measured mile method is the most easily understood and accepted method of analysing disruption. The object of the method is to compare the productivity in an undisrupted section of works (the measured mile) with the productivity in a

Section IV

Table 16.1   Causes of disruption and loss of efficiency.

• Late design	• Changes and variations in work scope
• Inaccurate detailed drawings	• Changes in working conditions (e.g. restricted working hours)
• Rework/corrective work	• Discovery of hazards
• Ripple effect of multiple changes	• Premature moves between activities
• Delayed or hindered access	• Work carried out in less than ideal conditions
• Adverse weather (usually severe)	• Double handling of materials
• Environmental conditions	• Constructive changes
• Crew overloading/crowding	• Over-inspection
• Learning and 'un-learning' curves (learning curves repeated)	• Works undertaken by others
• Fatigue (overtime/shift working)	• Fatigue
• Dilution of supervision	• Joint occupancy
• Stacking of trades in confined space	• Beneficial occupancy
• Repeated learning cycles or curves	• Morale and attitude
• Out-of-sequence access to work faces	• Reassignment of manpower
• Congestion at the work face (confined space, confusion, safety hazards)	• Crew size efficiency
• Stacking of trades (activities accomplished concurrently)	• Interruption of job rhythm
• Increase in labour gangs or labour force (above optimum levels)	• Overtime (physical fatigue and depressed mental attitude
• Increase in shifts	• Acceleration
• Out-of-sequence working or changes in the sequence of works (based upon industry standards and practice)	• Revisits or re-doing work (morale issue)
	• Excessive rework

disrupted section of the works. The reduction in productivity between the two is attributed to disrupting effects.

For the comparison to be of value, it is necessary to determine the productivity in an undisrupted section of the same type of work in comparable parts of the works when productivity is optimum, that is, not during the learning stages of the undisrupted work. One of the major problems with this approach is that in seriously disrupted projects, there may not have been a time or a section of similar works that were not disrupted. In those cases, it might be acceptable to use for comparison a measured mile from another similar project. Minimising the differences between the comparisons in different projects, for instance, the use of the same plant, gang sizes and subcontractors/main contractors at the same time of year and conditions, will reduce the unintended errors in the determination of the productivity. Alternatively, comparisons with work of a similar type rather than the same type might be acceptable with adjustments made to take account of the minor differences, for instance, laying of engineering bricks in comparison to laying facing bricks. In some instances, especially in mechanistic operations, industry norms and productivity rates might have been established. The establishment of such data requires a statistical approach processing data from many projects, which, if acceptable, will demonstrate a very small

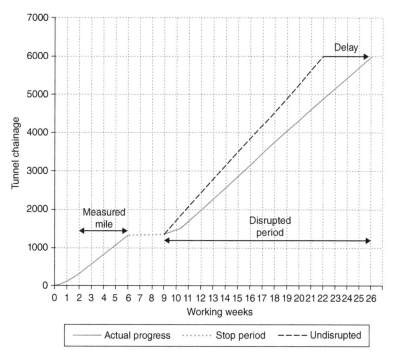

Figure 16.1   An example of a measured mile calculation.

variance in productivity. Whilst this is theoretically possible, the wide variation in construction projects and the conditions in which works are undertaken makes such norms unlikely to be reliable.

The use of the rates used in the tender is not considered acceptable as measured mile rates or 'norm' productivity rates. Without actual evidence of undisrupted productivity, the accuracy of the tender rates cannot be verified. In some cases, tender rates are adjusted for commercial reasons or may be overly optimistic or pessimistic.

In any case, the application of the measured mile method depends on detailed contemporaneous records that measured the productivity throughout the relevant section of the project (from undisrupted periods through to heavily disrupted periods). In addition, records of the causes of disruptions must be maintained to demonstrate the link between the disrupting event and the disruption caused.

Figure 16.1 shows the progress of boring 6000 m of tunnel. After 2 weeks (the 'learning curve'), the output became established at 5 m/h, working a 10 h day, 5 days a week; the weekly progress was around 250 m/week. However, a review of the tunnel route resulted in an instruction from the employer to stop work temporarily whilst the alignment data were verified. Works were stopped for 3 weeks before the instruction to restart was given. At the same time, the employer instructed the contractor to accelerate the tunnel boring to recover the lost time. The best the contractor could do was to increase the working day to 12 h and to work 6 days a week. There was another 'learning curve' over 2 weeks restarting the operations and adapting to the new shift pattern, and from then on, weekly progress of around 290 m was established; this equates to a just over 4 m/h. The efficiency of the new regime was only about 80% (4 m/h ÷ 5 m/h). Examination of the records showed that during the period when working 72 h/week compared to 50 h/week, there were more minor breakdowns of the tunnel boring machine and absenteeism of the crew increased especially at the beginning and ends of the shifts. The disruption due to stopping and then accelerating the works manifested themselves in two ways, an additional 'learning curve' following the restart and reduced productivity during the period of increased working hours. The 'undisrupted' line on the graph shows what the progress would have been if there had not been an additional 'learning curve' and that productivity had stayed at 5 m/h for the 72-h week. This demonstrates that the resulting delay due to a reduction in efficiency was around 4 weeks.

Leonard/Ibbs curves

Where it is not possible to use the measure mile approach, most usually because there are no periods of undisrupted work, it might be possible to use one of the statistical models that have been developed.

One of the earliest and most cited studies is that of Charles Leonard (1998). Leonard's study gathered data from 90 construction disputes in 57 U.S. projects in the years preceding its publication in 1988, and these were divided into two groups: 'civil and architectural projects' and 'electrical and mechanical work'. The study focuses on the effects of 'percentage change orders', which is the total of 'change order hours' divided by the actual hours spent on the original contract work. The actual hours being the project's total labour hours less the change order hours less any hours expended due to contractor mistakes and errors (such as reworking). The loss of productivity was calculated, where possible, using the 'measured mile' approach and the correlation graph drawn.

Each of Leonard's graphs has three curves (actually they are straight lines). Type 1 represents projects affected by change orders, Type 2 by change orders and one major cause and Type 3 by change orders and two major causes. The major causes include:

- Acceleration
- Out-of-sequence work
- Over-stacking of trades
- Late supply of plant and/or materials
- Late provision of information
- Increased complexity of the work
- Inadequate planning and coordination

Some of these 'major causes' are often the result of the impact of the change orders themselves so it is not always straightforward to separate the impacts into Type 1, 2 or 3.

Leonard's graphs are at Figure 16.2 and Figure 16.3.

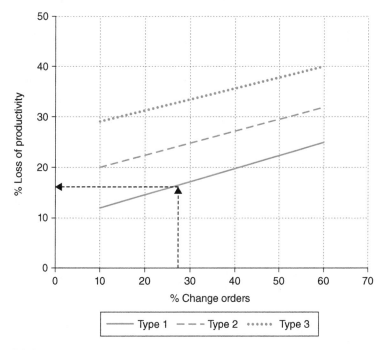

Figure 16.2   Leonard curve – civil and architectural projects.

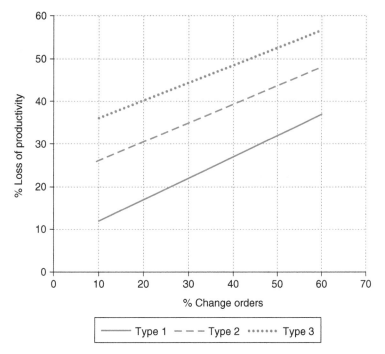

Figure 16.3   Leonard curve – mechanical and electrical work.

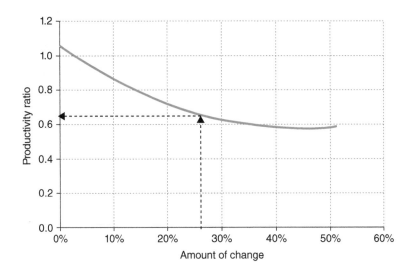

Figure 16.4   Ibbs curve.

For example, the records for a civil engineering project show that 30000 h were actually expended, and of those, 6000 h were worked on change orders and 2000 h were spent by the contractor inefficiently. Removing the hours expended on change orders and inefficiency hours leaves 22 000 h worked on the original contract works (30000 – 6000 – 2000 = 22 000), which includes the loss of productivity due to the change orders. The 'percentage change orders' is 27.3% (6000 change order hours ÷ 22000 base hours) on the x-axis, giving a percentage loss of productivity of 16.5% on the y-axis. As such, 3630 h were lost productivity out of a total of 22 000 h actually worked on the base contract hours (16.5% of 22 000).

Ibbs (Ibbs and Allen, 1995) study improved on the work of Leonard by analysing change on 104 projects from 35 contractors rising to 162 disputed and non-disputed projects from 93 worldwide contractors in his 2005 study (Ibbs, 2005). Ibbs does not differentiate between civil and architectural projects and electrical and mechanical work saying as he found this distinction was unnecessary as this variable did not make a significant difference in the impact of the change. Ibbs study also shows that at very low levels of change, around 3% and less, productivity is better than planned suggesting that contractors might make allowances in their bid for small levels of change. Ibbs graph is at Figure 16.4.

Following the previous example, 27.3% 'amount of change' results in a 0.65 productivity ratio suggesting 7694 h were lost productivity out of a total of 22000 h actually worked on the base contract hours (35% of 22000).

Ibbs also found that the timing of the change affected productivity differently; changes later in the project created lower levels of productivity (Ibb, 2005). Ibbs graph showing the effect of the timing of change is at Figure 16.5.

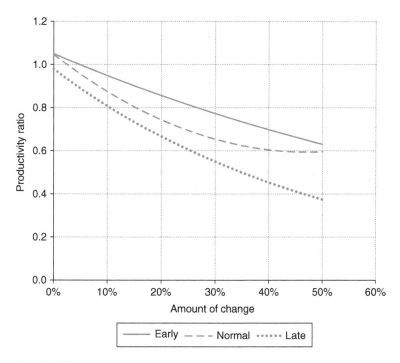

Figure 16.5   Ibbs curve – timing of change.

Following the previous example of 27.3 'amount of change', the early productivity ratio is 79.5% giving 4510 h lost productivity; the normal productivity ratio is 67.4% giving 7167 h lost productivity, and the late productivity ratio is 57.9% giving 9254 h lost productivity.

Indices and statistics

Various industry bodies, contractors and employers have developed models, guidelines and tables to assist in the calculation of productivity losses. In most cases, there is no real statistical basis for the indices, but they have been adopted by trade bodies and the like and by agreement between employers and contractors as a short-cut method of resolving disruption disputes. It is unlikely that, unless there is an agreement between the parties, that indices produced by one group or body would be applicable or acceptable in any other situation.

One of the most commonly referred to studies is from the Mechanical Contractors Association of America (2014, pp. 77–78). The manual addresses 16 factors, which can impact labour productivity. It presents a range of losses, expressed in percentages; for minor, average and severe cases, see Table 16.2.

Table 16.2   MCAA, labour estimating manual (Appendix B).

		Minor (%)	Average (%)	Severe (%)
1	Stacking of trades	10	20	30
2	Morale and attitude	5	15	30
3	Reassignment of manpower	5	10	15
4	Crew size inefficiency	10	20	30
5	Concurrent operations	5	15	25
6	Dilution of supervision	10	15	25
7	Learning curve	5	15	30
8	Errors and omissions	1	3	6
9	Beneficial occupancy	15	25	40
10	Joint occupancy	5	12	20
11	Site access	5	12	30
12	Logistics	10	25	50
13	Fatigue	8	10	12
14	Ripple	10	15	20
15	Overtime	10	15	20
16	Season and weather change	10	20	30

Key points

- Disruption is disturbance, hindrance or interruption to a contractor's normal working methods, resulting in lower efficiency.
- If caused by the employer, the disruption may give rise to compensation, either under the contract, or as a breach of contract.
- The 'measured mile' method for analysis disruption compares productivity in an undisputed sector of the works (the recognised mile) with productivity in a disrupted sector.
- While it is not possible to use the measured mile approach, it may be possible to use statistical models of productivity to assess disruption.

Chapter 17
Other Issues

There are a number of issues that can make what seems like a straightforward analysis into something that needs more consideration. It is rare that one of the previously described methods can be adopted without there being some problems that need addressing and extra steps or adjustments made to the basic principles. Ignoring or not being aware of some of these complications can render an analysis worthless or at least provide the opposition with chances to discredit or challenge the results.

Out-of-Sequence progress

A schedule represents one of the many ways in which the work can be undertaken; it is important because it shows the contractor's (and others' in some instances) intent as far as timing and sequence of activities is concerned. It is rare however that works go according to plan; work typically goes faster or slower than expected and activities are often carried out in a different sequence to that planned.

Out-of-sequence progress can drastically affect the validity of a delay analysis, in particular those methods that rely upon updating the schedule with progress; time impact analysis or time slicing/windows analysis. Most software used for critical path analysis of networks have two options for dealing with out-of-sequence progress: 'progress override' and 'retained logic'. If a project is generally running according to plan and activities are not progressed out of sequence, there will be virtually no difference between the two options. Significant differences occur when activities are progressed out of sequence. In most cases, the chosen option is applied to the whole of the network; it is not usually possible to select some activities that are to be scheduled with progress override and some with retained logic.

A Handbook for Construction Planning and Scheduling, First Edition. Andrew Baldwin and David Bordoli.
© 2014 John Wiley & Sons, Ltd. Published 2014 by John Wiley & Sons, Ltd.

Progress Override

Using the progress override option, the software allows the progress achieved to override the logic of the schedule. Consider the schedule in Figure 17.1. When the as-planned schedule is updated with schedule and the network reanalysed, it is apparent that activity 2 has started before activity 1 has finished. This is out of sequence.

When using the progress override option, the dependency between activity 1 and activity 2 is removed. The reasoning being that activity 2 actually started before activity 1 had finished the logic linking was incorrect or unnecessary.

Progress override invariably produces a shorter critical path than retained logic. It is also likely that there will be some illogical lack of dependencies resulting from activities having been progressed out of sequence and the dependency links being removed. After re-analysis, activity 1 in Figure 17.1 has no successor link, and it has free-float to the end of the project and therefore has no influence on the completion of the project. Ultimately, it does not have to finish for milestone 4 to be complete.

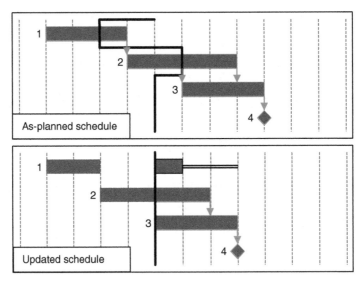

Figure 17.1   Reanalysis using progress override.

Retained Logic

Section IV

Using the retained logic option, the software reinstates logic that would otherwise be lost in progress override. The existing logic is not actually retained; it is replaced by another set of logic that is approximately equivalent to the missing logic. Consider the schedule shown in Figure 17.2. Again, when the as-planned schedule is updated with schedule and the network reanalysed, activity 2 has started before activity 1 has finished. This is out of sequence.

When using the retained logic option, the dependency between activity 1 and activity 2 is reinstated. The reasoning is that activity 2 should not have been able to start before activity 1 had finished, so the reinstated logic linking maintains the relationship that, at least, the remaining part of activity 2 cannot start until activity 1 is complete.

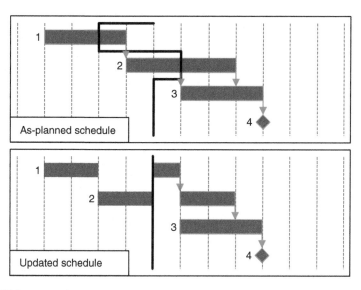

Figure 17.2   Reanalysis using retained logic.

Retained logic invariably produces the longest critical path. When activities have been progressed out of sequence, there could be some resulting illogical dependencies remaining. Figure 17.3 shows an extreme example where lag durations on finish-to-finish links are retained resulting in an extended completion date.

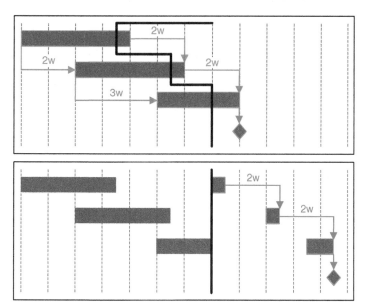

Figure 17.3   Retained logic resulting in an extended completion date.

In most cases, the retained logic ought to be adjusted to give a more realistic schedule to completion as shown in Figure 17.4.

When using critical path methods for analysing delays in disputes, the contesting parties will often select the opposing methods to analyse their schedules – thus producing wildly differing results from what seems like the same base data. As

Section IV

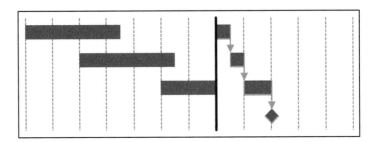

Figure 17.4   Adjusted logic schedule.

part of their preliminary discussions, the parties should agree what method best represents the situation. However, neither option is wholly correct. Retained logic often needs some dependencies to be deleted or their durations modified, and progress override often needs dependencies to be added to maintain logical sequences. In reality, using either option and adding, modifying or deleting logic of the out-of-sequence and progressed activities should result in the same result. Whatever method is used, a 'sense check' is required to ensure the resulting situation accurately models reality or intent.

Omissions

For one reason or another, the employer might wish to reduce the scope of the works or omit some items of work. If these omissions are significant, the period required to complete the works is liable to be shorter. Depending on the particulars of the specific contract, the employer may have the right to reduce the period for the works. In most cases, this does not permit the completion date to be earlier than the original completion date, but omissions could be taken into account when assessing extensions of time due to other delaying events or when considering what is fair and reasonable periods for the works taking into account all delaying and other events following completion of the works.

If the works to be omitted are on the critical path, their removal from the works is likely to reduce the overall planned period. However, it is often the case that omitting sections of the works do not have the dramatic effect that might initially have been considered. There are two main reasons for this: that there are other works that are on parallel critical paths or that have minimal float, they will then become the critical or driving activities controlling the duration of the works, and secondly, it might not be possible to reschedule subsequent works to take account of the apparent reduction in the time period resulting from the omission. Subcontractors and suppliers responsible for future critical works may not be able to reschedule their work to take account of the earlier access dates resulting from the omitted works.

The method of modelling omissions is the reverse procedure of that used to model the effects of delaying events. Primarily, it will be a combination of reducing the duration of an activity (the reverse of the 'Extended' delay type), removing activities from the programme and re-linking/adjusting the logic to suit (the reverse of the 'Additional' delay type) and adjusting the logic to reflect the new sequence of works (similar to the 'Sequence' delay type).

Calendars

Schedules for simple projects often have just one work calendar. For instance, it might reflect an eight hours a day, five days a week, working Monday–Friday from 08:00 to 12:00 and 13:00 to 17:00. It is also likely to include nonworking days for public holidays and other industry-specific, religious and customary holidays. As projects become more complex, the number of calendars in the schedule is also likely to increase; these might be:

- For specific trades or subcontractors
- To take account of different work patterns
- To take account of the breeding and migratory patterns of endangered or other fauna
- To take account of the growing seasons of endangered or other flora
- To take account of the seasonal effects of daylight
- To take account of the seasonal effects of temperature
- To take account of the seasonal effects of weather
- To take account of shutdown or possession periods when work is permitted

The differences in the calendars can include:

- The number of hours worked each day
- The start and finish times and breaktimes
- The number of days worked per week
- The number and duration of nonworking periods

Nonworking periods are not only weekends and holidays but might include, for instance, winter months when there are likely to be rough/high sea conditions preventing work at sea or a bird nesting season preventing work in or clearing wooded areas.

Problems with calendars are caused when activities with different calendars are linked. What seems logical on the as-planned schedule when delays and progress are taken into account in the schedule may become unreasonable. In Figure 17.5, the ground clearance works use an eight-hour-a-day, five-day-a-week calendar, whereas the piling works use a ten-hour-a-day, six-day-a-week calendar.

Table 17.1 lists the schedule data showing the start and finish dates for the activities, the progress on 21 February and the start and finish dates and delay to the completion of the piling resulting from progress made on 21 February.

Figure 17.6 shows the reanalysed schedule taking account of the progress on 21 February.

Figure 17.7 shows the planned finish of the ground clearance activity and the planned start and finish of the piling operation in greater detail.

The different calendars mean that the piling is scheduled to start at 17:00 on Friday evening. It is highly unlikely that the piling subcontractor would start the

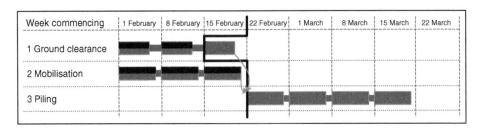

Figure 17.5   Planned schedule with progress on 21 February.

Table 17.1   Planned, progress and rescheduled activity data.

Planned schedule					
Activity	**Calendar**	**Planned start**	**Original duration**	**Planned finish**	**Successor**
1 Ground clearance	5 days at 8 h	Monday, 1 February, 08:00	15 days	Friday, 19 February, 17:00	F-S 0 link to 3
2 Mobilisation	6 days at 10 h	Monday, 1 February, 08:00	18 days	Saturday, 20 February, 19:00	F-S 0 link to 3
3 Piling	6 days at 10 h	Monday, 22 February, 08:00	24 days	Saturday, 20 March, 19:00	

Progress at end of third week (Sunday, 21 February)

Activity	**Calendar**	**% complete**	**Original duration**	**Remaining duration**
1 Ground clearance	5 days at 8 h	67% (10 days)	15 days	5 days
2 Mobilisation	6 days at 10 h	100% (18 days)	0 day	0 day
3 Piling	6 days at 10 h	0% (0 day)	24 days	24 days

Reschedule/delay on 21 February

Activity	**Calendar**	**Actual/ planned start**	**Remaining duration**	**Actual/ planned finish**	**Delay**
1 Ground clearance	5 days at 8 h	Monday, 1 February, 08:00 (actual)	5 days	Friday, 26 February, 17:00 (planned)	7 days
2 Mobilisation	6 days at 10 h	Monday, 1 February, 08:00 (actual)	0 day	Saturday, 20 February, 19:00 (actual)	0 day
3 Piling	6 days at 10 h	Friday, 26 February, 17:00 (planned)	24 days	Friday, 26 March, 17:00 (planned)	5 days 8 h

work late on Friday and would probably wait until Monday morning to make a start. Analysis without common sense would suggest then that the delay to the start of the piling operations was because the subcontractor was not ready to start and might then draw the conclusion that the delay to the start of the piling was

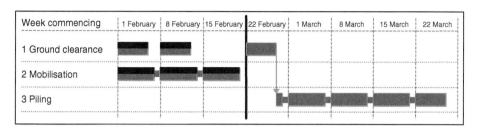

Figure 17.6   Reanalysed schedule piling staring at 17:00 on Friday, 26 February.

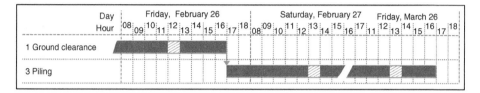

Figure 17.7   Detail of start and finish of piling.

the critical delay and was not as a result of the delay to the completion of the ground clearance.

Calendars can have a dramatic effect on the delay to project completion even from an event with a relatively small delaying effect, for example:

▪ On a project for the refurbishment of some airport buildings, included in the works was the replacement of the runway lights. The new runway lights were to be provided by the employer, and the work was scheduled to be completed over a weekend. (The airport was due to be closed to allow the works to be carried out.) The schedule for the works included a specific calendar for the installation of the new runway lights; mostly, it was made up of nonworking time but with two ten-hour working days on the Saturday and Sunday of the weekend runway possession. As it turned out, the employer was late in providing the new lights. This meant the contractor missed the weekend possession, and the next available weekend was some 6 months later. Using a prospective analysis, the contractor demonstrated that in receiving the lights from the employer a few days later than anticipated caused a delay to the completion of the works of 6 months.

▪ The construction of a new unloading jetty included the sinking of sheet piles to form the jetty walls. Once the coffer dam of sheet piles was in place, work could progress on filling in and constructing the jetty. The piling was to be undertaken from a barge in the sea, which would move around the perimeter of the proposed jetty. The schedule for the works included a specific calendar for the installation of the sheet piles. No work was scheduled between November 1 and March 31 due to the likelihood of high winds and rough seas that would make the piling operations too dangerous. Due to delays directly attributable to the contractor prior to the start of installing the piles, the contractor had used up all the float between the scheduled completion of the piling and 1 November – the sheet piling works were now on the critical path.

Section IV

Not long after the piling had started, the employer issued a change of work order and the jetty required extending a further 6 m out to sea. This required an additional 12 m of sheet piling which, at the planned output, would take 2 days to complete, going beyond 1 November. The contractor applied for an extension of time of 5 months and 7 days, arguing that the additional 2 days of piling could not be started until 1 April but a further week was required to mobilise and establish the barge and piling rig after 31 March before the two additional days of piling could start.

When compiling an as-built schedule, the basis information that is required is the start and finish dates of the activities. Very detailed examination of the records is required to determine the actual working and nonworking days for each activity. What might appear as a nonworking day might just be a day when no work was carried out on the project as opposed to a holiday, weekend or other such time when work would not normally have been carried out. Likewise, work might have actually been undertaken on what was normally a nonworking day. There are many reasons for this: to accelerate the works, to pick up lost progress, to allow the operatives to earn 'overtime' or that the actual working calendar was different to the scheduled working calendar. When compiling an as-built schedule, it is not uncommon to find that activities have started or completed on Saturdays or Sundays when the project calendar shows Monday–Friday 5-day working.

An absolute accuracy for an as-built schedule would require a different calendar for each activity. This would be disproportionally time-consuming and expensive and would also result in the problems associated with disparate calendars between linked activities. The general convention therefore when compiling an as-built schedule is to state activity durations in calendar days and to assign a seven-working-day, no-holiday calendar to all activities. The number of working hours each day should also be standard, say 10 h each day for ease of calculation, as as-built data rarely specifies the precise start of finish time of an activity.

This is particularly important when the as-built schedule is to be used as the basis of the 'collapsed as-built' method of analysis. The Recommended Practice for Forensic Schedule Analysis (AACEI, 2011, paragraph 2a, p. 86) endorses the seven-day calendar approach but also warns of possible anomalous results:

"This system may sometimes produce anomalous results. For example, if work started on Friday and completed on the next Monday, the duration assignment will be four days although only two were actually worked. Then in the collapse, if the same activity happens to start on the first day of a four-day holiday weekend, it will show to continue through the holiday weekend and complete on the last day of the holiday. However, the system tends to balance itself out because it is equally likely that an activity which started on a Friday and finished on the following Monday (a 2 workday activity taking 4 calendar days) would show up as occupying four workdays from a Monday through Thursday in the collapsed as-built. The counterbalancing rule is applicable to both work activities and no-work durations. Hence, the 7-day calendar is often used initially for assigning actual durations to both types of activities."

There are instances where the seven-day, no-holiday calendar is unsuitable. This is particularly where there are significant shutdown periods or periods of non-working in the project or for particular activities, for instance, as described earlier, where work is not permitted for significant periods for weather, environmental or other reasons. Similarly, many construction and engineering sites are closed for part of December and January for religious and winter/summer holidays. In these instances, the seven-day calendar should include nonworking time for the specified periods to ensure that 'collapsed' activities are not scheduled during shutdown periods.

Weather

Most contracts include specific clauses relating to the delays caused by the weather. The contract will usually specify:

- What constitutes a weather event
- What the weather thresholds are
- The category of the delay (excusable, neutral or culpable)

In addition, some contracts (such as the NEC suite of contracts) also specify where weather records (details of past weather) and weather data (details of the current weather in question) are to be acquired.

Weather events typically relate to rain and other precipitation, high winds and extremes of temperature (usually extremes of low temperatures). In virtually all cases, the contractor is obliged to take account of normal or average adverse weather, and extensions of time are only granted, either on an excusable or a neutral basis depending on the contract, for weather events that are *exceptionally* adverse. In many cases, there is no clear guidance on what constitutes 'exceptionally adverse weather conditions' or over what period the adversity should be measured and assessed. Some contracts however do take a more objective and measurable approach (see NEC, 2013):

"A weather measurement is recorded

- within a calendar month,
- before the Completion Date for the whole of the works and
- at the place stated in the Contract
  the value of which, by comparison with the weather data, is shown to occur on average less frequently than once in ten years. Only the difference between the weather measurement and the weather which the weather data show to occur on average less frequently than once in ten years is taken into account in assessing a compensation event."

For a weather delay claim to be successful, the contractor must be able to show, in accordance with the terms of the contract, that the actual weather experienced during the project was exceptionally adverse, that is, exceptionally more adverse than could have been contemplated.

As there is no authority or clear guidance on what constitutes 'exceptionally adverse weather conditions', it is open to interpretation. The generally established meaning is when, during a calendar month, it exceeds the long-term average, for instance, the average over the previous 10 years. The accepted definition of 'exceptional', however, is something that exceeds the average by a significant amount. It then becomes a subjective judgement when that level is reached, and unless the contract says anything to the contrary or is more specific, the test will be what is fair and reasonable. The NEC contract, for instance, does specify when the weather becomes sufficiently adverse to trigger a compensation event and when the weather occurs on average more frequently than 1 in 10 years.

To succeed in a weather claim, the contractor must show that the weather has been in excess of what the particular contract deems allowance for and that performance was actually affected by the weather, which resulted in a delay (or a likely delay) to the progress of the works that then impacted on the completion date for the project. This generally means it is only those activities on, or near, the critical path which, if delayed, will have a knock-on effect on the completion of the project.

Once it is established that the weather is exceptional, then the delays arising from it can be determined. The sequence of establishing a weather-based delay is as follows: was there exceptionally adverse weather, and if so, did it delay the actual progress, and if it did, how much did it delay the project completion?

Ideally, the contractor and others on-site (the Clerk of Works or Resident Engineer, for instance, if one is retained) should keep daily records of the weather – maximum and minimum temperatures, rainfall, wind speed and so on – and also a contemporaneous log of what areas of the works have been affected by adverse weather: how the work was affected, if the whole or part of the day was lost, if work continued but at a slower rate, what steps were taken to minimise the effects of the weather and the trades and operatives involved.

Although most contractors do keep some daily records, they tend not to be sufficiently robust to provide sufficient proof of the severity or extent of the weather. Few sites are furnished with the specialist equipment to make accurate climate recordings. Most contractors (and employers or contract administrators assessing claims) rely on weather measurements recorded at an independent weather station suitably close to the site. This information is available some time after the end of each month; it is then only at that point can the contractor determine if the weather has been exceptionally adverse.

When it has been established that the weather has been exceptionally adverse (has exceeded the contractual threshold for a weather claim) and that the progress of the works has been delayed, it is then necessary to determine which of the adverse weather days are the exceptional ones or what parts of the total rainfall are the exceptional parts. The contractor will generally want the exceptionally adverse days to be those that affect activities on the critical path, thus resulting in an extension of time. The employer will generally want the exceptionally adverse days to affect noncritical activities so that the risk of delay is the contractor's. A method of selecting the exceptional days that is transparent, fair and reasonable is required. There are four principal methods:

- **Random selection:** Perhaps the most logical selection would be made by random selection of the exceptional days. Although this is a fair and reasonable approach, it is far from transparent unless it is carried out jointly between the parties or by an agreed independent person. If a truly random selection were made, it is unlikely that opposing parties would arrive at the same exceptional days, and as any particular selection has the possibility of being a random selection, it is likely that accusations of bias would be made if the selection favoured one party or the other. This selection can only be made after the event so the contractor and employer would not know at the time if the day of adverse weather was an exceptional one or not.

- **Maximum measurement**: This method selects the exceptional days as those that have the greatest rainfall, lowest temperature, highest wind velocity and so on. This sounds feasible, perhaps because selecting the days of the most extreme weather measurements has some verbal correlation with the term 'exceptionally adverse'. Conversely, it is illogical as most of the weather records relate to the frequency of the events, not the quantity of the events.
- **Proportional**: This method uses a simple calculation to spread out regularly the normal adverse days and the exceptional adverse days.[1] Starting with the first day lost numbered 1 and subsequent days numbered in order, the exceptional days are to be selected as occurring every nth day lost, where n is given by

$$n = \text{INTEGER} \frac{\text{exceptional day} \times \text{total days lost}}{\text{exceptional days}}$$

where

Exceptional days = total days lost – threshold days

For example:

If 8 days were lost due to adverse weather where the contractual threshold was 5 days, the number of exceptional days would be three (8 days total lost – 5 threshold days = 3 exceptional days).

The first exceptional day will be

$$\text{INTEGER} \frac{(1 \times 8)}{3}$$

INTEGER [2.67] = 2

The second exceptional day will be

$$\text{INTEGER} \frac{(2 \times 8)}{3}$$

INTEGER [5.33] = 5

The third exceptional day will be

$$\text{INTEGER} \frac{(3 \times 8)}{3}$$

INTEGER [8.00] = 8

- **Last days**: The exceptionally adverse days are those that occur once the threshold days have passed. In the example earlier, the exceptional days are days 6, 7 and 8.

---

[1]   Also known a 'Scott's Method' as described in Scott (1993).

Using the random, maximum and proportional methods, the contractor does not know until the end of the month (or when weather records become available), when all the actual days are known, which days are just adverse and which are exceptional. As the contractor usually holds the risk for the adverse days, it might be in his interest to take steps to overcome the delaying effects of the weather to minimise any liability for liquidated and ascertained damages should the effects of the normal weather cause him to be in culpable delay. In order to minimise his risk, the contractor may well be involved in additional cost: heaters, rain shelters, working additional hours and so on. The additional cost would be balanced against the cost of overrun, damages, loss of reputation and so on.

Whilst the contractor is usually obliged to mitigate the effects of delays no matter whose responsibility, this does not generally mean that he should spend significant amounts to do so. The contractor is thus in a difficult position – to spend money to mitigate the delays which might eventually not be his responsibility or to not spend money and risk an overrun and the additional cost of damages. The rational solution would be for the contractor to make time contingency allowances in the schedule to take account of the risk of delay due to adverse weather below the delay event threshold.

Using the last days method, the prudent contractor, if he invests in the necessary historical records, will know month by month what the normal adverse weather will be and so can count down the days for which he should spend to mitigate his own risks and subsequently request of the employer instructions to spend to mitigate the employer's risks. Practically, it is unlikely, except on the largest of projects, that there will be a weather station on-site or locally that will be able to provide real-time data to allow the contractor to make informed decisions contemporaneously, so in all cases, the contractor and employer have to adopt a wait-and-see approach (until the weather records are available) to weather-based claims.

Concurrent delay

The SCL Protocol (2002, paragraph 1.4.4, p. 16), says of concurrent delay:

> "True concurrent delay is the occurrence of two or more delay events at the same time, one an Employer Risk Event, the other a Contractor Risk event, and the effects of which are felt at the same time. True concurrent delay will be a rare occurrence. A time when it can occur is at the commencement date (where for example, the Employer fails to give access to the site, but the Contactor has no resources mobilised to carry out any work), but it can arise at any time."

A similar definition (in content) has received approval in the English courts and elsewhere[2]:

> "… the expression 'concurrent delay' is used to denote a period of project overrun which is caused by two or more effective causes of delay which are of approximately equal causative potency."

---

[2] *Concurrent Delay*, John Marrin QC, SCL paper 100 (2002), and *Concurrent Delay*' John Marrin QC, 18 Const LJ 436 (2002). The definition was approved in *Adyard Abu Dhabi v SD Marine Services* [2011] EWHC848 (Comm), paragraph 277, and has been adopted by *Keating on Construction Contracts, idem* at para 8–025.

Whilst the definition of concurrency (or 'true concurrency') appears to be well established, there is casual reference to concurrency when the *effects* of two or more delaying events are felt at the same time even though the effective causes of the delay or delay events happen at different times. Despite Marrin's definition not being specific about the causes of the delay happening at the same time, he says[3]:

> "...It has already been observed that, in cases of supposed concurrent delay, the fact finding exercise often reveals that it is in reality one event only which can be regarded as a true cause of delay. In such circumstances that cause of delay is not one of concurrent delay at all. That kind of case is to be distinguished from those relatively rare cases where the fact-finding exercise leads to the conclusion that, as a matter of fact, two events are to be regarded as having independently caused delay to the contractor's progress during the same period."

The most common misinterpretation of concurrency is when two or more delaying events happen at different times but their effects are felt at the same time. Examination of the facts usually tends to show that the event that happened first caused the delay that would have resulted in a delay to the completion of the project if analysed contemporaneously. The second delaying event, which happens some time later, does not cause further delay to the completion of the project, and this would have been more than apparent if the delays had been analysed contemporaneously.

As in all matters, the specific obligations of the contract and the jurisdictions will affect how concurrent delays are analysed. Some contracts have clauses such that in concurrent delay situations, the contractor cannot be awarded an extension of time if one or more of the delaying events are his responsibility. Time bar provisions such that the contractor has a limited period in which to bring forward a delay claim may also diminish his rights in concurrent and other delay situations if the claims are not made timeously. Precedents and statutes in differing jurisdictions will also influence the method of analysis for concurrent delays – for instance, the Scottish Courts in *City Inn v Shepherd Construction* (2007) and (2010) found in favour of using the apportionment approach, as do Canada and likely New Zealand and Hong Kong[4] but not in England and Wales[5].

There are three principle methods of analysing concurrent delay, and the most acceptable will be that which takes into account the contract conditions, the prevailing precedents or statutes and the facts of the matter. To illustrate the differences in the methods, the following examples consider three overlapping delays with responsibility for those delays being a weather delay (neutral), an information delay (excusable) and a strike delay (culpable):

- **First in line**: Each delay is treated chronologically with the first to occur taking precedence and the next delaying event not taking effect until the previous delay has expired as shown in Figure 17.8.

---

[3]   *Concurrent Delay Revisited.* John Marrin QC, SCL paper 179 (2013).
[4]   See *Concurrent Delay Revisited.* John Marrin QC, SCL paper 179, p. 10.
[5]   See Walter Lilly & Co Ltd v Mackay [2012] EWCH 1773 (TCC) and *Keating on Construction Contracts*, (Furst et al., 2012, at paragraph 8–027).

Section IV

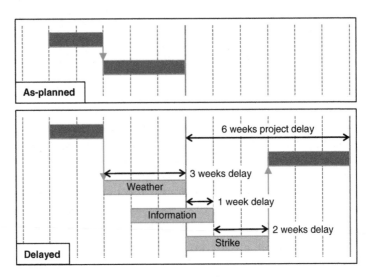

Figure 17.8   Concurrent delay: first in line.

Using the first-in-line method, the delays are apportioned:

Weather	3 weeks' delay
Information	1 week's delay
Strike	2 weeks' delay

Following the general rule for extensions of time, liquidated damages and direct loss and expense, the contractor would be granted an extension of time of 4 weeks (weather and information) and direct loss and expense for 1 week (information) and be charged 2 weeks' liquidated damages for the culpable delay due to the strike.

- **Dominant cause**: The contract administrator has to establish as a matter of fact which is the dominant cause of the delay, and this takes precedent over the other causes of delay as shown in Figure 17.9.

Using the dominant cause method, the delays are apportioned:

Weather	1 week's delay
Information	3 weeks' delay
Strike	2 weeks' delay

Following the general rule for extensions of time, liquidated damages and direct loss and expense, the contractor would be granted an extension of time of 4 weeks (weather and information) and direct loss and expense for 3 weeks (information) and be charged 2 weeks' liquidated damages for the culpable delay due to the strike.

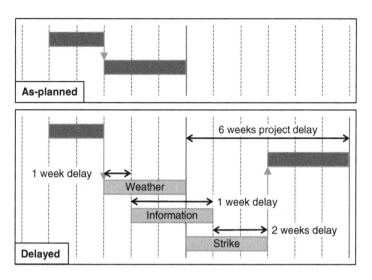

Figure 17.9   Concurrent delay: dominant cause.

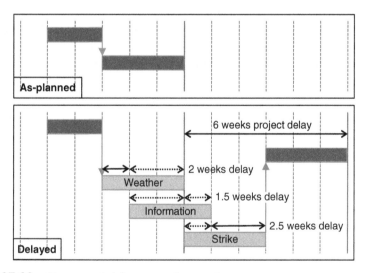

Figure 17.10   Concurrent delay: apportionment.

- **Apportionment:** Where there are concurrent delays, the responsibility for the delays is apportioned between the causes, generally on a 50:50 basis (if there are two concurrent causes) as shown in Figure 17.10.

Using the apportionment method, the delays are apportioned:

Weather	2 weeks' delay
Information	1.5 weeks' delay
Strike	2.5 weeks' delay

Following the general rule for extensions of time, liquidated damages and direct loss and expense, the contractor would be granted an extension of time of 3.5 weeks (weather and information) and direct loss and expense for 1.5 weeks (information) and be charged 2.5 weeks' liquidated damages for the culpable delay due to the strike.

Pacing

Pacing is a relatively new concept that allows one party, when it is aware that others' works are in critical delay, to slow down their works to a more efficient pace, to keep pace with the other party's critical delay so that there is 'no need to run to stand still later'. Pacing is inherently risky as there is no absolute certainty of the speed at which the delayed critical work will progress. Misjudgement or unexpected delays in the paced work could lead to further delays as a result of the paced work subsequently going slower than required. The concept is based on the premise that float would be generated on the non-delayed path and that the duration of the parallel non-delayed work can be slowed down to absorb the float.

When viewing delays retrospectively, if pacing was successful, it will appear that the paced work was concurrently delayed with the critically delayed work. In fact, most claims of pacing are proffered in retrospective analysis in an attempt to avoid culpability in concurrent delays.

When pacing is contended retrospectively, it should be treated with scepticism and the burden of proof lies with those claiming pacing. Pacing must be a forethought and not an afterthought, so there should be evidence that the party claiming pacing knew about and the extent of the critical delay to the works by the other party and made positive plans and decisions to pace the works. Whilst it is not usually contractually required that the pacing party informs the others about its intentions prior to instigating pacing, it would certainly be helpful in justifying its actions retrospectively. Notwithstanding evidence of prior consideration of pacing, the party must also be able to demonstrate by its actions that, after the pacing episode, it was able to pick up productivity to the as-planned levels; otherwise, it might be considered that the 'evidence' was in fact a calculated smoke screen for the contractor's inability to progress as planned.

Mitigation

In most jurisdictions, and often as a contract obligation, the parties are required to mitigate the effects of a delaying event.[6] A detailed discussion regarding the requirements to mitigate delay is outside the scope of this book; the SCL Protocol offers some guidance (paragraph 1.51, p. 18):

---

[6] Each contract should be examined individually to determine the extent of the party's obligations to mitigate delay, for instance, 'to do what is reasonable', 'to take all reasonable steps' or 'to take all measures'.

"The Contractor has a general duty to mitigate the effect on its works of Employer Risk Events. Subject to express contract wording or agreement to the contrary, the duty to mitigate does not extend to requiring the Contractor to add extra resources or to work outside its planned working hours."

Generally, this means that the contractor is not obliged to undertake mitigating action that is not commercially viable to him. However, it may decide to add extra resources and work extended hours to ameliorate its own delays, for instance, if the costs of mitigation are less than the liquidated damages that would be payable if the completion of the project was delayed.

In addition, there are actions that the contractor should take which would reduce the effects of delays and costs to the project. In relation to the calendar delays described in the Calendar section of this chapter.

- Having missed the weekend possession for installing the replacement runway lights, the project faced at least a 6-month delay to completion until the next weekend possession was scheduled. Following discussions with the employer, the contractor proposed that the replacement runway lights be installed between 23:00 and 05:30 over five consecutive nights.

  The cost of installing the new lights was greater than the original quotation to allow for night shift work, flood lighting and some temporary wiring and circuitry to allow a mixture of replacement and existing lights. The increase in cost was minimal to the disruption and additional cost that would result from delaying the project completion for 6 months. Allowing for time to mobilise, the night shift workers and the additional equipment, the installation of the runway lights was completed less than a month later than originally scheduled.

- Although the sheet piling works were not scheduled to be carried out between 1 November and 31 March due to the likelihood of high winds and rough seas which would make the piling operations too dangerous, the piling contractor estimated that there was only 2 days' work remaining when the 1 November deadline was hit. In discussions with the employer, the piling contractor agreed to keep the barge and piling rig at sea to complete the works. Whilst the weather did deteriorate, there was not a sudden change in the weather, and by closely monitoring the weather forecasts and the actual sea conditions, the additional piling works were completed by 3 November which allowed the filling of the coffer dam and construction of the jetty to start and to continue through November–March. There was a marginal increase in cost resulting from additional monitoring and the slightly lower installation rates. The saving in time was 5 months and the saving in cost not just from preventing the delay to the completion of the jetty but also from not having to remobilise the barge and piling rig the following year was immense.

When carrying out an analysis of delays after the work is complete and using a 'prospective' technique (such as time impact analysis or a windows time slice-type analysis), towards the end of the analysis, it will often show that there is a significant likely delay to the completion of the project beyond the completion date that actually occurred. As the delay analysis progresses for the final months of the project, the likely delay to the completion of the project rapidly reduces so that in the final period the likely delay of the analysis matches the actual delayed completion date. In some cases, the month-on-month reduction in likely delay can show that improvements or 'savings' of more than a month each month will be demonstrated. The analyst will say this demonstrates mitigating action taken by the contractor to reduce delays – this is almost always not the case.

As described previously, activities frequently progress to around 90% complete generally in accordance with the schedule, but the final 10% takes a disproportionate amount of time to complete. What actually happens is, towards the end of the project, *all* the outstanding small portions of the activities must be and are finally completed to allow the project as a whole to be completed. (Figure 17.3 illustrates the potential scheduling problems associated with unfinished activities and retained logic scheduling.) As each insignificant remaining part of each activity is completed, an *apparent* rapid improvement in performance results. This unrealistic phenomenon can be avoided by examining the schedule after each analysis and adding, modifying or deleting logic of the out-of-sequence and progressed activities (see the Out-of-Sequence Progress section of this Chapter).

Mitigation action, much like pacing, does not happen by accident; it has to be planned in advance. Within the records, there ought to be evidence that the steps to mitigate delays were planned and the steps to be taken and the envisaged outcomes. Without planning mitigating action, any reduction in apparent delays is usually down to good luck or accidents of critical path scheduling.

Acceleration

Mitigation of delays can include acceleration of the works, especially if the mitigation is being carried out by the contractor to recover delays that are his responsibility, and is therefore not entitled to compensation. Acceleration in this context is as a result of an employer's instruction for the contractor to increase productivity above that envisaged in the as-planned schedule.

Acceleration is defined as (SCL 2002, p. 52)

> "The execution of the planned scope of work in a shorter time than anticipated or the execution of an increased scope of work within the originally planned duration."

The right of the employer to instruct acceleration is dependent on the specific contract obligations, and where such clauses are not present, the parties are free to negotiate a collateral agreement (an associated, second or side but independent and separate contract made between the parties).

An employer may wish the contractor to accelerate the works to bring the original completion date forward, to maintain the original completion date following earlier delays or to maintain the original completion date when incorporating additional works. Acceleration measures in themselves, particularly increases in labour strength or working hours, as described in Chapter 16, can result in loss of productivity, so detailed plans should be formulated to ensure that the acceleration measures proposed are feasible along with additional supervision, plant, material testing, quality inspections and other resources associated with an increased workforce.

The acceleration instruction or agreement includes for the costs of the contractor instigating the acceleration measures which may include amounts to compensate for the additional risk or contingency against not completing by the revised or confirmed date for completion. The cost of the acceleration measures from the contractor must be weighed by the employer against the benefits gained from the new date for completion, for instance, the additional rental from commercial properties or increased sales from retail properties.

Employer/contractor/subcontractor schedules

In almost all cases, the contractor is obliged to produce a schedule showing how it intends to complete the works. This schedule is also used to manage the works and to report progress to the employer and for the employer to understand its schedule obligations to the contractor. It is also used extensively to demonstrate cause and effect of delay claims during the project and in post contract disputes. In this context, the schedule is referred to as the employer schedule; it is the one that the contractor uses as an interface with the employer.

Many contractors will also produce what they refer to as a 'working schedule'. This is essentially the same as the employer schedule, but the duration and sequence of the activities are slightly tweaked to provide a faster programme, giving the contractor some terminal float as a contingency against unforeseen culpable delays and as a target which will result in early completion, potentially earlier handover of the project and reduced overhead and preliminaries charges. In this context, the schedule is referred to as the contractor schedule as it is that schedule that the contractor is actually using to manage his works.

In an attempt to ensure the contractor can keep to its plans, further schedules are produced for the major subcontractors that incorporate even shorter activity durations and sequences. In this context, the schedule is referred to as the subcontractor schedule; it is the one that the contractor uses to interface with and manage its subcontractors.

So long as the subcontractor keeps to its schedule, the contractor will be ahead of its schedule and the project will be ahead of the employer schedule. However, keeping the three different schedules aligned is like juggling, and sooner or later, a ball will be dropped and there will be a delaying event that upsets the equilibrium of the three loosely related schedules. For example:

- As far as the employer is concerned, the project is progressing ahead of schedule, so when an information release date is missed by a couple of weeks, the project completion date is not jeopardised as the delay is absorbed by the terminal float resulting from the works being ahead of schedule.
- When it comes to the contractor schedule, there is less scope to absorb the effects of the late information as it shows a later planned completion date than the employer schedule, although because there is less terminal float, the late information date could consume it all, resulting in the contractor schedule becoming critical.

Section IV

- The subcontractor is only concerned with its work measured against its schedule, so when the late information issue filters down to the subcontractor, it is critical as far as the subcontractor schedule is concerned and results in a delayed start and finish to its work.

The situation is compounded in that subsequent trades and subcontractors are delayed by the delayed completion of the first subcontractor. The first subcontractor makes a claim against the contractor for delay caused by late information, and the subsequent subcontractors make claims against the contractor for the knock-on delay to the start of their work. Unfortunately, the contractor cannot make a delay claim against the employer for the late information as the employer's programme does not show the late information caused a critical delay to the completion of the works.

Whilst the employer/contractor/subcontractor schedule scenario appears to be a well-meaning method of ensuring timely completion of the works, it increases the contractor's risk and workload, attempting to keep the schedules aligned. In addition, if there comes a formal dispute at the end of the project, the documents will be subject to disclosure, and it may prove difficult to demonstrate delaying events and their effects when considering all three schedules and which one was actually operative during the project.

The alternative and transparent method of reducing risk and ensuring timely completion is to incorporate into the employer's programme adequate time risk contingency which will allow shortened activity durations countered by the contingency periods.

Key points

- Delay analysis is rarely straightforward. It is rare that construction works go according to plan. Out-of-sequence progress can drastically affect the validity of a delay analysis. Most software used for critical path analysis of networks have two options for dealing with out-of-sequence progress: 'progress override' and 'retained logic'.
- The employer might wish to reduce the scope of the works or omit some items of work. If the works to be omitted are on the critical path, their removal from the works is likely to reduce the overall planned period. However, it is often the case that omitting sections of the works does not have the dramatic effect that might initially have been considered.
- As projects become more complex, the number of calendars required to produce a truly accurate schedule is likely to increase. Different calendars may have different working days, different working hours, different holidays, etc. Problems arise when these calendars are linked. Calendars can have a dramatic effect on the delay to project completion.
- Most contracts include specific clauses relating to the delays caused by the weather, but extensions of time are usually only granted for 'exceptionally adverse weather'. In many cases, there is no clear guidance as to what constitutes 'exceptionally adverse weather'.
- Detailed weather data are essential if the contractor is to show that the weather has been in excess of that anticipated and that construction performance was affected.

- Concurrent delay is the occurrence of two or more delay events at the same time, one an employer risk event and the other a contractor risk event. True concurrent delay will be a rare occurrence. The most common misinterpretation of concurrency is when two or more delaying events happen at different times but their effects are felt at the same time.
- Pacing is a relatively new concept that allows one party, when it is aware that others' works are in critical delay, to slow down their works to a more efficient pace, to keep pace with the other party's critical delay so that there is 'no need to run to stand still later'. Pacing is inherently risky as there is no absolute certainty of the speed at which the delayed critical work will progress.
- In most jurisdictions, and often as a contract obligation, the parties to a construction project are required to mitigate the effects of a delaying event. Each contract should be examined individually to determine the extent of the party's obligations to mitigate delay, for instance, ' to do what is reasonable', 'to take all reasonable steps' or 'to take all measures'.
- Mitigation action, much like pacing, does not happen by accident; it has to be planned in advance. Within the records for the construction work, there ought to be evidence that the steps to mitigate delays were planned and the steps to be taken and the envisaged outcomes.
- Acceleration (in this context) is the right of the employer to request the contractor to bring the original completion date forward. The acceleration instruction or agreement normally includes provision for the costs of the contractor when instigating the acceleration measures. These costs may include amounts to compensate for the additional risk or contingency against not completing by the revised or confirmed date for completion.
- Contractors will normally be required under the contract to provide and update the employer with a schedule for the works. Many contractors will also produce what is called a 'working schedule' to be used for internal purposes and for monitoring the work of subcontractors. This approach may cause problems as it may prove difficult in a formal dispute to demonstrate the impact of delaying events as the argument will arise as to which schedule was actually operative during the project.

Appendix 1
BIM Case Study: One Island East

Introduction

One Island East is the name of a 70-storey, Grade A, commercial building located at North Point on Hong Kong Island. Completed in 2008, it is one of the first commercial buildings that was designed and constructed using building information modelling (BIM) including virtual construction. The adoption of these technologies enabled improvements in quality and significant savings in time and cost. The adoption of these technologies changed design and construction organisation and processes. Using BIM for the production of this building has formed the basis for efficient facilities management. One Island East comprises one of the most substantial BIM implementations for a commercial building. The building was awarded the American Institute of Architects' 2008 BIM Award for design delivery and process innovation. This appendix outlines the production of the model and then focuses on the use of the model for planning and scheduling the construction work.

Building the model

The BIM for One Island East was produced using Gehry Technologies Digital Project® software which is based on Dassault Systèmes® software. It comprises two main products: CATIA® (for product modelling) and DEMIA® (for process modelling). CATIA® enables integration and interoperability with Microsoft Project® and Primavera® planning and scheduling software.

The project had already reached the completion of the schematic stage of design when it was decided to adopt BIM for the detailed design stage onwards. Modelling was undertaken on the basis of a single 3D BIM project database. All the elements of the building were modelled by the creation of rule-based parametric object data including the comprehensive modelling of the mechanical, electrical and plumbing systems.

A Handbook for Construction Planning and Scheduling, First Edition. Andrew Baldwin and David Bordoli.
© 2014 John Wiley & Sons, Ltd. Published 2014 by John Wiley & Sons, Ltd.

The BIM model was used to facilitate the detailed design of the building, the drawing production, design coordination and cost implication of design decisions. To do this, it was necessary to manage the design team and manage the design model.

Gehry Technologies were awarded the contract for managing the design process and the production of the BIM model. The Architect (Wong and Ouyang (HK) Ltd), the Structural Engineer (Ove Arup and Partners (HK) Ltd), M&E consultants (Meinhardt (M&E) Ltd) and the Quantity Surveyors (Levett and Bailey) formed the key members of the design team. In addition to staff in these organisations' head offices, it was decided to co-locate design team members in a single project design office close to the construction site. Here, between 25 and 50 designers worked collaboratively on the design. Key to ensuring fast construction was clash detection and resolution before construction commenced. The design team using the BIM model resolved a total of more than 2000 clashes and design errors prior to tendering. The completeness of the BIM model based on the building elements and the components enabled accurate cost estimates to be prepared and maintained whenever design changes were made.

Figure A1.1 shows a schematic diagram of the organisation of the design team.

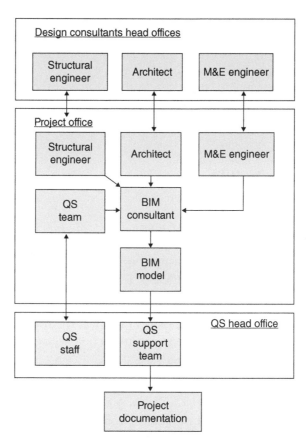

Figure A1.1   A schematic diagram of the design team.

The tendering process

The contract for the construction of the building was awarded following a selective tendering process. As part of this process, all the construction companies who tendered for the work were required to prepare a simulation model of how they proposed to undertake the construction process. To do this, they were provided with a copy of the BIM model and assistance with training on how to extract and use the data. Additional training and guidance was provided by BIM experts at Hong Kong Polytechnic University, HKPU.

The successful contractor (Gammon Construction) was required under the contract to take responsibility for the BIM, to maintain the model during the construction process and to hand over the updated model with the completed building, thus ensuring that on completion the client (Swire Properties) was in possession of a comprehensive, accurate model of the building that could be used for facilities management.

At the time of tendering, the data contained in all the main building elements were complete together with their main attributes: size, volume, weight, etc. Tendering was therefore on the basis of precise quantities. The BIM tool enabled many of the quantities to be measured automatically. There was no need for the contractors to measure the quantities of work. (However, the building model did not include all the reinforcing steel bars in the structural elements. For accuracy, the contractors bidding decided to calculate the quantity and price of the reinforcement in the structure manually using the ratio of reinforcement to concrete for individual elements.)

Using the model, Gammon were able to analyse the construction process and be confident that a four-day construction cycle could be achieved. The detail in the model enabled accurate cost estimates to be prepared.

Modelling the building process: Planning, scheduling and visualisation

During the construction of the building, the building model generated virtual construction models that became the central management tool for the production process, identifying construction problems and assisting the construction team to identify and solve construction problems before they impacted production on-site.

The BIM elements became a key part of the process of determining the construction sequence. (An example of this was the re-fabrication of the floor table formwork across the stages of construction where the outriggers changed the profile of the building.)

The visualisation and demonstration of construction sequences to the construction team (including subcontractor gangs of workers) enabled all concerned to contribute to ensuring the optimum construction process. The use of visualisation enabled production to be examined from a number of different aspects, for example, spatial requirements when moving temporary works and site safety, ensuring a fast safe production schedule.

The integration of Primavera scheduling software with Digital Project enabled planning engineers to link the BIM components to the activities in the Primavera construction model and produce updated schedules and 4D simulations of the construction sequence. (This integration was achieved by incorporating Primavera activity data in the building model 'tree structure'.)

This integration was two-way, enabling modelling staff to import project (and sub-project) data, work breakdown structures and construction activities from Primavera into the Digital Project modelling environment. The impact of changes in the construction schedule on the building product could therefore be assessed. This way of working ensured that construction proceeded to schedule. Gammon estimate that the use of the visualisation saved 20 days off the construction schedule. (The building was completed in March 2008 ahead of the 24-month construction schedule.) It is estimated that there was a 20% saving on the cost of construction.

Appendix 1

Appendix 2
The Shepherd Way and Collaborative Planning

Shepherd Construction have reviewed the principles and methods of Last Planner® and Lean Construction to produce a 'company-wide' approach to Collaborative Planning. This is a four-stage process involving the whole construction 'team' who work together to create and agree a 'lean' production schedule, identify risks to production performance and agree an implement solutions. They work to deliver the agreed schedule by identifying and removing constraints, producing weekly plans and measuring and monitoring performance against the agreed schedule. Examination of performance and investigation as to what has gone well and where problems have arisen identifies issues and areas for improvement throughout the programme.

Collaborative Planning

The Shepherd Way requires the team to establish a schedule collaboratively. This is a schedule that is agreed by all; it is a stable schedule, one fixed by a 'team' comprising representatives from all the trades involved. Working collaboratively, uncertainty is removed. The agreed way of working seeks to fully utilise the team's skills and experience. Opportunities for improved ways of working and increased production are explored, incorporated in the way of working or disregarded. Potential risks are evaluated and measures taken to mitigate the risk. Production issues are examined together with common interface problems. Forward planning identifies mobilisation needs and ensures that production may commence when anticipated.

Collaborative behaviours

Collaborative working requires changes in behaviour and changes in business culture. The Shepherd Way requires mutual respect between all of the team members, openness, honesty and consideration and understanding of everyone's point of view. It is only by such an approach that potential risks may be evaluated and commitment to an agreed schedule can be produced. Collaborative working requires regular communication and the provision of data, not opinions, to understand problems and focus on improvements. It is necessary to develop and maintain a working environment where problems are shared and solved as a team for the benefit of all parties involved. The aim must be to achieve success for the whole team, not just meeting individual goals.

The Collaborative Planning Process

Figure A2.1 shows the stages in the Shepherd Collaborative Planning Process.

Programme planning

The objective of the programme planning stage is the production of a master schedule created by the people who will actually carry out the work. The aim is to challenge the tender/contract programme with the whole team, contractor, subcontractor, design team, specialist consultants and specialist suppliers in order to create a 'lean' schedule. This must be a schedule which the whole supply chain understands and has been involved in creating a robust, workable, waste-free schedule of activities. All interface issues between trades must be examined together with opportunities for overall production improvement.

This planning stage is undertaken by mapping the current schedule using paper and post-it notes on the wall(s) of a room so that it can be fully examined by all involved. Key project milestones need to be established or identified and transferred from the existing contract documents. Goals are established with respect to the 'future' or 'new' schedule. It is imperative to consider the work sequence and the impact of each trade upon the next in the production process. It is important to recognise that the end user/client for the facility is not the only customer. You need to treat each trade in the production process as 'a customer' of the previous phase and a supplier to the next.

When assessing the duration of individual construction activities, it is important not to add buffers to individual tasks and to allocate reasonable, achievable task durations. Any 'buffer time' included should be planned and managed. It should be put at the end of the schedule or used as a series of stop checks throughout the production process. Buffer time should be controlled by the project leader and shared by the whole team.

Forward planning

The second stage of the planning process is forward planning, ensuring that the scheduled tasks are ready to happen. Identify from the schedule the tasks that should be starting in the next 6 weeks and review all possible constraints – what might stop the team from carrying out the tasks scheduled. What are the actions necessary to remove the constraints so that the task can start on time? (Typical

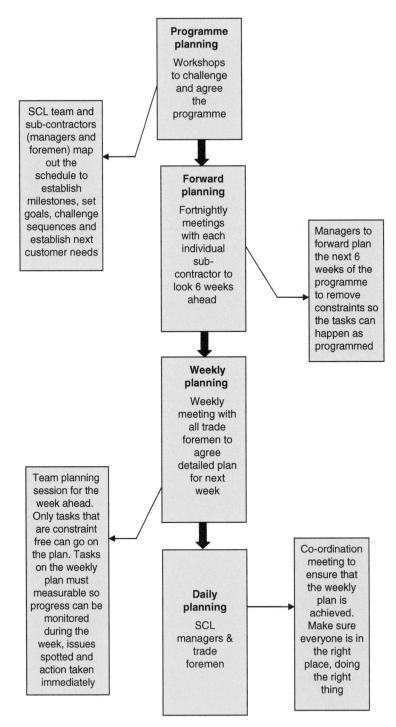

Figure A2.1   The Shepherd Collaborative Planning Process.

constraints include the supply of design information, the supply of materials, availability of labour, availability of plant and equipment and access to the works.)

This forward planning is achieved by agreeing with each trade the forward planning 'window', for example, six rolling weeks, and holding weekly/ fortnightly meetings with each subcontract foreman to review tasks within the window. This gives sufficient time to take actions to remove the constraints in time for the task to start. Checklists and action lists need to be produced for monitoring and control.

The main benefit of forward planning is that it ensures that all trades are ready to carry out the tasks as and when required, thereby maintaining production flow. It maintains a buffer of activities that can be carried out, and you are able to see which tasks can be pulled forward, if required, to maintain the flow of production and the buffers built into the schedule. Issues are encountered and addressed before work commences rather than as a task commences, preventing delays and stoppages. Overall, there is better plan predictability which generates confidence throughout all the production team and senior management.

Weekly Planning

Weekly Planning takes place at a weekly meeting with all trade site foremen. This meeting produces a detailed collaborative schedule of what can be done in the coming week. A review is made of the previous week's performance, the status of the coming week's plan is made, and the week's work is planned collaboratively based on what *can* be done.

Weekly Planning is undertaken to ensure that only work that can be done is scheduled, as opposed to what perhaps should be done but can't. At this stage, all constraints should have been removed and the final production schedule, what must be done, when and by whom, can be finalised. At this stage, individual trades are better able to plan and manage resources. Work is planned collaboratively, trades can support each other's work, the 'next customer' is satisfied, and the team committed.

Only tasks that are constraint-free are included in the weekly plan. Tasks are broken down so that they are no longer than 4 days in duration. The batch size is challenged to ensure that it is achievable. Progress is monitored throughout the week. If problems arise, the sooner that they are known, the sooner they can be rectified. Data on performance can be collected and analysed, allowing appropriate improvement actions to be taken.

The Daily Huddle

The Daily Huddle is a short stand-up meeting attended by the Site Manager and all trade foremen held at the start of each day where day-to-day issues are addressed and managed. This ensures that the weekly plan can be achieved. A visual board provides the focus for the meeting. This board can maintain details such as the areas where trades are working, drop of points for materials, and crane and hoist time slots. The Daily Huddle ensures that any issues can be quickly shared and appropriate action taken.

Focused improvement

Information captured from the weekly work plan can be used to identify areas for production improvement. This is particularly important for the types of construction that are repetitive, for example, the construction and fitting out of a hotel or a block of apartments. Shepherd use a range of appropriate tools to identify how to improve production. These include work observation, visual management, problem-solving methods and standardised work processes.

Appendix 3
Building Information Modelling (BIM) and English Law

Stacy Sinclair

Fenwick Elliott LLP, London, UK

The government's construction strategy requires that all government projects utilise a fully collaborative 3D computer model (Level 2) by 2016, with all project and asset information, documentation and data being electronic. Accordingly, BIM is the construction industry's buzzword at the moment – if you want to secure that next project, be it for public or private works, BIM certainly should be at the forefront of your bid.

BIM, as defined by the Construction Project Information Committee and supported by the RIBA, is the *'digital representation of physical and functional characteristics of a facility creating a shared knowledge resource for information about it forming a reliable basis for decisions during its life cycle, from earliest conception to demolition'*.

In other words, BIM is a way of approaching the design and documentation of a project, utilising 3D computer technology which is shared amongst the design and construction teams, incorporating cost, programme, design, physical performance and other information regarding the entire life cycle of the building in the construction information/building model. BIM is not simply the use of 3D technology – it is a way of design and construction.

As the use of BIM spreads throughout the construction industry, concerns regarding its legal and contractual implications rise. In particular:

- Does the use of BIM alter the traditional allocation of responsibilities as between the client, contractors, designers and suppliers?
- How (if at all) should standard form appointments and building contracts be altered to account for the use of BIM?
- Does the party managing the model assume additional liabilities and risk?

Before considering the answers to these questions, it is imperative that both the parties and their advisors understand what BIM environment they are

A Handbook for Construction Planning and Scheduling, First Edition. Andrew Baldwin and David Bordoli.
© 2014 John Wiley & Sons, Ltd. Published 2014 by John Wiley & Sons, Ltd.

working in. In 2008, Mark Bew of building Smart and Mervyn Richards of CPIC developed the BIM Maturity Diagram – a now well-known diagram which acknowledges the impact of both data and process management on BIM and defines various levels of maturity for BIM. In short, Level 2 BIM provides data and information in a 3D environment, with each member of the design (and possibly construction) team creating and maintaining their own model. These models and databases then 'fit' or work together with the use of proprietary technology (this consolidated model, comprised of the individual models prepared by each disciple, is often referred to as a 'federated model in a Common Data Environment (CDE)'). On the other hand, Level 3 BIM utilises a single project model, accessible by all team members. Any change in the traditional legal position? The consensus now appears to be that the use of BIM at Level 2 does not require wholesale changes to the traditional forms of contract or the allocation of responsibilities as between the parties. However, as BIM moves towards Level 3 in the future, changes to building contracts may well be required as the traditional legal position and relationships between the parties are likely to change. In March 2011, the Government Construction Client Group (GCCG) concluded in its Strategy Paper that '…*little change is required in the fundamental building blocks of copyright law, contracts or insurance to facilitate working at Level 2 of BIM maturity. Some essential investment is required in simple, standard protocols and services schedules to define BIM-specific roles, ways of working and desired outputs. Looking forward to the achievement of Level 3 integrated working, there are limited actions related to contracts, appointments and insurance that could be taken in advance to facilitate early adoption of integrated working*'.

With the government's current focus on Level 2 BIM, establishing a BIM Protocol and its associated services schedules on each project is clearly of utmost importance in structuring the project's design and development processes as well as addressing any legal concerns.

The BIM Protocol

At present, there is not a UK standard BIM Protocol as such – though they are likely to be available in the near future. Appendix 20 of the GCCG's Strategy Paper did provide an illustrative draft of a BIM Protocol for discussion; however, this protocol is likely to be considered too brief. By way of example, it does not fully address intellectual property issues.

The United States on the other hand has developed several standard BIM Protocols and execution plans: the AIA E202, ConsensusDOCS 301 BIM Addendum and the Penn State BIM Execution Guide. These documents are not intended to restructure contractual relationships or stand as a substitute for a complete building contract. They are simply addendums to be appended to the building contract and consultant appointments. These documents address important design, data and process issues which must be determined at the outset of a project: intellectual property rights, level of development (level of definition) of the model, model management, allocation of risk, ownership, permissible uses of the model, schedule of BIM deliverables, etc.

Whether or not the BIM Protocol is a contractual document and to what extent can the contractor and design professionals rely on each other's models are important issues to confront and address in the BIM Protocol.

To have any legal recourse, parties are likely to require that the BIM Protocol is indeed a contractual document. Most recently, the JCT contract amendments, introduced in December 2011 in the Public Sector Supplement (Fair Payment, Transparency and Building Information Modelling), require any BIM Protocol to be a Contract Document. The Public Sector Supplement amends the definition of Contract Documents at clause 1.1 of many of the JCT standard forms to include '*any agreed Building Information Modelling protocol*'.

As regards to the level of reliance on the parties' models (Level 2 BIM), the ConsensusDOCS 301 BIM Addendum allows parties to choose whether:

- Each party represents that the dimensions in their model are accurate and take precedence over the dimensions called out in the drawings
- Each party represents that the dimensions in their model are accurate to the extent that the BIM Execution Plan specifies dimensions to be accurate, and all other dimensions must be retrieved from the drawings or
- The parties make no representations with respect to the dimensional accuracy of their models and they are to be used for reference only – all dimensions must therefore be retrieved from the drawings.

In order to avoid complicated and potentially expensive disputes in the future, any BIM Protocol should address this dimensional accuracy/level of reliance issue along with the scope of the models created (often referred to as Level of Definition).

The Information Manager

Finally, a further legal issue which must be considered at the outset is the role, responsibilities and liabilities of the Information Manager – a key member of the design and construction team required for the successful implementation of BIM. If each party is responsible for their own model, to what extent is the Information Manager liable when clashes are not detected or the design is not coordinated?

The PAS 1192-2 requires the Information Manager to '*provide a focal point for all information modelling issues in the project; ensure that the constituent parts of the Project Information Model is compliant with the MIDP [Master Information Delivery Plan]; [and] ensuring that the constituent parts of the Project Information Model have been approved and authorized as "suitable for purpose" before sharing and before issuing for approval*'.

In March 2011, the Government Construction Client Group (GCCG) concluded in its Strategy Paper that '…little change is required in the fundamental building blocks of copyright law, contracts or insurance to facilitate working at Level 2 of BIM maturity. Some essential investment is required in simple, standard protocols and services schedules to define BIM-specific roles, ways of working and desired outputs. Looking forward to the achievement of Level 3 integrated working, there are limited actions related to contracts, appointments and insurance that could be taken in advance to facilitate early adoption of integrated working' (BIM Industry Working Group, 2011).

The specification goes on to state that the Lead Designer shall be responsible for the coordinated delivery of all design information. As such, the role of the Information Manager is not meant to be that of the Lead Designer – the Information Manager is responsible for the management of information, information processes and compliance with agreed procedures, not the coordination of design.

If the parties agree that this is the role of the Information Manager, clearly this needs to be identified and dealt with in the BIM Protocol – otherwise, a potential conflict arises as regards to design and design coordination roles.

Conclusion

The key to the successful implementation of BIM is not in the legal nuances – its success depends on close collaboration at the outset with the client, contractors, consultants and suppliers and the establishment of a well-developed BIM Protocol. Level 2 BIM is unlikely to change the current legal landscape of the construction industry, provided the BIM Protocol addresses risk allocation and other elements touched on in Chapter 9. As we see a move towards Level 3 BIM, contractual relationships and risk are likely to change – therefore resulting in more sophisticated contractual arrangements.

Glossary

In this glossary, we have adopted the term 'schedule' in preference to 'programme'. In the United Kingdom and current and former Commonwealth countries, 'programme' was generally the preferred term. However, increasingly, the original American term 'schedule' is being adopted throughout the world.

It should be noted that in the United Kingdom, the term 'schedule' may also refer to a tabular list of information. For example, an 'information required schedule' is a tabular list of information items and dates by which the information is required by the project team.

A

Acceleration   Taking or planning active measures to complete work ahead of **Project Schedule** or to recover delays. Such action usually increases the overall cost of the project. See also **Mitigation**.

Activity   A process in a project that consumes time and also usually consumes or uses other resources (e.g. people, money, materials and equipment). An activity is the smallest unit of work on a **Schedule** but, depending upon the hierarchy or level of detail of the schedule, may be divisible into smaller or more detailed activities. See also **Task**.

Activity Duration   The time calculated or estimated to carry out an activity, generally taking into account a specific level of resources, constraints and methods of working.

Activity File   The computer file (generally) that contains all the information relating to a particular activity.

Activity-on-Arc   A **Project Network Technique** that uses arced lines to represent activities. Preceding and succeeding activities join at nodes or events. See also **Activity-on-Line** and **Activity-on-Arrow**.

Activity-on-Arrow   A **Project Network Technique** that uses arrows to represent activities. Preceding and succeeding activities join at nodes or events. See also **Activity-on-Line** and **Activity-on-Arc**.

Activity-on-Line   A **Project Network Technique** that uses lines to represent activities. Preceding and succeeding activities join at nodes or events. See also **Activity-on-Arrow** and **Activity-on-Arc**.

Activity-on-Node   A **Project Network Technique** that uses nodes (generally 'boxes') to represent activities. See also **Precedence Diagram** and **PERT Chart**.

Activity-Orientated Scheduling   The method of developing a schedule that determines the sequence and timing of activities based on the logical work process only and does not take account of any potential limitations of resources. See also **Resource-Orientated Scheduling** and **Time-Limited Project**.

A Handbook for Construction Planning and Scheduling, First Edition. Andrew Baldwin and David Bordoli.
© 2014 John Wiley & Sons, Ltd. Published 2014 by John Wiley & Sons, Ltd.

Actual   The amount, relating to an activity that has been done (e.g. actual cost, actual labour, actual start and actual finish).

Actual Dates   The dates, relating to an activity, when it started and/or finished. Often compared to the **Planned Duration**.

Actual Duration   The duration that an activity or a group of activities took to complete, often compared to the **Planned Duration**.

Actual Finish   The date, relating to activity, when it finished.

Actual Progress   The amount of work that has been completed at a given point in time. Often compared to **Planned Progress**.

Actual Start   The date, relating to an activity, when it started.

Analytical Design Planning Technique   A technique that identifies where iteration is likely to be present within a process and provides methodologies for controlling, redoing or eliminating it, depending on the needs of the overall project.

Arrow
1. Represents an activity in an **Arrow-on-Arrow** network
2. Represents a dependency between two activities in an **Activity-on-Node** network

Arrow Diagram   See **Activity-on-Arrow**.

As-Built Bar Chart   An **As-Built Schedule** in **Bar Chart** format.

As-Built Dates   The actual start and finish for an activity.

As-Built Network   A **Network** such that the activity durations, sequence and start dates and finish dates reflect the actual **Durations, Start Dates** and **Finish Dates. Dependencies** and other Constraints in the as-built network should be carefully considered to represent the actual dependencies and constraints encountered in the project such that they result in the actual durations, start dates and finish dates of the activities.

As-Built Schedule   A schedule that represents the actual sequence, duration and timing of activities in a project.

As Late as Possible   Timing or positioning of an activity in a schedule at its **Latest Start/Latest Finish** dates such that there is no **Free-Float** on the activity and the timing of other activities in the schedule and overall duration of the schedule is not affected.

As-Performed Schedule   See **As-Built Schedule**.

As Soon as Possible   An option to position an activity in the earliest possible position. This is the normal setting for activities.

Assumptions   Suppositions that are used when data are unavailable or ambiguous to enable a schedule to be compiled.

B

Backward Press   The procedure whereby the **Latest Dates** of activities in a network are calculated.

Bar   A line on a bar chart that represents the timing and duration of an activity.

Bar Chart   A graphical chart on which activities are represented as bars drawn to a common time scale. Typically, a date scale is drawn across the top of the page and a list of activities down the left hand side of the page. Activity timing and durations are represented by horizontal bars. Additional information, such as resources, costs and dependencies, are also often shown on the chart.

Baseline   A record of all or any part of a project at a particular point in time against which current or future activities are referenced. Often taken to mean the first or original plan.

Baseline Dates   The earliest and latest start and finish dates of activities on the **Baseline Schedule**.

Baseline Schedule   A fixed or record schedule against which current or future activity is referenced. Often taken to mean the first or original plan but can be reset (for instance, following a change to the project scope), at which point the reset schedule becomes the (new) baseline schedule.

BIM   See **Building Information Modelling**.

BIM Maturity Levels   The level of information management, data sharing and collaborative working that is being adopted by the project or an organisation. (The UK Government BIM Industry Working Group identifies four different maturity levels, levels 0–3 inclusive.)

Branch A discrete part of a schedule generally represented by a single activity that is broken down by a project hierarchy and comprised further detail at the sub-activity level. See also **Child**, **Parent** and **Work Breakdown Structure**.

BREEAM The British Research Establishment's Environmental Assessment Method.

Buffer See **Float**.

Buffer Activity An activity in a schedule that acts as a **Contingency** or to artificially absorb **Float**.

Building Information Modelling (BIM) The digital representation of physical and functional characteristics of a facility creating a shared knowledge resource for information about it forming a reliable basis for decisions during its life cycle, from the earliest conception to demolition (see RIBA, 2012).

C

Calculate Schedule The mathematical analysis of a network, usually using a computer and **Project Management Software**, to determine the earliest and latest starts and finishes and float of the activities and the overall project duration. Often carried out following the addition of **Actual Progress** to determine the effect of progress on the network; primarily the completion date of the project (See also **Reschedule**).

Calendar A list of time intervals during which activities can be worked and/or resources used. Typical data includes working days/nonworking days, start and finish times for shifts, weekends, holiday periods and extra workdays. Each activity and/or resource will have a calendar attached to it. A project can contain many calendars, each with different working and non-working periods.

Calendar Days The **Elapsed Time** (measured in days or some other unit) between the start of an activity, or a group of activities, and its finish. For instance, an activity with a duration of 14 calendar days might have a **Working Time** of only 10 days, taking account of non-working weekends.

Child A subordinate activity in a schedule. See also **Branch**, **Parent** and **Work Breakdown Structures**.

Closeout The process or activity of completing the project.

Completed An activity or a group of activities that has finished. See also **Status**.

Completion Date The date at which the project is contractually obliged to finish. In most cases, the **Planned Completion Date** is the same as the Completion Date.

Concurrency The parallel timing of two or more activities or parts of a schedule. More often used to describe the effect of two or more discrete delaying events affecting or delaying the completion of a project in parallel. Had either of the delaying events happened in isolation, then the project would still have been delayed.

Concurrent Activity An activity whose timing coincides, in part or in total, with another activity (see also **Concurrency**).

Constraints Restrictions that affect the sequence or timing of an activity. These include **Predecessor** dependencies but more often refer to **Imposed Dates**.

Consumable Resource A resource which is used or diminished during an activity; once a consumable resource has been used on an activity, it is not available for use on other activities in the project. Bricks used in the construction of a wall are an example of consumable resource. See also **Permanent Resource**.

Contiguous Activity See **Continuous Activity**.

Contingency The planned allocation of time and/or resources to take account of unforeseen circumstances. See also **Buffer** and **Float**.

Contingency Planning The development of plans that involve courses of action to be taken to minimise the effects of unforeseen circumstances or predetermined risks.

Continuous Activity An activity that must be carried out on a continuous basis as opposed to an intermittent basis. See also **Discontinuous Activity**.

Cost Breakdown Structure A **Hierarchy** that results from breaking down or ordering the project on the basis of cost elements.

Cost Curve A graph illustrating cumulative costs over time. See also **S-Curve**.

Crashing Taking steps to reduce the overall project period by reducing the duration of individual activities. The activities selected from reduction in duration are selected on a combination of float (those with **Zero Float** or **Negative Float**), the sensitivity of duration to increasing resources, the cost of increasing resources and the availability of additional resources.

Critical Activity   An activity with **Zero Float**; if a critical activity is delayed or extended, it will delay or extend the completion of the project, and generally, if a critical activity is advanced or reduced, it will advance or reduce the completion of a project.

Critical Chain Project Management (CCPM)   Critical Chain Project Management is an approach to project management that is based on the concept of establishing a project 'buffer time' and then monitoring this time available and allocating resources to critical chain tasks.

Critical Delay   A delay that results in an extension to the **Critical Path** of a project and delays the project completion or increases the overall duration of a project.

Critical Path   A sequence of critical activities that form the longest sequence in a project, generally spanning from the start to the finish. The sum of the activity durations, taking into account **Leads** and **Lags**, determines the overall project duration. There may be more than one critical path.

Critical Path Analysis   The calculation of the **Critical Path** and **Float** in a network. Often used as synonym for **Critical Path Method**.

Critical Path Method   A method of calculating the overall duration of a project by taking account of the sequence and duration of activities that makes up the project and the interdependencies between the activities; the method is a mathematical model of the project which calculates the earliest and latest start and finish dates for activities, their float and the critical group of activities that must be completed on time.

Critically Index   A derivative of risk analysis that represents the frequency of occurrence or percentage that an activity was critical from a number of simulations. See also **Risk Analysis**.

Cut-Off Date   A date by which an action has to have taken place. For instance, the start or finish of a particular activity or the achievement of a **Milestone**.

Cycle Time   The time it takes to complete a repetitive group of activities in a project. For instance, the time taken between the completion of sequential floors of a multi-storey building.

D

Dangle   An activity that has either no **Predecessors** or no **Successors**. Unless an activity is a **Start Activity**, **End Activity** or **Isolated Activity**, it should have at least one predecessor and at least one successor.

Date for Completion   See **Completion Date**.

Delay
1. An event that results in an activity, or a group of activities, completing later than planned
2. How late an activity, or a group of activities, completes compared to when planned

Delay Analysis   The methodological investigation of the causes and effects of activities, or sequences of activities, completing later than planned.

Delaying Resource   In **Resource Scheduling**, the unavailability of one or more resources may result in the completion of an activity being delayed beyond the date on which it could otherwise be completed. See also **Earliest Feasible Date**.

Dependency   The term 'Dependency' refers to the logical interrelationships between activities. In a network, there can be one or more dependency between any two activities. There are four types of dependency, 'finish to finish', 'finish to start', 'start to start' and 'start to finish'. See also **Predecessor** and **Successor**. The dependencies dictate the sequence in which activities can be carried out.

Dependency Diagram   A diagram that shows the dependencies between activities of a project but may not show other data usually associated with a network diagram (for instance, activity durations and resources).

Dependency Free-Float   The amount by which the lag on the **Dependency** would have to be increased to cause a delay to the **Successor** activity. See also **Total Float**.

Dependency Link   See **Dependency**.

Dependency Total Float   The amount by which the lag on the **Dependency** would have to be increased to cause a delay to the completion of the **Schedule End Date**. See also **Total Float**.

Deterministic Network   A network that has only one set of dependencies between the activities and one fixed set of activity durations. This term is used to differentiate between the most common form of networks and **Probabilistic** networks.

Discontinuous Activity   An activity that can be carried out on an intermittent basis. This allows activities that are primarily constrained by finish dependencies to start at an earlier date. See also **Continuous Activity**.

Disruption   The effect on the project of events that affect the planned schedule of activities but may not affect the completion of the project. Typically, this will be as a result of delays to activities not on the critical path.

Driving   Where there are a number of dependencies to an activity, the driving dependency is the dependency, or dependencies, that results in the timing of the start and/or finish of that activity. The implication is that there is **Free- Float** on non-driving dependencies.

Drop Line   A method of indication progress of activities on a bar chart. A vertical line starting at **Time Now** links the ends of the bars at the point representing the progress achieved for that activity. Where the drop line deviates to the left, the activity is behind schedule; where the drop line is vertical, the activity is on schedule; and where the drop line deviates to the right, the activity is ahead of schedule. See also **Jagged Line**.

Dummy

An activity that represents no work but is required in a network to:
1. Represent a time delay between two activities
2. Maintain the logical group of activities
3. Ensure the uniqueness of nomenclature in an **Activity-on-Line** network

Duration   The estimated or actual time required for the completion of an activity, or a group of activities, based upon a particular resource allocation and method of working.

Duration Compression   The shortening of the duration of an activity, or a group of activities, perhaps to reduce the overall project period or to reduce the effects of delays. This often also results in an increase in cost and/or a reduction in quality and increased risk.

Duration Remaining   The working time remaining to complete an activity, or a group of activities, that has not started or is partially complete at a specific time. See also **Progress**.

E

Earliest Feasible Date   The earliest time that an activity, or a group of activities, can start within the constraints and logic of the network but ignoring any constraints imposed by resources.

Earliest Finish   The earliest time that an activity, or a group of activities, can finish within the constraints, resources and logic of the network.

Earliest Possible Date   See **Earliest Feasible Date**.

Earliest Start   The earliest time that an activity, or a group of activities, can start within the constraints, resources and logic of the network.

Early Dates   The dates calculated by the **Forward Pass** of network analysis. Within the constraints, resources and logic of the network, these dates represent the earliest start and finish dates for activities or sequences of activities.

Early Finish   See **Earliest Finish**

Earned Value Analysis (EVA)   A technique that compares the budgeted project costs, actual project costs and value of the work achieved to determine, *inter alia*, the status of the project, the likely completion of the project and the out-turn cost of the project.

Earned Value Management (EVM)   A management methodology for controlling a project based on measuring the performance of work and the Earned Value Analysis technique.

Earned Value Management System (EVMS)   The process, procedures, tools and templates used by an organisation to undertake Earned Value Management (Lukas, 2008).

Effort   The amount of labour time that is required to complete an activity or a group of activities. Usually expressed in, say, labour days and should not be confused with the duration of an activity which is dependent on the amount of labour allocated to it.

Effort Remaining   The amount of **Effort** remaining to complete an activity or a group of activities.

Elapsed Time   Durations that include both working and nonworking time, in other words **Calendar Days**. For example, a task might have a duration of 2 elapsed weeks, while the task's working duration is 10 **Working Days**.

End Activity   An activity that has no **Successors**. Usually the last activity or **Milestone** in a network.

End Float   See **Finish Float**.

Event

1. A point in a project, generally represented by a **Milestone**, that signifies the completion of all preceding activities prior to the start of any succeeding activities. Often used to represent a decision point.
2. The start or end of an important activity, or a group of activities, that signifies a defined point in the project. See also **Goal**.
3. A happening, an activity, or a group of activities that did not form part of the project that has an effect on the timing and/or resources of the project. Often used to describe something that delays or disrupts the project. See also **Relevant Event**.

Event-on-Node   See **Activity-on-Node**.

Expected Time   The **Activity Duration** calculated using **PERT**.

Extension of Time   A contractual award of additional time to complete the project usually as a result of a **Relevant Event**.

F

4D Planning and Scheduling   The integration of schedule and graphics to produce a time-based visualisation of the development of a project. Predominantly carried out by linking **Project Management Software** with graphics/drawing software though integrated software is now available.

Fast Tracking   The reduction of overall project duration by overlapping activities or sequences of activities that were originally or are usually planned to be carried out consecutively. Whilst reducing the overall project duration, this often also results in an increase in cost and/or a reduction in quality and increased risk.

F-F Link   See **Finish to Finish**.

Filter   A set of conditions that defines a subsection of the project; thus, facilitation focused on control, understanding or analysis. For example, a filter of activities with a particular resource (say, a particular contractor) to illustrate only the activities of that resource or a filter of all activities between particular dates.

Finish Activity   See **End Activity**.

Finish Date   The planned or actual date of when an activity or a group of activities is complete.

Finish Float   The period by which the finish of an activity, or a group of activities, can be delayed, brought forward or extended without affecting the **Schedule End Date**. For an activity with finish float, only the start of the activity cannot be delayed, brought forward or extended without affecting the **Schedule End Date**.

Finish to Finish   A relationship between two activities where the successor activity (the second activity) cannot finish until the predecessor activity (the first activity) has finished.

Finish-to-Finish Lag   The minimum period of time between the finish of the predecessor activity (the first activity) and the finish of the successor activity (the second activity). The time represents the portion of the work of the successor activity to be completed following completion of the successor activity.

Finish to Start   A relationship between two activities where the successor activity (the second activity) cannot start until the predecessor activity (the first activity) has finished.

Finish-to-Start Lag   The minimum period of time between the finish of the predecessor activity (the first activity) and the start of the successor activity (the second activity).

Finished Activity   An activity that is complete. See also **Status**.

Finishing Activity   See **End Activity**

Fixed Date   A date for the start or finish of an activity or of a milestone that cannot be changed. See also **Imposed Date**.

Fixed Finished   See **Imposed Finish**.

Fixed Start   See **Imposed Start**.

Flag   See **Imposed Date**.

Float   The period by which a task can be delayed, brought forward or extended without affecting the Schedule End Date. See also **Total Float, Start Float, Finish Float, Free-Float** and **Negative Float**.

Flow Chart   See **Network Diagram**.

Forecast   A projection of when an activity, or a group of activities, will start or finish based upon past performance. Typically, forecasts are made of the **Project Completion Date**.

Forward Pass   The reschedule procedure whereby the **Early Dates** of a plan are calculated.

Fragnet   A small network, an abbreviation of 'fragment network'. Typically inserted into the overall project network to represent a happening, an activity, or a group of activities that did not form part of the project that has an effect on the timing and/or resources of the project.

Free-Float   See **Start Float**.

F-S Link   See **Finish to Start**.

G

Gantt Chart   A Gantt chart is a graphical representation of the duration of activities against the progression of time and is a particular type of Bar Chart though used as a synonym for bar charts in general. Named after Henry Laurence Gantt (1861–1919) who originally developed the format (first described in 'Work, Wages and Profit' by H.L. Gantt in The Engineering Magazine, NY, 1910).

Goal   An event in the project signifying a defined point, usually the completion of a particular event often accompanied by a succinct definition of what is to be accomplished.

Graphical Evaluation and Review Telephone   A **Network** that has variable activity durations and variable dependencies between activities. See also **Probabilistic Networks**.

H

Hammock   An activity that gives an overview in terms of timing and duration of two or more activities that may be dispersed throughout the project. The duration of a hammock is initially unspecified and is ultimately determined by the timing and duration of the specified activities.

Handover   The transfer of the project, usually upon completion, to the property owner or operator.

Hazard   The potential to cause harm, including ill health and injury; damage to assets, products or the environment; production losses or increased liabilities (See ECI, 2013).

Hazard Effect Management Process   The structured hazard analysis methodology involving hazard identification, assessment control and recovery and comparison with objectives and acceptance criteria (ECI, 2013).

Hierarchy
1. Relating to **Plans** and networks, refers to showing different levels of detail, from high-level summary down to detailed working levels and the relationship between the plans
2. Relating to coding systems, such as **Work Breakdown Structure, Cost Breakdown Structure** and **Organisational Breakdown Structure**, refers to multilevel structures to aid classification and analysis of project components

Histogram   A graphical display of planned and/or actual resource usage and costs over a period of time. The format is of a vertical bar chart with the height of the bars representing the quantity of resource or cost for a specified time unit and the horizontal axis representing the period or project duration.

Holiday   A type of nonworking period. Can be for all project activities or for a calendar assigned to a particular resource.

Hypercritical Activity   Activity with **Negative Float**. An activity on which time must be recovered in order to meet the **Schedule Completion Date**.

I

I–J Number   The notation for the start and end nomenclature use in **Activity-on-Line** networks. An activity can be described by its start event number (I) and end event number (J). See also **Activity ID**.

Impact   The effect of an **Event** on a project.

Impact Analysis   The assessment of the effect of **Events** on a project. Specifically used in **Delay Analysis** to describe techniques that use **Fragnets** and the like to simulate the effect of Events and their effect on the **Project Completion Date**. See also **What-If Analysis**.

Imposed Date   A date imposed on an activity or Milestone by external constraints. Imposed dates can be:

1. Start on
2. Start no earlier than
3. Start no later than
4. Finish on
5. Finish no earlier than
6. Finish no later than

Imposed Finish   A **Finish Date** imposed on an activity by external constraints.

Imposed Start   A **Start Date** imposed on an activity or a milestone by external constraints. Often used, for instance, to set the start date of a project.

In Progress   An activity, a group of activities, or the project as a whole that has started but is not finished. See also **Status**.

Independent Float   See Free-Float.

Interfering Float   The period by which an activity can be delayed, brought forward or extended without affecting the **Schedule End Date** but will affect the successor activities. See also **Free-Float**.

Interruptible Activity   See **Discontinuous Activity**.

Interrupting Float   See **Interfering Float**.

Isolated Activity   An **Activity** or a **Milestone** that has no predecessors or successors. Usually time located by means of an **Imposed Date** and representing an **Activity** or **Event** that is associated with the project but has no direct influence on the timing or duration of any of the activities in the project.

J

Jagged Line   See **Drop Line**.

Job Safety Analysis (JSA)   A procedure used to review job methods and uncover hazards that may have been overlooked in the layout or design of the equipment, tools, processes or work areas; that may have developed after work started; and that may have resulted from changes in work procedures or personnel (ECI, 2013).

Just in Time   Associated with the philosophy of 'Lean Manufacturing'. Generally relates to scheduling raw materials and the like to arrive no earlier than required, thereby reducing stocks held at the point use (just-in-time supply). Can also refer to just-in-time production. There techniques are often seen as scheduling activities to be carried as **As Late As Possible**.

K

Key Date   See **Milestone**.

Key Event Schedule   A tabular report (usually) that extracts from the project plan the description and planned, forecast and actual dates of major events that are deemed to be crucial to the successful completion of the project.

Key Performance Indicator   A summarised measurement of indicators that can be used to assess the performance of the project.

L

Ladder A group of activities that are connected by both **Start-to-Start** and **Finish-to-Finish** dependencies to their predecessor. This produces an overlapping of a string of activities. Named as such because of the drawn resemblance to a ladder of a number of activities connected by **Dummies** in an **Activity-on-Line** network.

Lag
1. The minimum time between the end of one activity and the start of its successor. See also **Finish-to-Start Lag**.
2. The minimum time between the end of one activity and the end of its successor activity. See also **Finish-to-Finish Lag**.
3. The minimum time between the start of one activity and the start of its successor activity. See also **Start-to-Start Lag** and **Start-to-Start Lead**.

Last Planner Last Planner® (sometimes referred to as the Last Planner System) is a production planning system designed to produce predictable work flow and rapid learning in programming, design, construction and commissioning of projects. It is an approach to planning that is based on scheduling 'what should be done' on the basis of 'what can be done', that is, construction activities should only be commenced when all the constraints to production have been removed.

Late Dates The dates calculated by the **Backward Pass** of network analysis. Within the constraints, resources and logic of the network, these dates represent the latest start and latest dates for activities or sequences of activities.

Late Finish See **Latest Finish**.

Late Start See **Latest Start**.

Latest Finish The latest possible time by which an activity must end, within the logical and imposed constraints of the plan, without affecting the total project duration.

Latest Start The latest possible time by which an activity must start, within the logical and imposed constraints of the plan, without affecting the total project duration.

Latin Hypercube A statistical method used in **Risk Analysis** to select data from a risk profile (especially activity duration). Particularly used where a small number of iterations are required, for instance, because of a large number of activities or slow processing speeds. See also **Monte Carlo Simulation**.

Lead The minimum time between the start of one activity and the start of its successor activity. See also **Start-to-Start Lag** and **Start-to-Start Lead**.

Lead-In The period of time or activity, or group of activities, that is required before a schedule activity can commence. For instance, to represent the manufacturing period required once the design has been completed and an order placed before materials are delivered for incorporation in the works.

Levelling See **Resource Levelling**.

Line of Balance A graphical technique particularly suited for projects that comprise multiple and similar units, such as residential housing. The x-axis represents time and the y-axis the number of units. Sloping lines represent the activities of the project, the gradient of the line indicating the rate of production.

Link See **Dependency**.

Link Loop See **Logic Loop**.

Linked Bar Chart A **Bar Chart** that shows the **Dependencies** between the activities by drawn 'links' between the bars.

Logic The **Dependencies** linking activities in a network.

Logic Diagram A diagram that shows the **Logical Relationships** between activities, typically a **Network Diagram**.

Logic Link A **Dependency** between two activities and/or milestones.

Logic Loop An illogical circular configuration of dependencies, such that an activity is only able to start when a successor activity has finished.

Logical Relationship A logical interrelationship between two activities and/or two milestones.

Loop See **Logic Loop**.

M

Master Network   A high-level network showing the whole project and from which more detailed networks are derived.

Master Schedule   The principal schedule for a project, in particular that produced to comply with contractual requirements of the project. See also **Schedule** and **Project Plan**.

Microsoft Project Exchange   A standard file format, originating from Microsoft Project™, which allows project data to be transferred between software that support the MPX file format.

Milestone   A zero duration activity used to identify or highlight key points of **Events** in the project. Milestones are often used to identify the start or completion of sections of the project and therefore useful for **Monitoring** performance.

Milestone Plan   A plan that contains only **Milestones**. This is usually a **Filtered** version of the **Project Schedule**.

Mitigation   Making or planning measures to reduce the risk of delays or to recover delays. Mitigating action should not unreasonably increase the overall cost of the project. Action that does increase the overall cost of the project is considered to be **Acceleration**.

Mobilisation   The organisation and bringing together of **Resources** in readiness for the project start or subsequent activity starts. See also **Lead-In**.

Monitoring   The recording, analysing and reporting of project performance and comparing it to the Baseline Schedule.

Monte Carlo Simulation   A statistical method used in **Risk Analysis** to select data from a risk profile (especially activity duration) and to estimate a range of outcomes. See also **Latin Hypercube**.

Most Likely time   One of the three time estimates, the most frequently occurring duration, used in PERT.

Must Finish   See Imposed Start.

Must Start   See Imposed Finish.

N

Near Critical   An **Activity** or a **Milestone** that has float within a minimum defined limit. For instance, an activity with less than 3 days' float is to be deemed 'near critical' and treated as a **Critical Activity**.

Negative Float   Time that an activity must make up to stay on schedule. An activity with negative float is said to be **Supercritical**. Negative float is usually caused by conflicting **Constraints** on activities.

Network   See **Network Diagram**

Network Analysis   See Critical Path Analysis.

Network Diagram   A pictorial or graphical model of the project which includes **Activities** and their interrelated **Logical Links**. Also known as a **Flow Chart**, **Log Diagram**, **Network** or **PERT Chart** and so on.

Network Logic   See Logic.

Network Path   A series of **Activities** and their **Logic Links** within the Network. The **Critical Path** is a network path in which the activities have **Zero Float**.

Node
1. The start and end points of an activity in the **Activity-on-Line** network format
2. An activity in the **Activity-on-Node** network format

Nonworking period   A period in a **Calendar** when no work is carried out. Typically, amongst these are **Holidays** and **Weekends**.

Not Earlier Than   An **Imposed Date** with the restriction that the activity cannot start or cannot end earlier than the specified date.

Not Later Than   An **Imposed Date** with the restriction that the activity cannot start or cannot end later than the specified date.

O

Optimistic Time   One of the three time estimates, everything goes according to plan, used in **PERT**.

Organisational Breakdown Structure   A **Hierarchy** that results from breaking down the organisation into management levels and work groups for planning and control purposes.

Out-of-Sequence Progress   Activities that have been progressed even though **Predecessor** activities in the **Network** have not been completed.

Output Rate   The productivity of **Permanent Resources** and labour. Used to estimate **Activity Duration**: activity duration = quantity/(output rate × number of operatives).

Overtime   An irregular or additional working period in the **Calendar**.

P

Parallel Activities   Two or more activities that can be carried on at the same time. See also **Concurrent Activity**.

Parent   A superior activity in a schedule, one that is a product of one or more subordinate activities. See also **Parent** and **Work Breakdown Structure**.

Path   See **Network Path**.

Path Convergence   A point in a **Network** where two or more **Network Paths** join.

Percent Complete   The percentage of an activity, a group of activities or the project as a whole that is complete at a specific time. See also **Progress**.

Permanent Resource   A resource that is not depleted by its use in the project. Once a permanent resource has finished work on an activity, it is available to work on other activities in the project. Labour and plant are examples of permanent resources.

PERT Chart   Although not accurate, a synonym for a **Network Diagram** particularly when drawn in **Arrow-on-Line** format although latterly also used in relation to Activity-on-Node format.

Pessimistic Time   One of the three time estimates, everything goes wrong, used in **PERT**.

Plan   See **Schedule**.

Planned   An activity on the **Schedule** that has not yet started. See also **Status**.

Planned Completion Date   The finish date of the **End Activity** on the **Project Plan**. This date may be earlier than, the same as, or later than the **Completion Date**.

Planned Progress   The amount of **Progress** that is planned to be made at a specific time (as opposed to the amount that is actually made).

Planned Progress Monitoring   A systematic method of estimating the overall project **Position** employing a cumulative summation of all the work durations carried out compared to a summation of all the work durations planned. This technique was developed by the Property Services Agency (UK) 1988. See also **S-Curve**.

Planner   A member of the project team responsible for planning, scheduling and monitoring the progress of the project. Often used as a synonym for **Scheduler**.

Planning   The process of preparing for the commitment of resources in the most effective fashion. It aims to produce a workable schedule that will achieve project goals and serve as a standard against which actual progress can be measured. It defines:
1. What should be done (activities)
2. How should activities be performed (methods)
3. Who should perform each activity and with what means (resources)
4. When activities should be performed (sequence and timing)

Planning Engineer   See **Planner**.

Planning Engineer's Organisation   A professional body dedicated to serving planners and schedulers worldwide.

Planning Package   Project Management **Software** used by **Planners** and **Schedulers** to formulate and analyse **Project Plans**.

Planning Planet   A website networking planning, programming and control of professionals around the world at http://www.planningplanet.com/.

Position   The **Status** of an activity, a group of activities or the project as a whole. Position at a specified time can be ahead of schedule, on schedule or behind schedule.

Positive Float   See **Total Float** and to differentiate from **Negative Float**.

Precedence Diagram   A **Network Diagram** in the **Activity-on-Node** format.

Precedence Diagram Method   See **Precedence Diagram**.

Preceding Float   See **Start Float**.

Predecessor   An activity that must be partially finished or finished before a specified activity can start, that is, those activities that have outgoing **Start-to-Start, Finish-to-Start** or **Finish-to-Finish** dependencies. See also **Successor**.

Predecessor Activity   See **Predecessor**.

Probabilistic Network   A **Network** that has variable activity durations (the value dependent on the probability of it occurring) and/or variable dependencies between activities (the path dependent on the probability of it occurring). See also **Deterministic Networks** and **Graphical Evaluation and Review Technique**.

Program Evaluation Research Task   See **Programme Evaluation Review Technique**.

Programme   A group project or a portfolio of projects containing multiple projects generally with a common purpose or linkage.

Programme Evaluation Review Technique   A technique that uses three estimates of activity duration, **Optimistic Time** (o), **Most Likely Time** (m) and **Pessimistic Time** (p), to calculate the **Expected Time** (e). Expected Time $= (o + 4m + p)/6$. When applied to **Activity-on-Line** network methods, statistical techniques can be used to estimate the probability of the project completing within specified time periods.

Programme Management   The management of a project or programme containing multiple projects with a common purpose.

Progress   The measurement of the completeness of an activity or a group of activities or the project as a whole.

Progress Override   A method of calculating the **Critical Path** in partially completed projects taking account of **Out-of-Sequence Progress**. Where activities have been progressed out of sequence, **Predecessor** dependencies are ignored. See also **Retained Logic**.

Progress Period   A particular time period during which progress is reported. Progress periods usually coincide with producing **Progress Reports** and are, for example, at weekly or monthly intervals.

Progress Report   A regular report detailing the progress of the project, usually including a comparison of the actual progress achieved compared to that planned for the individual activities and for the project as a whole and a forecast of completion of key dates and the project as a whole. See also **Project Report**.

Progress Report Line   See **Time Now**.

Project   A project may be defined as a temporary endeavour undertaken to create a unique product. Projects are performed by people, constrained by limited resources, planned, executed and controlled.

   A typical construction project includes construction work incorporating planning, design, management or any other works involved until the end of the construction phase – that is it includes construction, alteration, conversion, maintenance, fitting out, commissioning, renovation, repair, upkeep, redecoration, decommissioning, demolition or dismantling. A full definition is available in the Temporary and Mobile Construction Sites Directive (92/57/EEC) and relevant national legislation.

Project   A temporary endeavour undertaken to create a unique product. Projects are performed by people, constrained by limited resources, planned, executed and controlled.

Project Calendar   The specific **Calendar** that denotes the default working and nonworking time for the activities in the **Schedule**.

Project Completion Date   See **Completion Date**.

Project Date   See **Time Now**.

Project Execution Plan   Includes plans, procedures and control processes for project implementation and for monitoring and reporting progress, providing a means of ensuring that everyone understands, accepts and carries out their responsibilities (see CIOB, 2010).

Project Logic   The **Dependencies** linking the activities in a **Network**.

Project Logic Drawing   See **Network**.

Project Management Software   Computer applications used by **Planners** and **Schedulers** to draw, develop, analyse and communicate **Project Schedules**.

Project Network Diagram   See **Network**.

Project Plan   A document for management and communication purposes that details the **Activities**, methods, **Resources**, sequence and timing.

Project Report   A regular report detailing all aspects of the project performance and management, costs and other issues and usually incorporating a **Progress Report**.

Project Schedule   The term Project Schedule may be interpreted in two ways.
1. The project activities and milestones and durations and planned sequence and timing
2. The physical document (**Bar Chart** or **Network**), for instance, that illustrates and communicates the aforementioned

Project Schedule   (US) See **Project Programme**.

R

Relationship   See **Dependency**.

Relationship Free-Float   See **Dependency Free-Float**.

Relationship Total Float   See **Dependency Total Float**.

Relevant Event   An **Event** that results in a delay to the **Project Completion Date** that, in accordance with the terms of the contract, is not the responsibility of the contractor.

Remaining Duration   The amount of time remaining until the activity is complete.

Replanning   Planning, carried out during the currency of the project, of activities that are not complete. This is usually carried out to recover delays or to incorporate changes in the scope of the project. See also **Acceleration** and **Mitigation**.

Reschedule   A mathematical calculation (**Critical Path Analysis**) performed on the tasks and links to ensure that the project is completed in the minimum possible time within the logical and imposed constraints of the plan and any progress that might have been achieved.

Resource   Any goods or services required to complete the work of an activity. For example, labour, materials, plant and money. See also **Consumable Resource** and **Permanent Resource**.

Resource Accumulation   See **Resource Aggregation**.

Resource Aggregation   A simple summation that allows calculation of each type of resource throughout the project duration. See also **Resource Analysis**.

Resource Analysis   Analysis and action with the objective of making the best use of **Resources**. See also **Resource Aggregations, Resource Smoothing** and **Resource Levelling**.

Resource Breakdown Structure   A **Hierarchy** that results from breaking down or ordering the project on the basis of the resources assigned to them.

Resource Calendar   The calendar to which a resource works, irrespective of the calendar on the task to which the resource is assigned.

Resource-Driven Activity Durations   An activity where the availability of resources determines the **Activity Duration**. See also **Resource-Limited Project**.

Resource Histogram   A graphical display of planned and/or actual resource usage over a period of time. The format is of a vertical bar chart with the height of the bars representing the quantity of resource for a specified time unit and the horizontal axis representing the period or project duration.

Resource Level   The quantity of **Resource** assigned to an activity or a group of activities per time unit.

Resource Levelling   The process of producing a schedule that reduces the variation between maximum and minimum values of resource requirements.

Resource-Limited Project   A project where the availability of resources determines the project duration. See also **Resource-Orientated Scheduling**.

Resource-Limited Resource Scheduling   See **Resource Levelling**.

Resource Link   A **Dependency** in a network that is in addition to those modelling the logical work process. These links simulate the transfer of a resource from one activity to another.

Resource-Orientated Scheduling   The method of developing a schedule that determines the sequence and timing of activities based on the logical work process and the availability and limitation of resources. Generally, this involves estimating activity durations based on available resources and the introduction of **Resource Links** to stimulate the transfer of resources from one activity to another. See also **Activity-Orientated Scheduling**.

Resource Plan   Part of the **Project Plan** that deals with how the project will be resourced.

Resource Planning   Determining what resources, when they are required and the quantity needed to complete the project.

Resource Profile   The level of resource requirement or availability over a period of time. This information is generally generated by **Resource Aggregation**.

Resource Scheduling   See **Resource Smoothing**.

Resource Smoothing   Smoothing out of peaks and troughs in the resource requirements without extending the planned project duration. The technique is to delay some activities within their floats to remove peaks and troughs in demand. The scope for smoothing is limited by the float available and constraints of the schedule.

Retained Logic   A method of calculating the **Critical Path** in partially completed projects taking account of **Out-of-Sequence Progress**. Where activities have been progressed out of sequence, **Predecessor Activities** are reduced to their **Remaining Duration** and predecessor dependencies are maintained. **See Progress Override**.

Risk   The product of the chance that a specified undesired event will occur and the severity of the consequences of the event.

Risk Analysis   Techniques to quantify the impact of uncertainty by analysing the probability of **Events** occurring and the consequence of their occurrence.

Risk Assessment   The process of identifying hazards, analysing and evaluating risk. It involves both causal and consequence analysis and requires determination of likelihood and risk. Assessment is not an end in itself, it must lead to risk management where the risk is eliminated, reduced or controlled.

Rolled-Up Activity   A **Summary Activity** representing a **Sub-Network**, a **Sub-Project** or a group of activities from a lower level in a **Hierarchy**.

S

Schedule
1. The timetable for a project. Showing how **Activities** and **Milestones** are planned to be carried out over a period of time
2. The physical document for communicating the **Plan**, especially timing and sequence

Schedule Compression   See **Crashing**.

Schedule Date   See **Imposed Date**.

Schedule End Date   See **Planned Completion Date**.

Schedule Performance Index   A parameter of **Earned Value Analysis**. The ratio of actual performance to planned performance. A ratio of less than 1 indicates the project is performing less than planned (i.e. it is behind schedule); a ratio of greater than 1 indicates the project is performing better than planned (i.e. it is ahead of schedule).

Schedule Variance
1. A parameter of **Earned Value Analysis**. The difference between the budgeted cost of work carried out and the budgeted cost of work planned.
2. The difference between the planned performance and actual performance. See also **Variance**.

Scheduled Finish   A 'finish on' or 'finish no earlier' than **Imposed Date**.

Scheduled Start   A 'start on' or 'start no earlier' than **Imposed Date**.

Scheduler   A member of the project team responsible for determining when the activities should be performed (sequence and timing), utilising methods such as **Bar Charts** and **Networks** which employ **Project Management Software**. Often used as a synonym for **Planner**.

Scheduling   The particular **Planning** duty determining when the activities should be performed (sequence and timing), utilising methods such as **Bar Charts** and **Networks** employ **Project Management Software**.

Scheduling Constraints   A general term for the variety of options that can constrain tasks. For example, **Imposed Dates** and **Resource** availability.

Sequence   The order in which activities are planned to be performed, a product of the **Dependencies** and other **Constraints** in the schedule.

S-F Link   See **Start to Finish**

SHE Plans Occupational safety, occupational health and construction environment plans (as defined by the European Construction Institute (ECI); see ECI, 2013).

Shutdown
An irregular nonworking period. For instance, a **Holiday** is a shutdown but a **Weekend** is not. See also **Calendar**.

Site Waste Management Plans (SWMP)   Plans for the management of construction site waste produced aimed at promoting waste reduction and material reuse and recycling.

Slack   See **Float**.

Slippage   Difference between **Planned Progress, Planned Start** or **Planned Finish** and **Actual Progress, Actual Start** or **Actual Finish** when the project is in **Delay**. See also **Variance**.

Smoothing   See **Resource Smoothing**.

Software   See **Project Management Software**.

S-S Link   See **Start to Start**

Stage   A section of the project. For instance, a **Subproject**, a subsection of the works or a defined time period.

Start Activity   An activity that has no **Predecessors**. Usually the first activity or **Milestone** in a network.

Start Date   The planned or actual date of when an activity or a group of activities commences.

Start Float   The period by which the start of an activity, or a group of activities, can be delayed, brought forward or extended without affecting the project end date. For an activity with start float only, the finish of the activity cannot be delayed, brought forward or extended without affecting the project end date.

Start to Finish   A relationship between two activities where the successor activity (the second activity) cannot finish until the predecessor activity (the first activity) has started. This is essentially an illogical dependency as it implies time running backwards. However, it is sometimes used to schedule the successor activity 'as late as possible' in relation to the predecessor activity.

Start to Start   A relationship between two activities where the successor activity (the second activity) cannot start until the predecessor activity (the first activity) has started.

Start-to-Start Lag   The minimum period of time between the start of the predecessor activity (the first activity) and the start of the successor activity (the second activity). The time represents the position of the work of the predecessor activity that must be completed before the successor activity can start. Also known as **Start-to-Start Lead**.

Start-to-Start Lead   See **Start-to-Start Lag**.

Started Activity   An activity that is in progress. See also **Status**.

Starting Activity   See **Start Activity**.

Status
1. See **Position**.
2. The status of an activity, a group of activities, or the project as a whole can be **Planned, Started** or **Finished**.

Status Date   See **Time Now**.

Status Report   See **Project Report**.

Subnet   See **Sub-network**.

Sub-critical Activity   See **Near Critical**.

Sub-network   Part of a **Network** representing a **Sub-project**. The activities of the subnetwork can be Rolled Up and represented by a **Summary Activity**.

Sub-project   A group of activities that is represented by a **Summary Activity** or **Summary Bar** higher in the **Hierarchy** of the schedule.

Succeeding Float   See **Finish Float**.

Successor   An activity that cannot start and/or finish before a specified activity has started and/or finished, that is, those activities that have incoming **Start-to-Start, Finish-to-Start** or **Finish-to-Finish** dependencies. See also **Predecessor**.

Summary Activity   See **Summary**.

Summary Bar   A **Summary Activity** on a **Bar Chart**.

Supercritical Activity   See **Hypercritical Activity**.

T

Target Completion Date   A date that the contractor aims to finish an activity or a group of activities. This may be earlier or later than the **Project Completion Date** depending on the circumstances. See also **Target Schedule**.

Target Date   See **Imposed Date**.

Target Finish Date   See **Target Completion Date** and **Target Schedule**.

Target Schedule   See **Target Schedule**.

Target Start Date   A date that the contractor aims to start an activity or a group of activities.

Task   The smallest measurable unit of work. **Activities** comprise one or more tasks.

Three Duration Technique   See **Program Evaluation Review Technique**.

Time Analysis   See **Critical Path Analysis**.

Time Chainage Chart   A graphical technique particularly suited for linear projects (such as roads, railways and pipelines). The x-axis represents distance (or chainage) and the y-axis the time. Sloping lines represent the activities of the project, indicating the rate of production and where, and at what time, the activity is taking place.

Time–Distance Chart   See **Time Chainage Chart**.

Time Impact Analysis   A method of **Delay Analysis** that analyses the effect of each **Event** incrementally and takes account of the status of the project at the time the delaying event occurred, the changing nature of the critical path and the effects of action taken or that should have reasonably been taken to **Mitigate** delays.

Time-Limited Project   A project that must be completed by a specific date with the assumption that whatever resources are required are available. See also **Activity-Orientated Scheduling**.

Time-Limited Resource Scheduling   See **Resource Smoothing**.

Time Location Chart   See **Time Chainage Chart**.

Time Now   The time at which all progress information entered for a project should be correct as of this date. The time at which all remaining work starts. This is the date when **Progress** is measured up to and generally coincides with the production of the **Progress Report**.

Timescaled Logic Drawing   A **Network Diagram** drawn such that the activities are positioned and dimensioned horizontally to represent, much as a **Bar Chart**, the planned timing of the activities.

Timescaled Network Diagram   See **Timescaled Logic Drawing**.

Time Slice Analysis   See **Windows Analysis**.

Time Unit   A unit of time in which activities are created, edited and examined. For instance, hours, days and weeks.

Today's Date   See **Time Now**.

Total Float   The period by which the start of an activity, or group of activities, can be delayed, brought forward or extended without affecting the project end date. Total Float comprises **Free-Float** and **Interrupting Float**. See also **Float**.

Tracking   See **Monitoring**.

Turnaround Document   A tabular report generated to enable those **Monitoring** to enter their activity **Progress** status against a list of activities that are schedule to be in progress during a particular time window.

V

Variance   The difference between planned and actual performance of an activity or a sequence of activities. Usually relating to differences between **Planned Progress, Planned Start** or **Planned Finish** and **Actual Progress, Actual Start** or **Actual Finish** but can also be applied to other project parameters such as costs. See also **Slippage**.

Virtual Construction   The ability to digitally model the construction process by modelling the building components within the construction process so that the project team can visualize different aspects of the construction of the facility throughout the production period.

W

Weekend   A regular **Nonworking Period**. See also **Calendar**.

What-If Analysis   The process of evaluating alternative strategies by changing the **Sequence, Durations and Constraints** on activities.

Windows Analysis   A method of **Delay Analysis** that analyses the project in discrete 'windows' of time. The **Status** of the project and forecast **Schedule Completion Date** is examined at the beginning of the time window and at the end of the time window. Any **Variance** (generally a **Delay**) is due to the activities, milestones and events on the **Critical Path** within the window period.

Work Breakdown Structure   A **Hierarchy** that results from breaking down or ordering the project so that the activities on any level are a subsection of the level above.

Work Rate   The rate at which a **Permanent Resource** can work. Usually expressed in quantity per time unit, for instance, m^2/h.

Working Time   The net time an activity is planned to be working or actually working. The working time does not take account of **Weekends, Holidays** and other **Nonworking Time**. For instance, an activity that spans 2 weeks would have an **Elapsed Time** of, say, 14 days but a working time of only 10 days taking account of the nonworking weekends.

WRAP   WRAP is a private, not-for-profit company and an authoritative body on waste and recycling and resource efficiency backed by funding from Defra, the Northern Ireland Executive, the Scottish Government, the Welsh Government and the European Union.

Z

Zero Float   Activities with zero float are said to be **Critical**. Any variance in their duration or start and finish dates will generally affect the **Schedule Completion Date**.

4D CAD   The term used to cover the use of computer based tools to visualize the construction plan in a 4-dimensional (3D + time) environment.

5D CAD   The term used to cover the use of computer based tools to visualize the construction plan in a 5-dimensional (3D + time + cost) environment.

Abbreviations

4D-PS	4D Planning and Scheduling
AdePT	Analytical Design Planning Techniques
ALAP	As late as possible
ASAP	As soon as possible
BIM	Building Information Modelling
CBS	Cost Breakdown Structure
CPA	Critical Path Analysis
CPM	Critical Path Method
EF	Earliest Finish
EoT	Extension of Time
ES	Earliest Start
EVA	Earned Value Analysis
EVM	Earned Value Management
F-F	Finish to Finish
F-S	Finish to Start
GERT	Graphical Evaluation and Review Technique
JIT	Just In Time
KPI	Key Performance Indicator
LF	Latest Finish
LoB	Line of Balance
LS	Latest Start
MPX	Microsoft Project Exchange
OBS	Organisational Breakdown Structure
PDM	Precedence Diagram Method
PEO	Planning Engineers Organisation
PERT	Programme Evaluation Review Technique
	Programme Evaluation Research Task
PPM	Planned Progress Monitoring
RBS	Resource Breakdown Structure
S-F	Start to Finish
SPI	Schedule Performance Index
S-S	Start to Start
WBS	Work Breakdown Structure

References

AACE International. (2011) *AACE International Recommended Practice No 29R-03 Forensic Schedule Analysis*, AACE International Inc. Available at http://www.aacei.org/non/rps/29R-03.pdf (accessed 30 October 2013).

ACA. (2008) PPC2000 The ACA Standard of Contract for Project Partnering, (amended 2008).

Ackoff, R. (1970) A concept of corporate planning, *Long Range Planning*, 3(1), pp. 2–8.

ADAPT. (2013) *Bridge Design Software with 4D Construction Sequencing and Time-Dependency*. Available at http://www.adaptsoft.com/specs-abi.php (accessed on 15 October 2013).

Addis, B. and Jenkins, O. (2008) Briefing: design for deconstruction, *Proceedings of the Institution of Civil Engineers*, 161(1), pp. 9–12.

ADePT. (2010) ADePT Management Ltd., Internal Report, produced by ADePT Management Ltd. Coventry, UK.

AIA. (2008) *AIA Document E202*, American Institute of Architects, Washington, DC, USA.

Alarcon, L. (Ed.) (1997) *Lean Construction*, A.A. Balkema Publishers, Rotterdam, The Netherlands.

Association for Project Management (APM. (2012) *APM Body of Knowledge* (Sixth Edition), Association for Project Management, Prices Risborough, Buckinghamshire, UK.

Austin, S.A., Baldwin, A.N., Li, B., and Waskett, P.R. (1999) Analytical design planning technique (ADePT): programming the building design process, *Proceedings of Institution of Civil Engineers*, 134, pp. 111–118.

Baldwin, A.N., Austin, S.A., Hassan, T.M., and Thorpe, A. (1999) Modelling information flow during the conceptual and schematic stages of building design, *Construction Management and Economics*, 17(2), pp. 155–167.

Ballard, G. (1999) Improving work flow reliability, *Proceeding of the 7th Annual Conference International Group for Lean Construction*, Berkeley, CA, April 17–19, available from the Lean Construction Institute website at www.leanconstruction.org.

Ballard, G. (2000) *The last planner system of production control*, PhD Thesis, School of Civil Engineering, University of Birmingham, Birmingham, UK.

Ballard, G. and Howell, G. (1998a) Shielding production: an essential step in production control, *Journal of Construction Engineering Management*, 124, pp. 11–17.

Ballard, G. and Howell, G. (1998b) What kind of production is construction, *Proceeding of the 6th Annual Conference International Group for Lean Construction*, Guarujá, Brazil. Available at http://www.leanconstruction.org/media/docs/BallardAndHowell.pdf (accessed on 30 October 2013).

Ballard, G., Hammond, J., and Nickerson, R. (2009) *Production Control Principles*. Available at http://www.adeptmanagement.com/amltechnologiesus/pubs/production_control_principles.pdf (accessed on 30 October 2013).

Barber P., Tomkins, C., and Graves, A. (1999) Decentralised site management – a case study, *International Journal Project Management*, 17, pp. 113–120.

Barry, D. (2009) *Beware the Dark Arts! Delay Analysis and the Problems with Reliance on Technology*. A paper presented to the Society of Construction Law International Conference held in London, 6th–7th October 2008 and published as Paper number: D095. Society of Construction Law, London.

BIM Industry Working Group. (2011) *BIM Management for Value, Cost and Carbon Improvement*. A report for the Government Construction Client Group commissioned by the Department of Business, Innovation and Skills dated March 2011 and first published July 2011 URN 11/948.

Bloch, A. (1977) *Murphy's Law, and Other Reasons Why Things Go Wrong!* Stern Sloan, Los Angeles, CA.

Blunck, F. (2006) *What is Competitiveness? British Research Establishment 2011, BREEAM New Construction, Nondomestic Buildings – Technical Manual 2011*, BRE, UK.

Booz, A.H. (2011) *Earned Value Management Tutorial Module 2: Work Breakdown Structure*, Office of Science, Tools & Resources for Project Management. Available at www.science.energy.gov (accessed on 15 October 2013).

Brandon, P. and Kocaturk, T. (2008) *Virtual Futures for Design, Construction and Procurement*, Blackwell Publishing Ltd., Oxford, UK.

Brook, M. (2008) *Estimating and Tendering for Construction Work* (Fourth Edition), Butterworth-Heinemann, Elsevier, Oxford, UK.

Brownhill, D. and Rao, S. (2002) *A Sustainability Checklist for Developments: A Common Framework for Developers and Local Authorities*, BRE Press, Garston, UK.

British Standard 3375-2:1993.

British Standard BS6079-1:2002.

British Standards BS6079-1:2000.

Building Research Establishment (BRE). (2008) *BRE Environmental and Sustainability Standard*, BRE Global, BRE, UK.

BRE. (2009) *SMARTWaste Plan: User Guide*, Version 3, BRE, Watford, UK.

BRE. (2011) *BREEAM New Construction, Non-Domestic Buildings – Technical Manual 2011*, BRE, UK.

Burke, R. (2003) *Project Management: Planning and Control Techniques* (Fourth Edition), John Wiley, Chichester, UK.

Burnett, J. (2007) City buildings – Eco-labels and shades of green, *Landscape and Urban Planning*, 83(1), pp. 29–38.

Cheshire, D., Attenborough, M., and Grant, Z. (2007) *Introduction to Sustainability*, CIBSE. Available at www.cibse.org (accessed on 25 October 2013).

Choo, H.J. (2003) *Distributed planning and coordination to support lean construction*, PhD Thesis, Civil and Environmental Engineering in the Graduate division of the University of California, Berkeley, CA.

CIOB. (2009) *Code of Estimating Practice* (Seventh Edition), Wiley-Blackwell, Oxford, UK.

CIOB. (2010) *Code of Practice for Project Management for Construction and Development* (Fourth Edition), Wiley-Blackwell, Oxford, UK.

CIOB. (2011) *Guide to Good Practice in the Management of Time in Complex Projects*, Wiley-Blackwell, Oxford, UK.

CIOB. (2012) We need to talk about BIM, *Construction Manager Newsletter*, Chartered Institute of Building, Englemere, UK, April 2012.

Computer Integrated Construction Research Program. (2011) *BIM Project Execution Planning Guide*, Version 2.1, Penn State University, University Park, PA.

Constructing Excellence (2011) *UK Industry Performance Report 2011-Based on UK Construction Industry Performance Indicators*, Published by Constructing Excellence, London, UK.

Construction Clients Group by the BIM Industry Working Group entitled, BIM management for value, cost and carbon improvement which was published in 2011.

Cooke, B. and Williams, P. (2009) *Construction Planning, Programming and Control* (Third Edition), Wiley-Blackwell, Oxford, UK.

Cole, R. (1998) Emerging trends in building environmental assessment methods, *Building Research and Information*, 26(1), p. 16.

ConsensusDOCS. (2008) ConsensusDOCS 301 – Building Information Modeling (BIM) Addendum ConsensusDocs, Arlington, VA.

Cooper R., Aouad, G., and Kagioglu, M. (1998) *Generic Guide to the Design and Construction Process Protocol*, University of Salford, Salford, UK.

Coventry, S., Woolveridge, C., and Patel, V. (1999) *Waste Minimisation and Recycling in Construction*, Special Publication 135, Boardroom Handbook, CIRIA, London.

Coventry, S., Shorter, B., and Kingsley, M. (2001) *Demonstrating Waste Minimisation Benefits in Construction*, CIRIA, (C536), London.

Covey, S. (2004) *The 7 Habits of Highly Effective People*, Simon & Schuster, London.

Covey, S. (2005) *The 7 Habits of Highly Effective People: Personal Workbook*, Simon & Schuster, London.

Crawley, D. and Aho, I. (1999) Building environmental assessment methods: applications and development trends, *Building Research and Information*, 27 (4–5), pp. 300–308.

Currie, R.M. (1977) *Work Study* (Fourth Edition) (reprinted 1987), Pitman Publishing, London.

Dainty, A. and Brooke, R. (2004) Towards improved construction waste minimisation: a need for improved supply chain integration? *Structural Survey*, 22(1), pp. 20–29.

Department of Defense. (2006) *Earned Value Management Implementation Guide*, USA Defense Contract Management Agency. Available at http://www.everyspec.com/DoD/DoD-PUBLICATIONS/DOD_EVMIG-OCT2006_2841/ (accessed on 15 October 2013).

Department for Trade and Industry. (2004) *Sustainable Construction Brief*, Department for Trade and Industry, London.

Dua, R. (1999) *Implementing Best Practice in Hospital Project Management Utilising EVPM methodology*, Micro Planning International, Pty Ltd. Available at http://www.microplanning.com.au/index.php/white-papers/by-raphael-dua/ (accessed on 15 October 2013).

Dudek, G. (2004) *Collaborative Planning in Supply Chains: A Negotiation-Based Approach*, Springer, New York.

D'Onofrio, M.F. and Hoshino, K.P. (2010) *AACEI Recommended Practice for Forensic Schedule Analysis*, American Bar Association.

Eastman, C., Teicholz, P., Sacks, R., and Liston, K. (2011) *BIM Handbook* (Second Edition), Wiley, NJ.

ECI. (2013) Safety Health & Environment (SHE) Management Guide published by the European Construction Institute (ECI), Loughborough University, Loughborough, UK.

Edwards, B. (1998) *Green Buildings* (First Edition), Pay, E & FN Spon, Routledge, London.

EEC Directive 92/57.(1992) EEC – temporary or mobile construction sites of 24 June 1992 on the implementation of minimum safety and health requirements at temporary or mobile construction sites (eighth individual Directive within the meaning of Article 16 (1) of Directive 89/391/EEC).

Egan, J. (1998) *Rethinking Construction*, Department of the Environment, Transport and the Regions, London.

Ekanayake, L. and Ofori, G. (2000) Construction material waste source evaluation, *Proceedings: Strategies for a Sustainable Built Environment*, Pretoria, South Africa, August 23–25.

Ekins, E.L.P. (2008) The impact of EU policies on energy use in and the evolution of the UK built environment, *Energy Policy*, 36 (12), pp. 4580–4583.

Electronic Industries Alliance. (2007) EIA Standard Earned Value Analysis Management Systems 748-B, an American Institute/Electronic Industries Alliance Standard, Electronic Industries Alliance, Arlington, VA.

Faniran, O. and Caban, G. (1998) Minimizing waste on construction project sites, *Engineering, Construction and Architectural Management*, 5(2), pp. 182–188.

Faniran, O.O., Love, P.E.D., and Li, H. (1999) Optimal allocation of construction planning resources, *Journal of Construction Engineering Management*, 125, pp. 311–319.

Fenwick, E.R. (1993) *Delay Analysis* (conference on Delay and Damages, London).

Fenwick, E. (2012) *BIM and the Law Fenwick Elliott Newsletter*, January 2012. Fenwick Elliott, LLB, London.

Fisk, W.J. (2000) Health and productivity gains from better indoor environments and their relationship with building energy efficiency, *Annual Review of Energy Environment*, 25, pp. 537–566.

Fischer, M. and Kam, C. (2001) 4D Modelling: technologies and research. Presentation given to the Workshop on 4D Modelling: experiences in UK and Overseas organised by The Network on Information Standardisation, Exchanges and Management in Construction, Milton Keynes, UK, October 17.

Furst, S., Ramsey, V., Hannaford, S., Williamson, A., Uff, J., and Keating, D. (2012) *Keating on Construction Contracts* (Ninth Edition), Sweet & Maxwell, London.

Goldacre, B. (2009) *Bad Science*, Fourth Estate, London.

Goldratt, E.M. (1997) *Critical Chain*, The North River Press, Great Barrington, MA.

Goldratt, E.M. and Cox, J. (2012) *The Goal*, The North River Press, Great Barrington, MA.

Gray, C. and Hughes, W. (2012) *Building Design Management* (First Edition 2001), Routledge, Oxford, UK.

Griffiths, N. (2004) *Site Waste Management Plans: Guidance for Construction Contractors and Clients*, DTI, London.

Heesom, D. and Mahdjoubi, L. (2004) Trends of 4D CAD applications for construction planning, *Construction Management and Economics*, 22, pp. 171–182.

Herrolen, W. and Leus, R. (2001) On the merits and pitfalls of critical chain scheduling, *Journal of Operations Management*, 19, 559–577.

Herroelen, W., Leus, R., Demeulemeester, E. (2002) Critical chain project scheduling-do not over simplify, *Project Management Journal*, 33(4), pp. 46–60. Published by John Wiley & Sons, Inc.

Hinze, J.W. (2012) *Construction Planning and Scheduling* (Fourth Edition), Pearson Prentice Hall, Upper Saddle River, NJ.

HMSO. (1964) *The Placing and Management of Contracts for Building and Civil Engineering Work*, The Banwell Report, a report produced by the Committee on the Placing and Management of contacts for building and civil engineering work, Chaired by Sir Harold Banwell and Ministry of Public Building and Works, UK.

HM Treasury. (2011) Autumn Statement 2011, presented to Parliament by the Chancellor of the Exchequer by Command of Her Majesty, November 2011. Report reference Cm 8231 HMSO, London.

Hoshino, K.P. Livengood, J.C., and Carson, C.W. (2011) *AACEI Recommended Practice No. 29R-03 – Forensic Schedule Analysis*, AACE International, Inc. Available at www.aacei.org/resources/rp (accessed on 15 October 2013).

Houlihan, J.B. (1985) International supply chain management, *International Journal of Physical Distribution & Materials Management*, 15(1), pp. 22–38.

Howell, G.A. (1999) What is lean construction, *Proceedings of the 7th Annual Conference International Group for Lean Construction*, 1999, Berkeley, CA. Available at www.leanconstruction.org (accessed on 25 October 2013).

Howell, G.A. and Koskela, L. (2000) Reforming project management: the role of lean construction, *Proceedings of the 8th Annual Conference International Group for Lean Construction*, 2000, Brighton, UK. Available at www.leanconstruction.org (accessed on 25 October 2013).

Hughes, V. (2011) *Breeam In-Use* published in FM World, the online magazine of the British Institute of Facilities Management, 24 March 2011. Available at http://www.fm-world.co.uk/good-practice-legal/legal-articles/breeam-in-use/ (accessed on 15 October 2013).

Ibbs, C.W. (2005) *Impact of change's timing on labor productivity, Journal of Construction Engineering and Management*, 131 (11), pp. 1219–1223.

Ibbs, C.W. and Allen, W.E. (1995) Quantitative Impacts of Project. Source Document 108, Construction Industry Institute, University of Texas, The Institute, Austin, TX.

INSAG (2002) *Key Practical issues in Strengthening Safety Culture- INSAG 15*, A report by the International Nuclear Safety Advisory Group published by the International Atomic Energy Agency, Vienna, 2002.

Johnston, R.B. and Brennan, M. (1996) Planning or organizing: the significance of theories of activity for the management of operations, *International Journal of Management Science*, 24(4), pp. 367–384.

Joint Contracts Tribunal. (2011) *PC/N 2011 Partnering Charter – Non-binding 2011*, Sweet & Maxwell, part of Thomson Reuters (Professional), UK Limited, London, EC3N 1DL.

Jones, M. (2008a) Site Waste Management Plans, *Construction Manager*, April, pp. 27–30, Chartered Institute of Building, CIOB, Englemere, Berkshire.

Jones, P. (2008b) Briefing: resource efficiency in construction – a perspective from the waste resources sector, *Waste and Resource Management*, 161(1), pp. 3–6.

Jones, M. (2009) *Turning Waste into Profit: Focus: Sustainability*, WRAP. Published by *Contract Journal*, 25th February, 2009. Available at http://www2.wrap.org.uk/downloads/contract_journal_25.02.09_turning_waste_into_profit.dcdabb5c.7051.pdf (accessed on October 30 2013).

Keane, P.J. and Caletka, A.F. (2008) *Delay Analysis in Construction Contracts*, Wiley-Blackwell, Oxford, UK.

Keys, A., Baldwin, A., and Austin, S. (2000) Designing to encourage waste minimisation in the construction industry, *Proceedings of CIBSE National Conference*, CIBSE 2000, Dublin, September 2000. Available at https://dspace.lboro.ac.uk/2134/4945 (accessed on 25 October 2013).

Kilger, C. and Reuter, B. (2002) Collaborative planning. In: *Supply Chain Management and Advanced Planning: Concepts, Models, Software and Case Studies* (Eds Stadler, H. and Kilger, C.), Springer, New York, pp. 223–237).

Kirkham, R. (2007) *Ferry and Brandon's Cost Planning of Buildings* (Eighth Edition), Wiley-Blackwell, Oxford, UK, ISBN 978-1-4501-3070-7.

Koskela, L. (1992) *Application of the New Production Philosophy to Construction*, Technical Report No.72, CIFE, Stanford University, Stanford, CA.

Koskela, L. (2000) *An exploration towards a production theory and its application to construction*, PhD Thesis, Technical Research Centre of Finland, Finland.

Koskela, L. and Howell, G. (2001) Reforming project management: the role of planning, execution and controlling, *Proceedings of the 9th Annual Conference International Group for Lean Construction*, 2001, Kent Ridge Crescent, Singapore. Available at www.leanconstruction.org (accessed on 25 October 2013).

Koskela, L. and Howell, G. (2002a) The theory of project management: explanation to novel methods, *Proceeding of the 10th Annual Conference International Group for Lean Construction*, 2000, Gramado, Brazil. Available at www.leanconstruction.org (accessed on 25 October 2013).

Koskela, L. and Howell, G. (2002b) The underlying theory of project management is obsolete, *Proceedings of PMI Research Conference*, 2002, Project Management Institute, Portland, OR.

Koskela, L. and Ballard, G. (2006) Should project management be based on theories of economics or production? *Building Research & Information*, 34(2), pp. 154–163.

Langdon, D. (2009) *Designing out Waste: a Design Team Guide for Buildings*, WRAP, Oxon, UK.

Latham, M. (1994) *Constructing The Team*, Final Report of the Government/Industry Review of Procurement and Contractual Arrangements In The UK Construction Industry, HMSO, London.

Latham, M. and Great Britain. (1994) *Constructing the Team*, Joint Review of Procurement and Contractual Arrangements in the United Kingdom Construction Industry, HMSO, London.

Latyea, S. and Hughes, W. (2008) How contractors price bids: theory and practice, *Construction Management and Economics*, 26(9), pp. 911–924.

Laufer, A. (1997) *Simultaneous Management: Managing Projects in a Dynamic Environment*, AMACOM, New York.

Laufer, A., Tucker, R.L., Shapira A., and Shenhar, A.J. (1994) The multiplicity concept in construction project planning, *Construction Management and Economics*, 12(10), pp. 53–65.

Laufer A., Denker, G.R., and Shenhar, A. J. (1996) Simultaneous management: the key to excellence in capital projects, *International Journal of Project Management*, 14(4), pp. 189–199.

Leach, L.P. (1999) Critical chain project management improves project performance, *Project Management Journal*, 30(2), pp. 39–51.

Lean Construction Institution. (2005) Available at http://www.lean construction.org (accessed on 15 October 2013).

Lee, W.L. and Yik, F.H.W. (2004) Regulatory and voluntary approaches for enhancing building energy efficiency, *Progress in Energy and Combustion Science*, 30(5), pp. 477–499.

Leonard, C.A. (1988) *The effects of change orders on productivity*, Concordia University, Montreal, Canada.

Lingard, H., Gilbert, G., and Graham, P. (2001) Improving solid waste reduction and recycling performance using goal setting and feedback, *Construction Management and Economics*, 19(8), pp. 809–817.

Liu, M., Li, B., and Yao, R. (2010) A generic model of Exergy Assessment for the Environmental Impact of Building Lifecycle, *Energy & Buildings*, 42(9), pp. 1482–1490.

Lockyer, K.G. (1974) *An Introduction to Critical Path Analysis* (Third Edition), Pitman, London.

Lord, W.E. (2008) Embracing a modern contract – progression since Latham? *A Paper Presented at COBRA 2008*, The construction and building research conference of the Royal Institution of Chartered Surveyors, Dublin, September 4–5.

Lukas, J.A. (2008) Earned Value Analysis – why it doesn't work, *American Association of Civil Engineers Transactions*. Available at www.icoste.org (accessed on 15 October 2013).

Mawdesley, M., Askew, W., and O'Reilly, M. (1997) *Planning and Controlling Construction Projects*, Addison Wesley Longman and The Chartered Institute of Building, Harlow, UK.

McDonald, B. and Smithers, M. (1998) Implementing a waste management plan during the construction phase of a project: a case study, *Construction Management and Economics*, 16(1), pp. 71–78.

McGrath, C. (2001) Waste minimisation in practice, *Resources, Conservation and Recycling*, 32(3–4), pp. 227–238.

Mechanical Contractors Association of America. (2014). Change Orders, Productivity, Overtime – A Primer for the Construction Industry (2014 Edition). *Factors Affecting Labor Productivity* at pages 77–78.

Milberg, C. and Tommelein, I. (2003) Role of tolerances and process capability data in product and process design integration. In: *Construction Research Congress* (Eds Molenaar, K.R. and Chinowsky, P.S.), pp. 1–8, ASCE. Available at http://dx.doi.org/10.1061/40671(2003)93 (accessed on 22 November 2013).

Moon, D. (2008) *Soft Option: Free Technology Tools to Help Firms with the Waste Management Task at Hand*, Building. Available at http://www.building.co.uk/story.asp?sectioncode=629&storycode=3127660 (accessed on 15 October 2013).

Moon, D. (2009) *Committing to waste reduction: Government business*, 15(12), pp. 25–27.

Moore, D. (2002) *Project Management: Designing Effective Organizational Structures in Construction*, Blackwell Science, London.

Morris, W.G.P. (1994) *The Managing of Projects*, Thomas Telford, London.

Morris, W.G.P. (2002) Science, objective knowledge and the theory of project management, *Proceedings of the ICE, Institution of Civil Engineering*, 150(2), pp. 82–90.

Mossmann, A. (2012) Last Planner – 5 + 1 Crucial and Collaborative Conversations for Predictable Design and Construction Delivery. Available at www.thechangebusiness.co.uk (accessed on 15 October 2013).

Mossmann, A. (2013) Last Planner – Collaborative Production Planning, Collaborative Programme Co-Ordination. Available at www.lci-uk.org (accessed on 15 October 2013).

Neale, R. and Neale, D.E. (1989) *Construction Planning*, Thomas Telford, London.

NEC (2013) *NEC3 Engineering and Construction Contract (ECC)*, Thomas Telford, London.

NEDO Report. (1983) *Faster Building for Industry*, HMSO, London.

NetRegs. (2007) *Site Waste – It's Criminal: A Simple Guide to Site Waste Management Plans*, NetRegs, Bristol, UK.

NetRegs. (2008) *Half of Construction Businesses Still Unaware that SWMPs are Law*. Available at http://www.environment-agency.gov.uk/static/documents (accessed on 15 October 2013).

Newton, A.J.N. (1996) *The planning and management of detailed building design*, PhD Thesis, Loughborough University, Loughborough, UK.

Oliver, R.K. and Webber, M.D. (1992) Supply-chain management: logistics catches up with strategy. In: *Logistics: The Strategic Issues* (Ed. Christopher, M.), Chapman & Hall, London.

Osmani, M. (2011) Construction waste. In: *Waste: A Handbook for Management* (Eds Letcher, T.M. and Vallero, D.A.), Academic Press, Burlington, MA, pp. 207–218.

Osmani, M., Glass, J., and Price, A. (2006) Architect and contractor attitudes to waste minimisation, *Waste and Resource Management*, 159(2), pp. 65–72.

Osmani, M., Glass, J., and Price, A. (2008) Architects' perspectives on construction waste reduction by design, *Waste Management*, 28(7), pp. 1147–1158.

Parkinson, C.N. (1955) Parkinsons Law. *The Economist*, 19 November.

Pearsall, J. and Trimble, B. (2002) *Oxford English Reference Dictionary*, Oxford University Press, Oxford, UK.

Pickavance, K. (2005) *Delay and Disruption in Construction Contracts* (Fourth Edition), Sweet & Maxwell, London.

Pilcher, R. (1992) *Principles of Modern Construction Management* (Third Edition), McGraw-Hill, Maidenhead, UK.

Poon, C., Yu, A., Wong, S., and Cheung, E. (2004) Management of construction waste in public housing projects in Hong Kong, *Construction Management and Economics*, 22(7), pp. 675–689.

Price, T., Wamuziri, S., and Gupta, N. (2009) Should SWMP2008 be integrated with CDM2007? *Construction Information Quarterly*, 11(1), pp. 12–17.

Prior, J.J. (2007) BREEAM – A step towards environmentally friendlier buildings, *Structural Survey*, 9(3), pp. 237–242.

Project Management Institute (PMI). (1996) *A Guide to the Project Management Body of Knowledge*, PMI Standards Committee, PMI, Newtown Square, PA.

PMI. (2004) *A Guide to the Project Management Body of Knowledge* (Third Edition), PMI Standards Committee, PMI, Newtown Square, PA.

PMI. (2013) *A Guide to the Project Management Body of Knowledge (PMBOK Guide)* (Fifth Edition), PMI, Newton Square, PA.

Reiss, G. (1995) *Project Management Demystified* (Second Edition), E and F N Spon, for Chapman Hall, London.

Reuter, B. (2002) *Collaborative planning*. In: *Supply Chain Management and Advanced Planning: Concepts, Models, Software and Case Studies* (Eds Kilger, C. and Stadler, H.), Springer, New York, pp. 223–237.

Royal Institute of British Architects (RIBA) (2013) *RIBA Plan of Work 2013*, RIBA, London. Available at www.ribaplanof work.com (accessed on 25 November 2013).

Rutkauskas, A.K. (2008) On the sustainability of regional competitiveness development considering risk, *Technological and Economic Development of the Economy*, 14(1), pp. 89–99.

Scott S. (1993) The nature and effects of construction delays, *Construction Management and Economics*, 11(6), pp. 358–369.

Shen, L., Tam, V., Tam, C., and Drew, D. (2004) Mapping approach for examining waste management on construction sites, *Journal of Construction Engineering and Management*, 130(4), pp. 472–481.

Shewhart, W.A. and Deming, W.E.(1939) *Statistical Method from the Viewpoint of Quality Control*, the Graduate School, the Department of Agriculture, Washington, DC.

Sinclair, D. (Ed.) (2012) *BIM Overlay to the RIBA Outline Plan of Work Royal Institute of British Architects*, RIBA, RIBA Publishing, London.

Stadtler, H. (2002) *Supply Chain Management and Advanced Planning: Concepts, Models, Software, and Case Studies*, Springer, New York.

Strategic Forum for Construction. (2002) Accelerating Change: A Report by the Strategic Forum for Construction. Chaired by Sir John Egan. Rethinking Construction, Construction Industry Council, London.

Steele, J.L. (2010) Three steps forward, two steps back: managing information-driven design and engineering processes, Published in *The Proceedings of the 2010 PMI Global Congress Proceedings, National Harbor*, MD.

Steward, D. (1981) *Systems Analysis and Management*, Petrocelli Books, New York.

Strategic Forum for Construction. (2002) *Accelerating Change: A Report by the Strategic Forum for Construction*. Chaired by Sir John Egan. Rethinking Construction, Construction Industry Council, London.

Tah, J.H.M., Thorpe, A., and McCaffer, R. (1993) Contractor project risks contingency allocation using linguistic approximation, *Journal of Computing Systems in Engineering*, 4(2–3), pp. 281–293.

Tam, V. (2007) On the effectiveness in implementing a waste-management-plan method in construction, *Waste Management*, 28(6), pp. 1072–1080.

Teo, M. and Loosemore, M. (2001) A theory of waste behaviour in the construction industry, *Construction Management and Economics*, 19(7), pp. 741–751.

The Society of Construction Law (SCL). (2001, November) *Protocol for Determining Extensions of Time and Compensation for Delay and Disruption – Consultation Copy*. At paragraph 3.9.4.2, p. 18.

The Society of Construction Law (SCL). (2002) *The Society of Construction Law Delay and Disruption Protocol*. (October 2002 – reprinted March 2003). The SCL. Available at www.eotprotocol.com (accessed on 15 October 2013).

Thompson, D. (1996) The Oxford Compact English Dictionary, published by Oxford University Press, Oxford, UK.

Tommelein, I.D. (1997) Models of lean construction processes: example of pipe-spool materials management, *Proceedings of Construction Congress 97*, Minneapolis, MN. Available at www.leanconstruction.org (accessed on 25 October 2013).

Tommelein, I.D. (2000) Impact of variability and uncertainty on product and process development. In: *Construction Congress VI* (Ed. Walsh, K.), ASCE, Springer, New York.

Trufil-Fulcher, G., Reeves, S., Myers, D., Reid, M., Tong, R., and Ferris, C. (2009) *Site Waste Management Plans Impacts Survey 2009*, TRL, Wokingham, UK.

Uher, T.E. and Zantis, A.S. (2011) *Programming and Scheduling Techniques* (Second Edition), Spon Press an imprint of Taylor & Francis, Abingdon, UK.

UK Government. (2011) *Autumn Statement 2011*. The UK Government's Autumn Statement deleivered to parliament, November 2011. Available at https://www.gov.uk/government/news/autumn-statement-2011--3 (accessed on 25 October 2013).

Ulrich, K.T. and Eppinger, S.D. (2011) *Product Design and Development* (Third Edition), McGraw-Hill, New York.

US Department of Defense. (1998) *Department of Defense Handbook 881 – Work Breakdown Structure*, US Department of Defense, Washington, DC. Available at http://www.srs.gov/general/EFCOG/03OtherAgencies/MilHdbk881.pdf (accessed on 25 October 2013).

Verheij, H. and Augenbroe, G. (2006) Collaborative planning of AEC projects and partnerships, *Automation in Construction*, 15(4), pp. 428–437.

Wall, M. (2006) Energy efficient terrace houses in Sweden. Simulations and measurement. *Energy and Buildings*, 38(6), pp. 627–634.

Waskett, P., Newton, A., Steele, J., Cahill M., and Beaumont, J. (2010) Achieving reliable delivery of design information for procurement and construction, *Proceedings of 3rd 'World of Construction Project Management' Conference*, Coventry University, UK, October 20–22.

Waste and Resources Action Programme (WRAP). (2007a) *Achieving Good Practice Waste Minimisation and Management: Guidance for Construction Clients, Design Teams and Contractors*, WRAP, Oxon, UK.

WRAP. (2007b) *Achieving Effective Waste Minimisation: Guidance for Construction Clients, Design Teams and Contractors*, WRAP, Oxon, UK.

WRAP. (2008) *Site Waste Management: Guidance and Templates for Effective Site Waste Management Plans*, NHBC Foundation, Bucks, PA.

WRAP. (2009a) *Stakeholder Briefing: Resource Efficiency*, WRAP, Oxon, UK.

WRAP. (2009b) *Reducing your Construction Waste: A Pocket Guide for SME Contractors*, WRAP, Oxon, UK.

WRAP. (2009c) *Review of Activities for Year Ended 31 March 2009*, WRAP, Oxon, UK.

WRAP. (2009d) *Site Waste Management Plans Impacts Survey 2009*, WRAP, Oxon, UK.

Weaver, P. (2012) Henry L Gantt, 1861–1919: debunking the myths a retrospective view of his work, *PM World Journal*, I(V). Available at www.pmworldjournal.net (accessed on 15 October 2013).

Weist, J.D. (1964) Some properties of schedules for large projects with limited resources, *Operations Research*, 12(3), 395–418.

Winch, G.M. (2002) *Managing Construction Project: An Information Process Approach*, Blackwell Science, London.

Winch, G.M. (2003) Models of manufacturing construction process: the genesis of re-engineering construction, *Building Research & Information*, 31(2), pp. 107–118.

Winch, G.M. (2006) Towards a theory of construction as production by projects, *Building Research & Information*, 34(2), pp. 154–163.

Winch, G.M. and Carr, B. (2001) Processes, maps and protocols: understanding the shape of the construction process, *Construction Management and Economics*, 19(5), pp. 519–531.

Winch, G.M. and Kelsey, J. (2005) What do construction project planners do? *International Journal of Project Management*, 23(2), pp. 141–149.

Winch, G.M., Courtney, R., and Allen, S. (2003) Re-valuing construction, *Building Research & Information*, 31(2), pp. 82–84.

Winograd, T. and Flores, F. (1986) *Understanding Computers and Cognition: A New Foundation for Design*, Ablex, Norwood, NJ. (Cited in Koskela, L. and Howell, G. (2002a) The theory of project management: explanation to novel methods, Proceeding of the 10th Annual Conference International Group for Lean Construction, Gramado, Brazil.)

Womack, J.P. and Jones, D.T. (1996) *Lean Thinking: Banish Waste and Create Wealth in Your Corporation*, Simon and Schuster, New York.

Womack, J.P., Jones, D.T., and Roos, D. (1990) *The Machine that Changed the World: The Story of Lean Production*, HarperCollins, New York.

Yeomans, S. (2006) *Collaboration, Construction Excellence*. Available at http://www.constructingexcellence.org.uk//wiki/article.jsp?id=203 (accessed on 15 October 2013).

Yourdon, E. (1989) *Modern Structured Analysis*, Yourdon Press/Prentice-Hall, New York.

Law reports

Adyard Abu Dhabi v SD Marine Services [2011] EWHC848 (Comm).

City Inn Ltd v Shepherd Construction Ltd [2007] ScotCS CSOH 190.

City Inn Ltd v Shepherd Construction Ltd [2010] CSIH 68.

Cleveland Bridge UK Ltd v Severfield – Rowen Structures Ltd [2012] EWHC 3652 (TCC).

Costain Ltd v Charles Haswell & Partners Ltd [2009] EWHC 3140 (TCC).

Great Eastern Hotel Company Ltd v John Laing Construction Ltd & Anor [2005] EWHC 181 (TCC).

John Barker Construction Ltd v London Portman Hotel Ltd. (1996) 83 BLR 31.

John Doyle Construction Ltd v Laing Management (Scotland) Ltd [2004] ScotCS 141.

London Underground Ltd v Citylink Telecommunications Ltd Rev 1 [2007] EWHC 1749 (TCC).

Mirant Asia-Pacific Construction (Hong Kong) Ltd v Ove Arup and Partners International Ltd & Anor [2007] EWHC 918 (TCC).

Skanska Construction UK Ltd v Egger (Barony) Ltd. [2004] EWHC 1748 (TCC).

Walter Lilly & Co Ltd v Mackay [2012] EWCH 1773 (TCC).

Wharf Properties Ltd v Eric Cumine Associates (1991) 52 BLR 1.

Index

Note: Page numbers in *italics* refer to Figures; those in **bold** to Tables.

A Handbook for Construction Planning and Scheduling, First Edition. Andrew Baldwin and David Bordoli.
© 2014 John Wiley & Sons, Ltd. Published 2014 by John Wiley & Sons, Ltd.

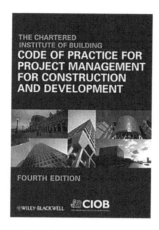
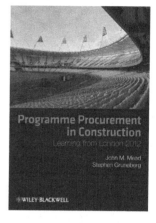
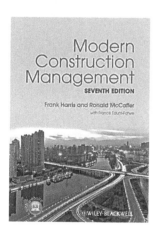

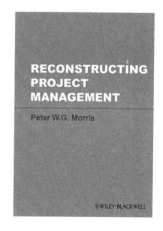

TJI84210-9780470670323-22-03-19